Basic Plumbing With Illustrations

Howard C. Massey

Craftsman Book Company
6058 Corte Del Cedro ● Carlsbad ● CA 92009

Acknowledgments

The author expresses his sincere thanks and appreciation to the following companies and organizations.

-To the Florida Energy Committee and the Environment Information Center of the Florida Conservation Foundation, Inc. for pertinent information necessary to authenticate the thermosyphon and pumped solar water heating systems

-To Amtrol Inc., West Warwick, R.I. (Illustrations used for the Well-X-Trol water systems)

-To Sta-Rite Industries, Inc., Delavan, Wisconsin (Technical information for troubleshooting private water systems)

-To Josam Manufacturing Co., Michigan City, Indiana (Fixture carrier illustrations)

-To American Standard, P. O. Box 2003, New Brunswick, New Jersey (Plumbing fixture photos and roughing-in measurements)

-To Mansfield Sanitary, Inc., Perrysville, Ohio (Brass items illustrated)

-To Montgomery Ward (Dishwasher installation illustrations)

-To Moen Faucet, 2701 Washington Boulevard, Bellwood, Illinois (Service repair illustrations)

-To Delta Faucet Company, Greensburg, Indiana (Service repair illustrations)

To my wife, Hilda, for her untiring assistance in typing this manuscript.

Library of Congress Cataloging in Publication Data

Massey, Howard C
　　Basic plumbing illustrated.

Illustrations by
Mike Aten and the author

　　Includes index.
　　1.　Plumbing. I. Title.
TH6122.M38　　696'.1　　　　　　　79-20008
ISBN 0-910460-72-8
© 1980 Craftsman Book Company
Eighth printing 1992

Cover/Gary Symington

Contents

Chapter 1 Introduction To Plumbing 5
  Plumbing Codes and Permits 5

Chapter 2 The Plumbing System 7
  The Drainage System 7
  The Water Supply System 7
  Plumbing Fixtures 9

Chapter 3 Plot Plans 11
  Sample Plans 11

Chapter 4 Isometric Drawings and
Definitions 16
  Naming the Pipes 16
  Public Sewers 22
  Private Sewers 22
  Fixture Branches and Drains 23
  Vents 24

Chapter 5 The Drainage System 26
  Pipe Sizing 26
  Trap Sizing 26
  Sizing From a Layout 27
  Sewer Sizing 27
  Building Drain Sizing 31
  Waste Pipe Sizing 31
  Vent Sizing 32

Chapter 6 The Vent System 33
  Sizing 33
  Venting Principles 33
  Stack Venting 34
  Wet Venting 34
  Horizontal Distance of Fixture Trap
  From Vent 38
  Isometric Drawings 40

Chapter 7 Traps 45
  Fixture Traps 45
  Code Restrictions. 45
  Drum Traps 50
  Broken Trap Seals 50
  House Traps. 51
  Installations Not Permitted 53

Chapter 8 Cleanouts 57
  Requirements 57
  Access 58
  Underground Piping. 58
  Clearance 59
  Vent Stacks 59
  Cleanout Plugs 60
  Sizing 61

Chapter 9 Floor Plans And Layouts 62
  Seven Bathroom Floor Plans 63
  Four House Layouts. 67

Chapter 10 DWV Materials And Installation . 72
  Materials 72
  Changes of Direction 73
  Cast Iron Soil Pipe and Fittings 77
  Wyes and Bends 79
  Plastic Pipe and Fittings 82
  Plastic Building Sewers. 86
  Cutting Cast-Iron Pipe 86
  Cutting Plastic Pipe 88
  Cast-Iron Hub and Spigot Joints 88
  Caulking 89
  Cast-Iron No Hub Joints 90
  Plastic Pipe Joints. 91
  DWV Piping Within a Building 93
  Supports 94
  Venting 96
  Air Conditioning Condensate Drains . . . 96

Chapter 11 Septic Tanks And Drainfields . . 101
  Size and Location of Septic Tanks . . . 102
  Design and Construction of Septic Tanks. 103
  Sizing Drainfields. 103
  Tile Drainfields. 105
  Reservoir Type Drainfields 108
  Corrugated Plastic Perforated Tubing
  Drainfields 108
  Drainfield Replacement 109
  Drainfield Restrictions. 109

Chapter 12 Water Supply And Distribution
System 113
  Sizing 113
  Maintenance. 118
  Reducing Corrosion 119
  Insulation. 120
  Pipe Covers 120

Chapter 13 Valves And Faucets 122
  Types of Valves 123
  Types of Faucets 124

Chapter 14 Hot Water Systems 127
  Design Objectives. 127
  Heater Capacities and Sizing. 128
  Safety Devices 129
  Solar Water Heater Components 131
  Solar Collector Installation 133
  Water Circulation. 134
  Materials and Installation of Solar Water
  Heater Systems 136

Chapter 15 Water Pipe And Fittings 141
Galvanized Steel Pipe 141
Threaded Fittings. 142
Copper Pipe and Tubing 144
Copper Fittings 145
Plastic Pipe 145
Plastic Fittings 146

Chapter 16 Installation Of Water Systems . . 148
Water Service Supply Pipe 146
Water Distributing Piping 150
Cross Connection 155
Measuring Pipe 156
Offsets 157
The Pipe Vise 157
Cutting and Joining Steel Pipe 158
Threaded Steel Pipe 161
Cutting and Joining Copper Pipe 162
Cutting and Joining Plastic Pipe 167

Chapter 17 Private Water Wells And
Sprinkler Systems. 172
Open End Casings. 173
Well Points 175
Well Requirements 175
Suction Lines and Sizing 176
Pressure Tank Water Systems 178
Lawn Sprinkler Systems 182
Sizing 182
Pumps 183
Materials 183
Troubleshooting Guide 186

Chapter 18 Roughing-In 194
Fixture Clearances 194
Fixture Carriers 195
Roughing-In Dimensions 199

Chapter 19 Plumbing Fixtures 203
Minimum Fixture Requirements 203
Fixture Overflows 205
Bathtubs 206
Waste and Overflow Installation 210
Showers 212
Lavatories 215
Water Closets 221
Urinals 226
Laundry Sinks 227
Food Waste Disposers 229
Dishwashers 229
Water Heaters 230

Chapter 20 Maintenance Of Plumbing
Systems 239
Types of Stoppages 239
Freeing Stoppages 240
Clogged Sinks 241
Clogged Drains 244
Clogged Water Closets 245
Water Heater Maintenance 247
Reduced Water Pressure 247
Water Hammer 249
Thawing Frozen Pipes 249
Cleaning 250

Chapter 21 General Plumbing Repairs . . . 253
Globe Valves 253
Compression Faucets 255
Diaphram Faucets 256
Mixing Faucets 257
Trap Washers 263
Basket Strainers 266
Washers 266
Water Closets 274
Locating Leaks 277

Chapter 22 Simple Gas Installations . . . 280
Sizing Gas Systems 281

Chapter 23 Materials For Simple Gas
Installations 285
Underground Piping 285
Piping in Walls 287
Drip Pipes 287
Shutoff Valves 287
Appliance Installation 290

Chapter 24 Private Swimming Pools . . . 292
Water Supply 292
Waste Water Disposal 293
Minimum Equipment 294
Filtration Requirements 297
Materials and Installation Methods . . . 298
Pool Heaters 299

Chapter 25 Definitions, Abbreviations, And
Symbols 301
Plumbing Symbol Legend 311
Fitting Symbols 312

Index 314

1

Introduction To Plumbing

This book is a guide to good plumbing practice. Even those with little or no experience with plumbing systems should have no difficulty understanding what is explained here. The text covers the basic principles required to plan, install and maintain common plumbing systems in residential and light commercial buildings. An understanding of the plumbing code is essential, and code requirements are discussed throughout this volume. Questions are included at the end of each chapter to help you test your understanding of the concepts presented.

The objective of this book is to help you select the materials, pipe sizes and methods of installation generally accepted as correct by most model codes. The basic, practical side of plumbing is covered fully here; the highly technical and specific language sections of the code and information on hydraulic and pneumatic systems that could lead to confusion are omitted.

Plumbing Codes

Every state has adopted plumbing codes (laws) to protect the health, safety, and welfare of its people. Specific powers, duties, and responsibilities have been delegated by the states to municipal authorities to enforce these regulations at the local level.

Everyone knows that contaminated water or leaking sewage pipes can be hazardous to health and can lead to serious illness or even death over a wide area. The purpose of the plumbing code is to establish plumbing standards that protect the health and welfare of everyone in the community.

Although it is difficult to learn plumbing from a code book, anyone who wishes to install plumbing (apprentice plumber, maintenance plumber, or home plumber) should own a copy of the local plumbing code. Your local building department will sell you a copy of the code they enforce.

Please note that a homeowner who installs plumbing must work within the same rules as the professional plumber. He is subject to the same penalties for not obtaining permits or inspections and for not complying with the code. Good plumbing does not depend on *who* does the work, but on *how* it is done.

Plumbing details vary from code to code in design, layout and installation methods. But the basic principles of sanitation and safety are the same regardless of the code that has been adopted in your area. What you learn in this book should be applicable almost anywhere in the United States.

However, this manual is *not* the plumbing code. You will have to refer to your local code from time to time. The minor differences between model plumbing codes are emphasized throughout this book. The differences will be apparent as you read and compare sections of this book with your local code.

Some codes require that only a licensed plumbing contractor do new or extensive remodeling work. However, most codes allow a homeowner to do all plumbing work in the home he owns and lives in. Some codes also permit a homeowner to install the plumbing in a new house he is building if he can show proof of ownership and satisfy the building official that he has the basic knowledge to do the work without outside help.

What About Plumbing Permits?

Codes require a permit for all plumbing work, with limited exceptions: no plumbing permit is necessary to repair leaks, clear obstructions in sewer lines or waste pipes, repair faucets or valves, or clean septic tanks. When a permit is required it must be available at the site for examination upon request of the plumbing inspector.

No portion of a plumbing installation can be covered or concealed in any way until the plumbing inspector has inspected the work for compliance with code requirements. It is the responsibility of the permit holder to notify the plumbing inspector when the work is ready for test and inspection.

The permit holder requests the necessary inspections when the work for which the permit was issued is completed. Inspections are classified into three categories for a one-story building: (1) the ground roughing-in, (2) the tub set and interior water piping, and (3) the final. For a two-story building a fourth inspection is required. This fourth inspection is referred to as the *topping out*. All rough piping must be inspected above the first floor up to and through the roof.

The request for final plumbing inspection should not be made more than 30 days after completion of the permitted work. The building or construction in which the work is done can not be used or occupied until a final inspection is made. These inspections will be discussed in detail in later chapters so you know how to prepare for each inspection.

2

The Plumbing System

Approximately ten percent of the construction cost of the average home is in the plumbing system. Nearly 300 feet of pipe are installed beneath the floor and in the partitions. Each foot of this pipe will function quickly and efficiently and last many years when properly installed and maintained.

There are three basic components to every plumbing system:

1) The drainage and vent system

2) The water service pipe and distributing pipes

3) The plumbing fixtures

Each component should be designed, installed, located and maintained to serve its intended use. These components and their uses will be discussed in detail in later chapters. In this chapter we will review briefly each of the three systems.

The Drainage System

Plumbing fixtures are the end of the potable (pure) water supply system and the beginning of the sewage system. The drainage system consists of three major classes of piping and fittings. In the *first class* are all the portions of the system necessary to receive and convey the used water generated within a building to a public sewer or other approved private disposal system.

The *second class* of drainage system pipes and fixtures consists of the fixture traps. The trap has a unique importance in protecting the health of a building's occupants. It is designed and constructed to provide a liquid seal to prevent drainage system odors, gases and even vermin from entering the building at fixture locations.

The *third class* of pipes and fittings in the drainage system consists of the vent pipes, which are designed to admit and give off air to and from all parts of the drainage system. Vent pipes are sized and arranged to relieve pressure that builds up as water is discharged into the sanitary drainage system by various types of plumbing fixtures. This free flow of air within the system keeps back pressure and siphoning from destroying fixture trap seals.

These three classes of drainage and vent pipes and fittings are commonly called the "drain-waste-vent" system and are referred to in the trade as the DWV system. The DWV piping is a very important part of the plumbing system and is the part least understood.

The sanitary drainage and vent arrangements are considered by most professional plumbers to be the heart of a plumbing system. This system is the first of the three basic components to be installed and inspected in any construction. Since a major portion of it is installed beneath the floor of a building, it is called "ground roughing-in" in the trade.

The three classes of pipes and fittings may be better understood by examining Figure 2-1, an isometric drawing of a sanitary system. The drainage system and the traps (the portion of the system that receives liquid waste) are illustrated by solid (continuous) lines. The dry portion (vents) is illustrated by dashed (broken) lines.

Become familiar with isometric drawings if you have had no previous training in plumbing practice. Construction blueprints that are acceptable to local authorities generally include isometric drawings of the systems to be installed. Simple jobs are easy to understand if you know how to read isometric drawings.

The Water Supply System

The second basic component of a plumbing system is the water service pipe and the distributing pipes. These pipes and fittings

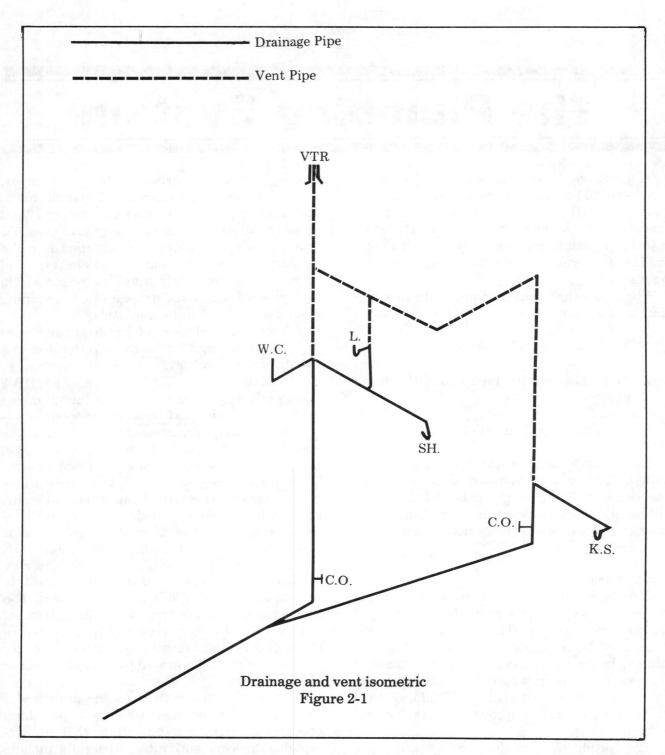

———————— Drainage Pipe

— — — — — — Vent Pipe

VTR

W.C.

L.

SH.

C.O.

K.S.

C.O.

Drainage and vent isometric
Figure 2-1

convey both hot and cold water through separate systems to plumbing fixtures and other water outlets on private property.

The water source for these pipes may be either a public water distribution system adjacent to each lot or parcel of land or a private system that depends on a well.

It is important to understand the symbols for hot and cold water as shown in piping diagrams.

Cold water is illustrated as a solid line broken with a single dot or dash: (————— ——). Hot water is shown as a solid line broken with two dots or dashes (————— — —). Where a plumbing fixture requires both hot and cold water, the hot water pipe is always on the left and the cold on the right as you face the fixture. Figure 2-2 illustrates a simple water piping diagram.

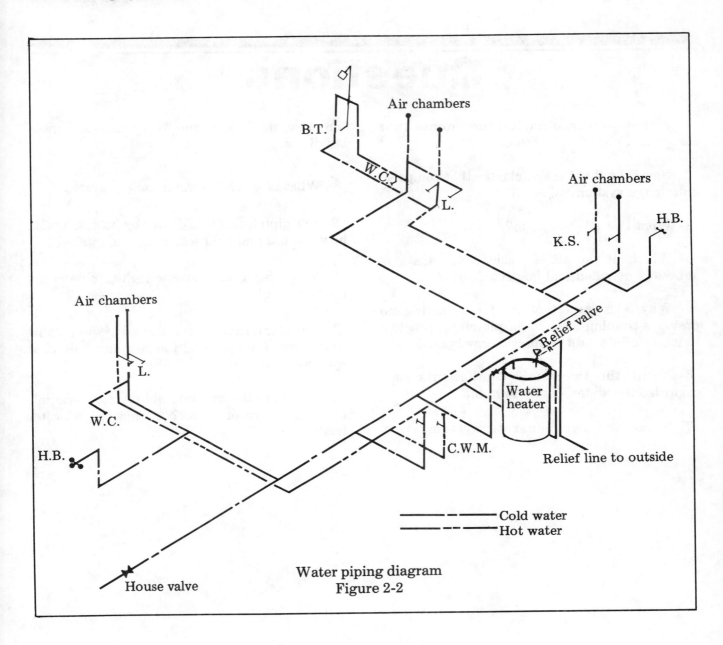

Water piping diagram
Figure 2-2

The Plumbing Fixtures

The third basic component of a plumbing system is the plumbing fixtures. These must be of the type and quantity to meet the needs of the occupants of the building.

Abbreviations are often used to identify various types of plumbing fixtures in isometric drawings and floor plans. For example, you will see the letter "L" used to designate a lavatory. Some plans may use "LAV." The following abbreviations will be used throughout this book to identify the plumbing fixtures in isometric drawings and floor plans.

Plumbing Fixtures	Abbreviated Symbols
water closet	W.C.
bathtub	B.T.
shower	SH.
lavatory	L.
kitchen sink	K.S.
clothes washing machine	C.W.M.
water heater	W.H.

Questions

1- Name the three basic components that comprise a plumbing system.

2- Name the three major classes that comprise a drainage system.

3- What is a DWV system?

4- Which of the three basic components is generally installed and inspected first?

5- Why is it important for those having no previous training or skills in plumbing practice to learn to interpret isometric drawings?

6- Name the two basic components that comprise the water supply system.

7- Name the two separate systems that comprise the water supply system for most buildings.

8- What is a private water supply system?

9- In a piping diagram drawn by professionals, how are hot and cold water lines identified?

10- When facing a fixture, on which side is the hot water outlet?

11- Plumbing fixtures are the end of the potable (pure) water supply and the beginning of what system?

12- What are the common abbreviated symbols used for a water closet? a shower? a water heater?

Plot Plans

The drawing of a lot on which a building stands or on which a building will be constructed is known as a *plot plan*. You should be familiar with plot plans and know how to interpret them. A plot plan shows:

- The shape and size of the lot

- The location and size of the building

- The setbacks from the property line

- All permanent outside construction above ground, including driveway and walkway

The electric service, easements (if any) and streets are also shown, plus other useful information. Below-ground information includes location and size of the gas service, source, location and size of the water facility, storm drainage and any disposal facility, type, location and size of the sewage disposal facility.

Plot plans are required by code and are an essential part of all construction plans. Plot plans are first approved by plans examiners; they are then called "as-built" plans. Plans examiners may ask for minor corrections on the plans before approval.

What is shown to the plans examiners and approved by them must be strictly adhered to. Making any changes to the plan after its approval (for example, relocating the septic tank to another area) would require submitting a revised plot plan for reapproval.

The three types of plot plans presented in Figures 3-1 to 3-3 show the three basic types of outside plumbing facilities: 1) a building having public water and sewers, 2) a building having public water and a septic tank, and 3) a building having a well and a septic tank.

The house in Figure 3-1 is a typical single-family dwelling. It is connected to a public sewer lateral located at approximately the center of the lot. The water service is connected to a water meter located on the right side of the lot. Also note that in this plot plan the utilities serving the dwelling are in the street and not in the utility easement at the rear of the property. The sewer and water service were properly sized and thus no corrections by the plumbing plans examiner are noted.

The plot plan in Figure 3-2 shows a house in a suburban area. It has public water, but it depends on a septic tank for its sewage disposal. The rapid spread of urban communities around large metropolitan areas has often outstripped the installation of sanitary sewage collection systems.

This plot plan shows a 1200-gallon septic tank with a 480-square-foot drainfield. The soil is sandy loam. It is located in the front yard of the dwelling and is sized for four bedrooms. The water service is connected to a water meter located on the right side of the lot on the plan. Also note that this service is for a three-bath house and is unsized. The plumbing plans examiner sized this service for the architect. The separation of five feet between the septic tank and water service is also noted on the plans, as required by code.

The house in Figure 3-3 is in a rural area. It depends on a domestic well for its source of water and on a septic tank for sewage disposal.

The well is located at the right rear of the property. The septic tank is located on the left side of the residence. The water service (suction line) from the well to the house is sized at 1¼ inches because its developed length exceeds 40 feet. If it had been less than 40 feet, a one-inch service (suction line) would have been sufficient. The septic tank and drainfield area is also sized properly for a two-bedroom home. The soil is fine sand. The 50-foot separation between the septic tank and the well is also noted on the plans, as required by the Department of Environmental Resource Management (DERM), or in some cases by the local health department. Since there are no notations by the plumbing plans examiner on this plot plan, it would be proper to install the facilities as shown.

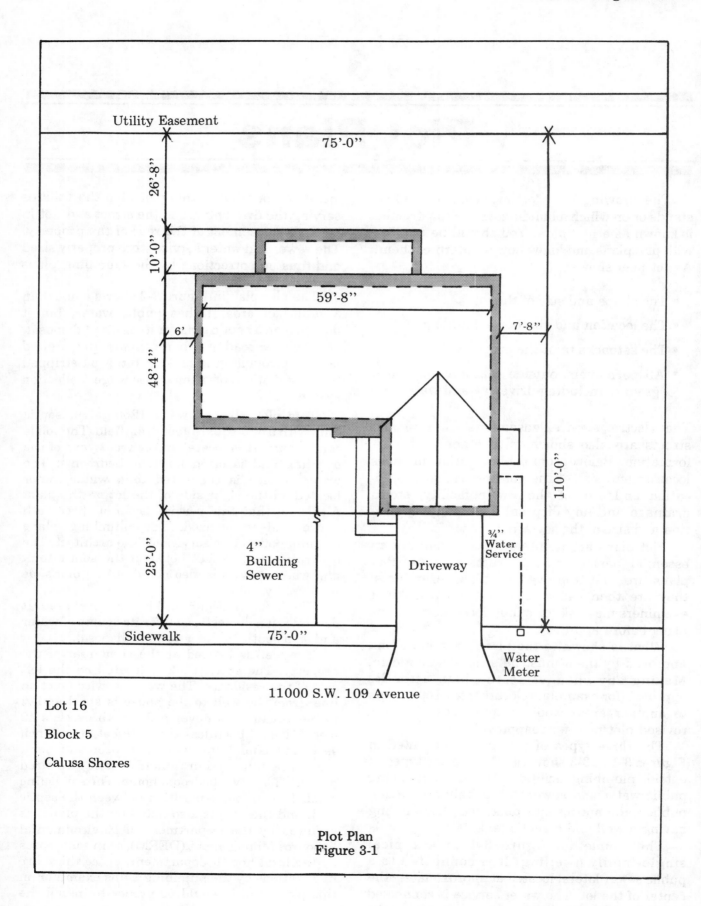

Utility Easement

75'-0"

26'-8"

10'-0"

59'-8"

6'

7'-8"

48'-4"

110'-0"

25'-0"

4"
Building
Sewer

Driveway

¾"
Water
Service

Sidewalk   75'-0"

Water
Meter

11000 S.W. 109 Avenue

Lot 16

Block 5

Calusa Shores

Plot Plan
Figure 3-1

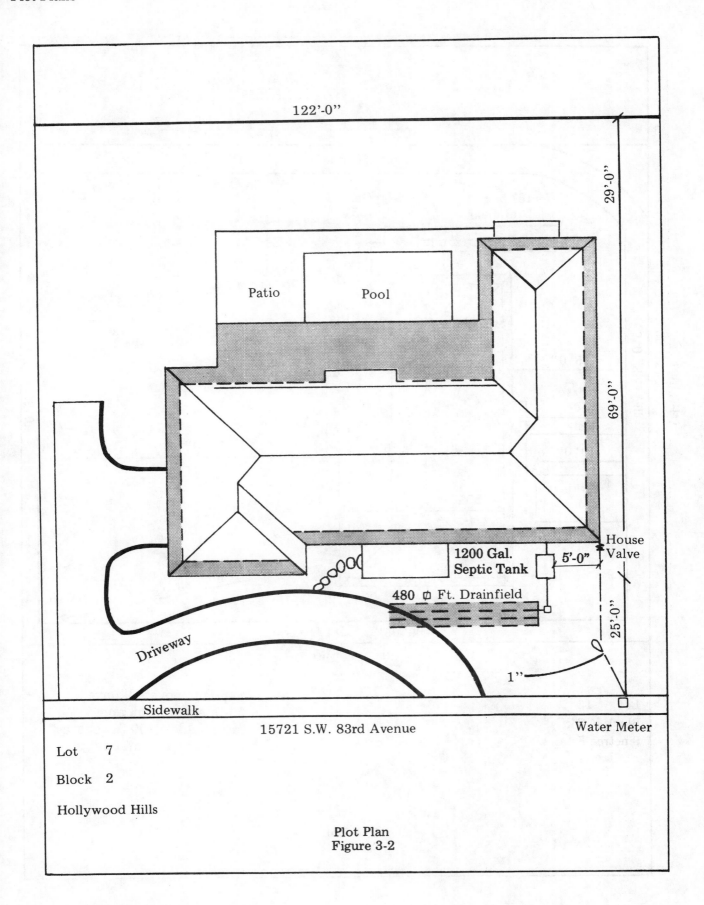

**Plot Plan
Figure 3-2**

122'-0"

29'-0"

69'-0"

25'-0"

Patio

Pool

House Valve

1200 Gal. Septic Tank

5'-0"

480 ∅ Ft. Drainfield

1"

Driveway

Sidewalk

15721 S.W. 83rd Avenue

Water Meter

Lot 7

Block 2

Hollywood Hills

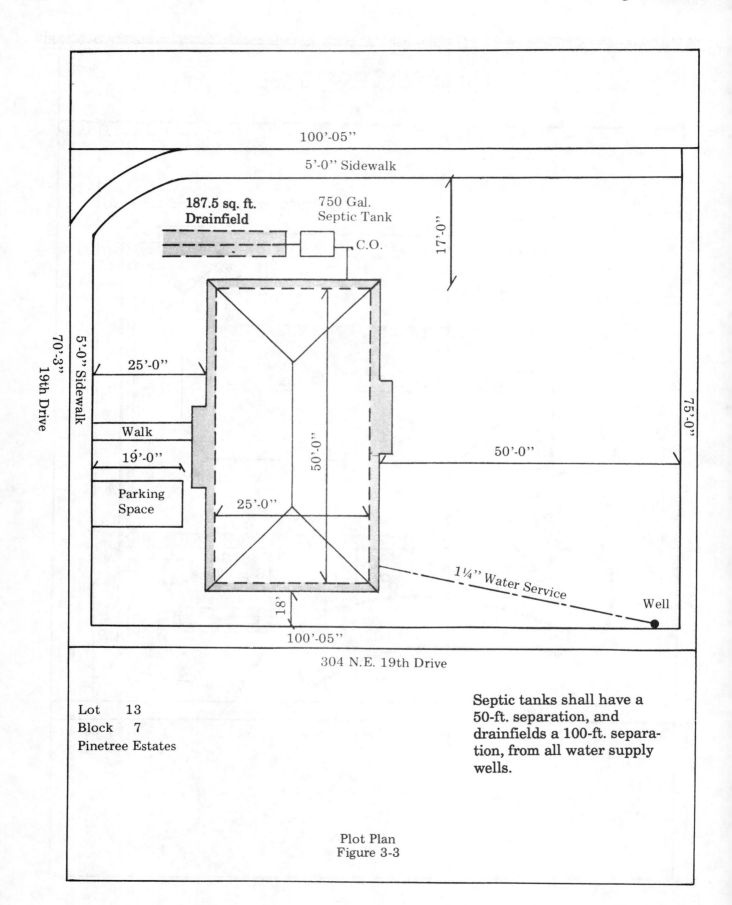

Plot Plan
Figure 3-3

Questions

1- What is the term used for approved plans?

2- If any changes are made in approved plans, what steps would have to be taken?

3- What is included in a plot plan?

4- What below-ground information is included in a plot plan?

5- When is it acceptable to install facilities as shown on a given plot plan?

6- In addition to the plumbing plans examiner, who else must approve the plot plans?

7- Why is it necessary that a plumber know how to read and interpret plot plans?

4

Isometric Drawings And Definitions

Everyone who installs or designs plumbing systems should know how to read and make isometric drawings. They are used by plumbing contractors to estimate the cost of new work and to show a job foreman how to rough in a particular job.

Reading and making isometric illustrations is a skill you can master quickly. Only three basic angles are needed to express the plumbing system: the horizontal pipe, the vertical pipe, and the forty-five degree angle pipe.

Figure 4-1 is a valuable guide to practicing these isometric angles. The only pieces of equipment necessary are a sharp no. 2 pencil and a 90-60 right triangle. By following the directions here and practicing the exercises in this chapter, you will soon be able to produce your own isometric projections or follow those made by others.

Making Isometric Drawings

Draw a circle with a dot in the exact center. Place the letter N at the top of the page to designate the direction of north. Look at Figure 4-1. (Do not let the dotted line confuse you; it is used only to determine the proper angle necessary to illustrate the sanitary system shown in the unbroken solid lines.)

Using a 90-60 right triangle (see Figure 4-2), square the short base with the right edge of your paper and draw line A through the center dot. This represents A, the north-south horizontal pipe.

Again using the 90-60 right triangle, square the short base with the left edge of your paper and draw line B through the center dot. This represents B, the east-west horizontal pipe.

To arrive at the C vertical line, square the short base of the 90-60 right triangle with the lower edge of the paper and draw line C using the long base as a straight edge. Connect line C with any of the horizontal lines, as desired.

To determine the placement of line E, divide the area equally between the horizontal line B and the dividing dotted line. Then draw line E east and west through the center dot. (This line is necessary to show the change in direction assumed by the 45-degree fittings of either a wye or 1/8 bend.) Repeat the same procedure, of course, for line D north and south. The lower portion of Figure 4-1 shows a simple isometric drawing of all three basic angles used in designing rough plumbing for any building.

Fittings Within an Isometric Drawing

The lines on isometric drawings represent pipe and fittings. Symbols are used to show the type and location of fixtures. The illustrations that follow show sixteen common fittings used with no-hub pipe. The symbols are the same, regardless of the type of pipe used.

Figures 4-3, 4-4, and 4-5 show typical isometric drawings. Each fitting in these drawings is numbered to correspond to a drawing of the same fitting shown below. Look at Figure 4-5. The horizontal twin tap sanitary tee (also known as an "owl fitting") is the same as fitting number 14 at the top of the page. This fitting connects two similar fixtures to the same waste and vent stack at the same level. In this case, it connects two lavatories.

To become proficient and knowledgeable about plumbing, learn how to make isometric drawings and know the symbols for each fitting.

Naming the Pipes

Basic to mastering plumbing work is learning the names and the various terms used. Master these plumbing terms and definitions now so that they are familiar when reviewing the details of installations later in this book. Without a good grasp of key terms, you won't understand how to design, lay out or install pipes and fittings.

Figures 4-6, 4-7, and 4-8 are three sanitary isometric drawings which depict many of the major terms of the basic drainage and vent

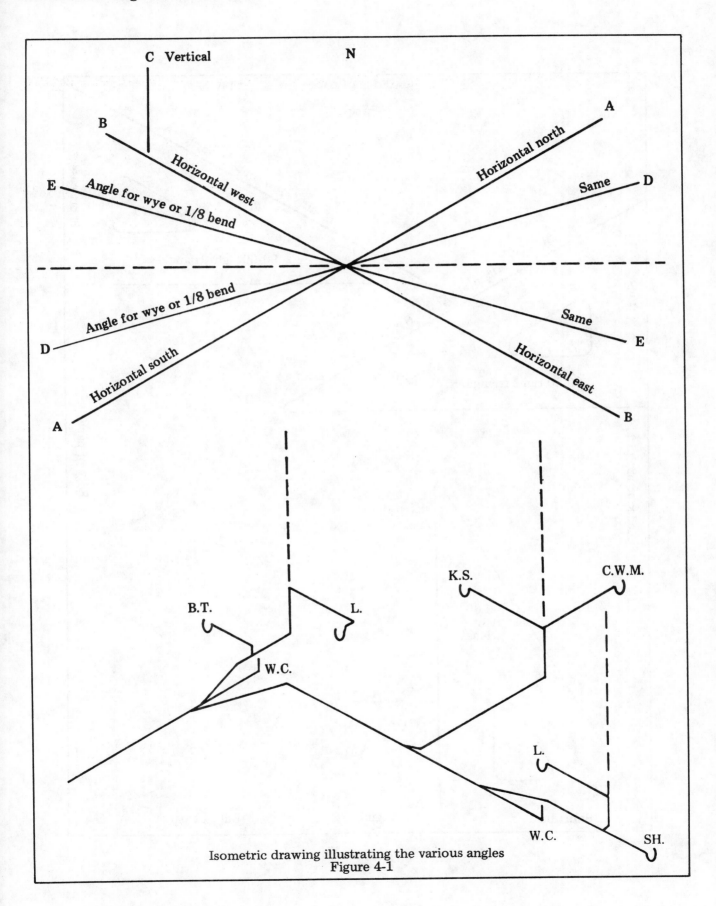

Isometric drawing illustrating the various angles
Figure 4-1

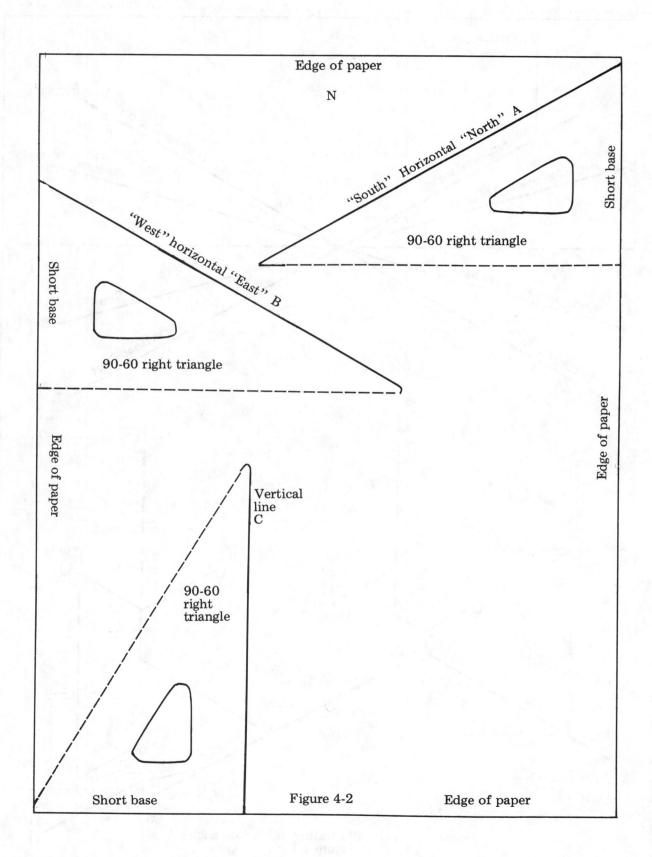

Figure 4-2

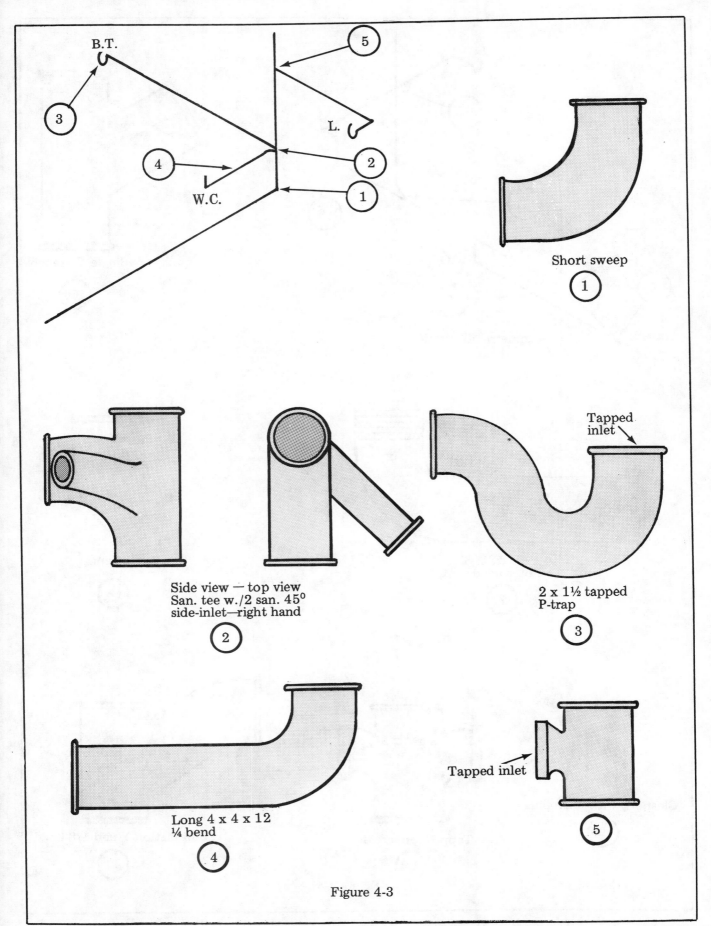

B.T.

③

④

W.C.

⑤

L.

②

①

Short sweep

①

Side view — top view
San. tee w./2 san. 45°
side-inlet—right hand

②

Tapped
inlet

2 x 1½ tapped
P-trap

③

Long 4 x 4 x 12
¼ bend

④

Tapped inlet

⑤

Figure 4-3

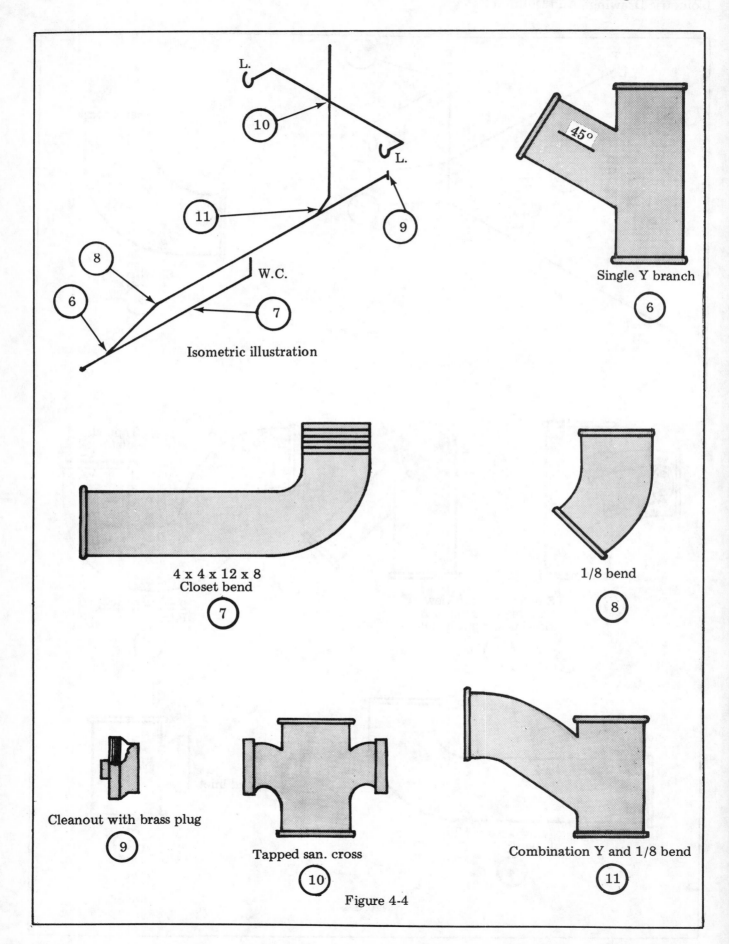

L.

10

11

L.

9

8

6

W.C.

7

Isometric illustration

45°

Single Y branch

6

4 x 4 x 12 x 8
Closet bend

7

1/8 bend

8

Cleanout with brass plug

9

Tapped san. cross

10

Combination Y and 1/8 bend

11

Figure 4-4

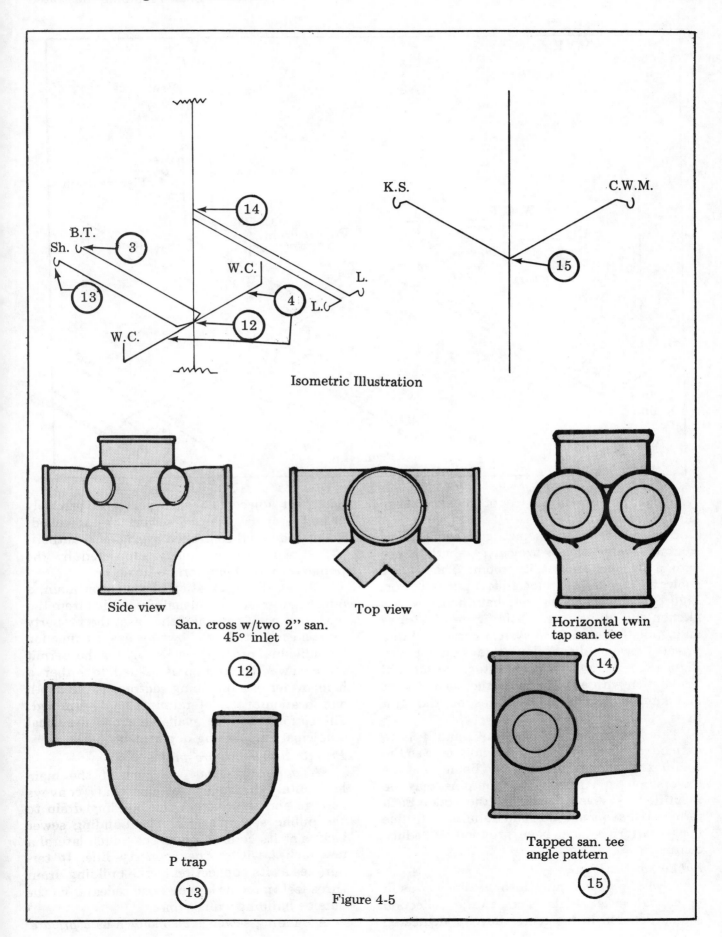

Isometric Illustration

Side view

Top view

San. cross w/two 2" san.
45° inlet

⑫

Horizontal twin
tap san. tee

P trap

⑬

Tapped san. tee
angle pattern

⑮

Figure 4-5

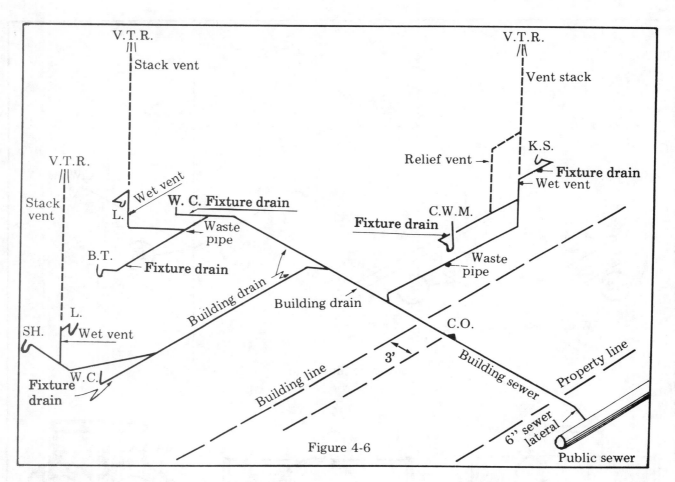

Figure 4-6

systems. The important parts of the system are labeled.

Figure 4-6 is an isometric drawing of a sanitary system for a typical two-bath house with a kitchen and utility room. This figure shows a *flat system* connected to a public sewer. Figure 4-7 shows a small one-bath house with a kitchen and utility room. This figure illustrates an installation on a *stack system* connected to a private sewage disposal unit (a septic tank). Figure 4-8 is a typical two-story residential installation with a full bath on the second floor and kitchen and utility facilities on the first floor.

The terms in Figures 4-6, 4-7, and 4-8 are defined in this chapter in simple language so the complex wording of the code can be more easily understood. A single section of pipe may be identified by several terms in the code. Each term may have a special code definition. In this book the terms have been grouped to reduce confusion.

Public Sewer (*Figure 4-6*)

A *public sewer* may also be known as a *municipal sewer*. This is a sewage collection system located in a street, alley, or dedicated easement adjacent to each parcel of **privately** owned property. Public sewers are **installed**, maintained, and controlled by the local **authori**ties. Their cost is usually supported by the public through some form of taxation.

During the installation of the sewer main, a 6-inch sewer lateral is usually extended from the main to a point several inches past the property line on each lot. This allows for easy connection by individual property owners. After the **permit** for a sewer connection is issued to either a homeowner or a plumbing contractor, the depth and location of the lateral must be determined. This information is available from the local municipal engineering department.

Private Sewer (*Figure 4-6*)

A *building sewer* is part of the **main** horizontal drainage system that conveys sewage and waste from the building drain to the public sewer lateral. The building sewer begins at its connection to the 6-inch lateral a few inches within the property line. It terminates at its connection to the building drain three feet (more or less in some codes) from the outside building wall or line.

A building sewer is also known as a *private*

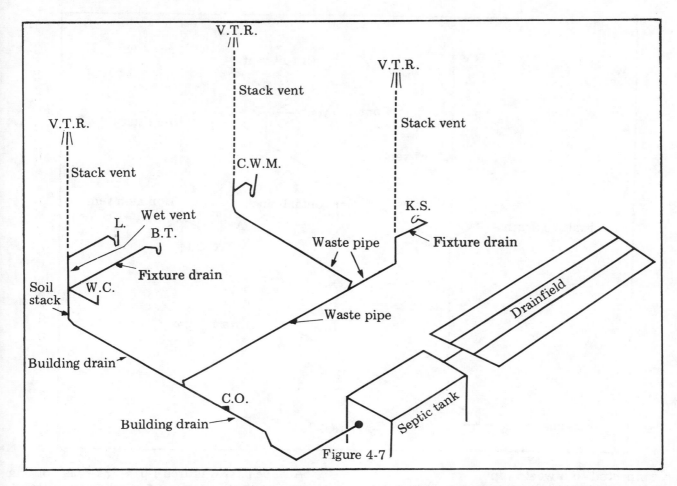

Figure 4-7

sewer, since it is not controlled directly by the public authority. Its installation and maintenance are the responsibility of the individual property owner.

Sanitary sewer is another name used for the building sewer in the plumbing code because it carries sewage that does not contain storm, surface, or ground water.

The building drain (Figures 4-6 and 4-7) is the main horizontal collection system, exclusive of the waste and vent stacks and fixture drains, and is located within the wall line of a building. It conveys all sewage and other liquid waste substances to the building sewer which begins three feet (more or less in some codes) outside the building wall or line.

It is also considered a main, as it acts as the principal artery to which other drainage branches of the sanitary system are connected.

The fixture drain (Figures 4-6, 4-7, 4-8) is the pipe from each fixture trap to the vent serving that fixture (referred to in some codes as fixture branch). It may connect directly to a vertical vent stack above the floor or, in the case of a shower or bathtub, to the horizontal wet vent section beneath the floor. See Figure 4-6. It may be referred to by plumbers as the "sink arm" or "lavatory arm" when it is located above the floor.

The fixture branch is the drainage pipe that conveys liquid waste from several fixtures to the junction of any other drain pipe (usually the building drain). Some codes refer to it as a fixture drain. In the code it is also called a waste pipe, as it does not carry the waste from water closets, bedpan washers, or fixtures having similar functions.

Another term for this pipe is the wet vent, because it often conveys liquid waste from plumbing fixtures, excluding water closets, to the building drain and also serves as a vent for these fixtures. Thus, there are waste pipes serving kitchen sinks, lavatories, bathtubs, showers, laundry trays, clothes washing machines, and similar fixtures.

Fixture drains can not carry body waste. Body waste as defined in the code includes only solid fecal matter; the code makes no mention of urine. Therefore, it is permissible for urinals to

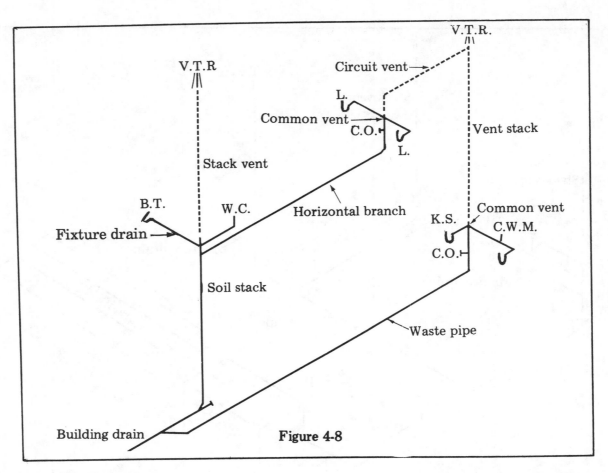

Figure 4-8

be connected to a waste pipe.

A *soil stack* (Figures 4-7 and 4-8) is the vertical section of pipe in a plumbing system that receives the discharge of water closets, with or without the discharge from other fixtures, and conveys this waste substance (usually) to the building drain. A building drain that receives waste from water closets is generally referred to as a *soil pipe*.

As the term is generally used, *stack* refers to any vertical pipe, including the soil, waste and vent piping of a plumbing system.

A *horizontal branch* (Figure 4-8) is the portion of a drain pipe extending laterally from a soil or waste stack that receives the discharge from one or more fixture drains.

A *common vent* (Figure 4-8) is a vertical vent that serves two fixture drains (fixture branches in some codes) that are installed at the same level. In Figure 4-8 this is the sink and clothes washing machine and the two lavatories installed in the second floor bathroom.

A *continuous vent* (Figure 6-8) is the vent vertical portion that is a continuation of the drain to which it is connected. It is also known as an *individual vent*.

A *stack vent* (Figures 4-6, 4-7, 4-8) is nothing more than the extension of a soil or waste stack (dry section) up and through the roof of a building.

A *vent stack* (Figure 4-8) is the vertical portion of a vent pipe. Its primary purpose is to provide circulation of air to and from all parts of a drainage system.

The *relief vent* (Figure 4-6) is sometimes referred to in the trade as a "re-vent." Its primary function is to provide circulation of air between the drainage system and the vent of a plumbing system.

The function of a *circuit vent* (Figure 4-8) is similar to that of a *branch vent*. A circuit vent serves two or more fixtures and rises vertically from between the last two fixture traps located on a horizontal branch drain. This vent must then connect to the vent stack.

A *vent system* consists of all the vent pipes of a building. It may be one or more pipes installed for the primary purpose of providing a free flow of air to and from a drainage system. This prevents back pressure or siphonage from breaking the water seals.

Questions

1- What is the purpose of isometric drawings?

2- What are the basic angles necessary to express the sanitary plumbing system in an isometric drawings?

3- What type of triangle is used to make an isometric drawing?

4- In an isometric drawing, what do the lines represent?

5- Why is it important to be able to name the various pipes in a drainage system?

6- How is a public sewer maintained?

7- How is a private sewer maintained?

8- What kind of waste can not be conveyed by a sanitary sewer?

9- The waste and vent stacks and fixtures drains are not a part of what system?

10- Name three fixtures in a residence which a waste pipe may serve.

11- Name three fixtures in a residence which a soil stack or soil pipe may serve.

12- In a drainage system, what constitutes a stack?

13- A common vent serves how many fixtures installed at the same level?

14- What name is given to the continuation of the waste and vent pipes through the roof?

15- What is the primary function of a vent stack?

16- A vent system is comprised of which pipes?

The Drainage System

Sizing Drain and Waste Pipes

In the drainage system, the maximum fixture unit load determines the size of the building sewers, building drains, and individual waste pipes. The *fixture unit load* is the total gallons expected to be discharged by the fixture per minute. Most lavatories, for instance, are rated at one fixture unit, which is equal to 7.5 gallons per minute or approximately one cubic foot per minute. This is generally accepted nationally as the standard flow rate for various types of plumbing fixtures.

The pipe sizes to use for building drain and waste pipes are the minimum sizes required by the code. Using larger waste pipes than required is not recommended. Here is the theory behind pipe sizing:

1) The normal operating capacity of waste piping is considered to be one-third full. The balance of the pipe's capacity is needed to prevent back-ups of waste into floor-mounted fixtures at peak volume.

2) Minimum sizes promote scouring action within the pipe and help avoid stoppages, especially in kitchen sink lines.

The most widely accepted model codes vary on sizing requirements in several respects:

1) Each code has its own fixture unit loads for the various plumbing fixtures.

2) Each code has a distinct way of applying these fixture units to waste pipe sizes.

Listed in Table 5-1 are several commonly used residential plumbing fixtures. Opposite each fixture is the fixture unit rating of each of three codes. Obviously, the codes reach about the same result. The difference of one F.U. is very minor and would not affect the sizing of waste pipes for most residential buildings.

	Fixture Unit Value		
Fixture Type	South Florida Building Code	Standard Plumbing Code	National Plumbing Code
Bathtub (with or without overhead shower)	2 F.U.	2 F.U.	2 F.U.
Lavatory	1 F.U.	1 F.U.	1 F.U.
Shower stall, domestic	2 F.U.	2 F.U.	2 F.U.
Kitchen sink, domestic	2 F.U.	2 F.U.	2 F.U.
Water closet, tank operated	4 F.U.	4 F.U.	4 F.U.
Clothes washing machine	4 F.U.	3 F.U.	3 F.U.
Laundry tray (1 or 2 compartments)	2 F.U.	2 F.U.	2 F.U.
Total Fixture Unit Values	17 F.U.	16 F.U.	16 F.U.

Table 5-1

Table 5-2 lists the trap sizes for various fixtures required by each model code. Again, with one exception, there is no difference between the three codes. This is important, as the trap size determines the fixture drain size in every case. A fixture drain cannot be smaller or larger than the fixture trap. For example, if a "shower stall, domestic" requires a 2-inch trap, the drain must also be 2 inches. It cannot be 1½ inches or 2½ inches.

	Trap Size In Inches		
Fixture Type	South Florida Building Code	Standard Plumbing Code	National Plumbing Code
Bathtub (with or without overhead shower)	1½ or 2	1½ or 2	1½ or 2
Lavatory	1¼	1¼	1¼
Shower stall, domestic	2	2	2
Kitchen sink, domestic	1½	1½	1½
Water closet, tank operated	3	3	3
Clothes washing machine	1½	2	2
Laundry tray (1 or 2 compartments)	1½	1½	1½

Table 5-2

Table 5-3 shows the sizing of waste pipes. The pipe sizes listed would normally be used in residential or light commercial jobs. The fall (pitch) of 1/8 inch per foot has been used, as this is the standard fall acceptable to most authorities.

Diameter of Pipe in inches	South Florida Building Code	Standard Plumbing Code	National Plumbing Code
1¼	1 F.U.	1 F.U.	1 F.U.
1½	4 F.U.	3 F.U.	3 F.U.
2	10 F.U.	6 F.U.	6 F.U.
3	28 F.U.	20 F.U.	20 F.U.
Total Fixture Unit Values	43 F.U.	30 F.U.	30 F.U.

Maximum number of fixture units that may be connected to any portion of the building drain or horizontal fixture drains, 1/8 inch fall per foot
Table 5-3

Note that the 1¼-, 1½-, and 2-inch fixture pipe drains listed in Table 5-3 for the Standard Plumbing and National Plumbing Codes apply to a minimum fall of ¼-inch per foot. Inspectors do not have precise measuring devices or the time to verify fall with total accuracy. Therefore, to avoid controversy with the installer, the inspector generally accepts a fall of from ⅛ to ½ inch per foot for all horizontal waste lines.

The same three model codes are compared in Table 5-3. The differences between the codes are minor and will usually have no effect on sizing waste pipes in most residential buildings.

Your code has more complete tables for many fixture types and larger drain pipe sizes. The tables in the code are generally used for complex computations for large commercial buildings. The beginning plumber is not concerned with these tables initially. Use the more technical portions of the code once you have mastered the basics and your knowledge of plumbing has grown.

Here is an example of how to size residential piping from a layout using the tables in this chapter. Figure 5-5 shows a simple two-bath house on a single level. The first step in sizing the pipe is to find the numbers of fixture units of required capacity. Use Table 5-1. Then size the drainage pipes using Tables 5-2 and 5-3.

Size the building drain at 3 inches from Table 5-3, as a 3-inch drain is sufficient under all three codes to carry the total fixture unit load generated (20 and 19 F.U.'s) to the building sewer. See Table 5-4.

	South Florida Building Code	Standard Plumbing Code	National Plumbing Code
1 - Bathtub	2 F.U.	2 F.U.	2 F.U.
1 - Shower	2 F.U.	2 F.U.	2 F.U.
2 - Lavatories	2 F.U.	2 F.U.	2 F.U.
2 - Water closets	8 F.U.	8 F.U.	8 F.U.
1 - Kitchen sink	2 F.U.	2 F.U.	2 F.U.
1 - Clothes washing machine	4 F.U.	3 F.U.	3 F.U.
Total Fixture Units	20 F.U.	19 F.U.	19 F.U.

Table 5-4

The horizontal fixture drains are also sufficient to meet all three model code requirements. See Table 5-3.

The minimum size of the building sewer is uniform in all codes: "the minimum size of a building sewer shall not be smaller than 4 inches."

The drainage and waste pipes for a two-bath house, as sized and illustrated in Figure 5-5, should meet the minimum model code requirements, regardless of the state or area where the house is constructed.

In Figure 5-6, the building sewer is shown connected to a septic tank. Although this could be a building sewer, it is not classified as such because its developed length does not exceed 10 feet. Therefore, the portion of the sewer pipe exceeding the 3-foot limit beyond the exterior wall of this particular building (more or less in some codes) may be considered part of the building drain and thus be sized at 3 inches.

Another perplexing situation which could cause some confusion is illustrated in Figure 5-7. A room addition and bathroom is to be installed at the side of an existing building. The plot plan shows the sewer from the new addition installed around the outside of the existing structure and connected to the existing sewer line in the front yard. The new sewer pipe is sized at only 3 inches. Normally it would be sized at the full 4 inches required for building sewers. The code has made an exception in this and similar cases, and the interpretation of the code reflects this:

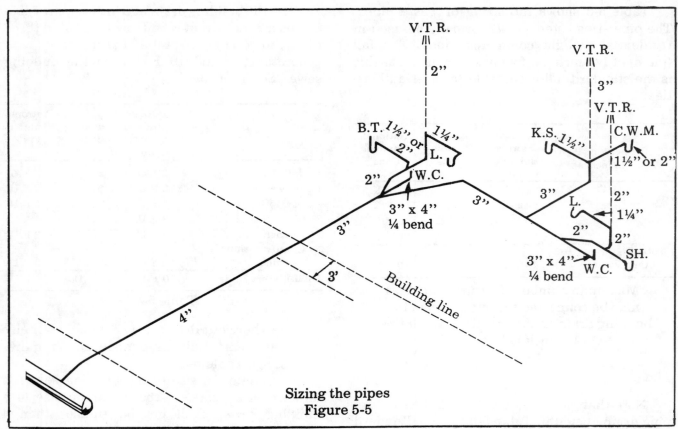

Sizing the pipes
Figure 5-5

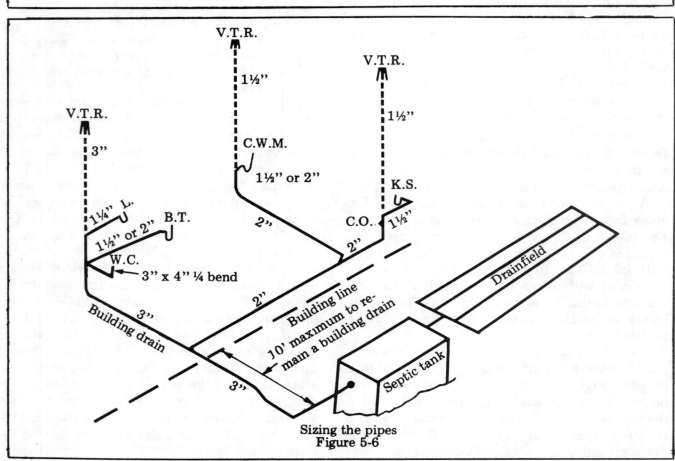

Sizing the pipes
Figure 5-6

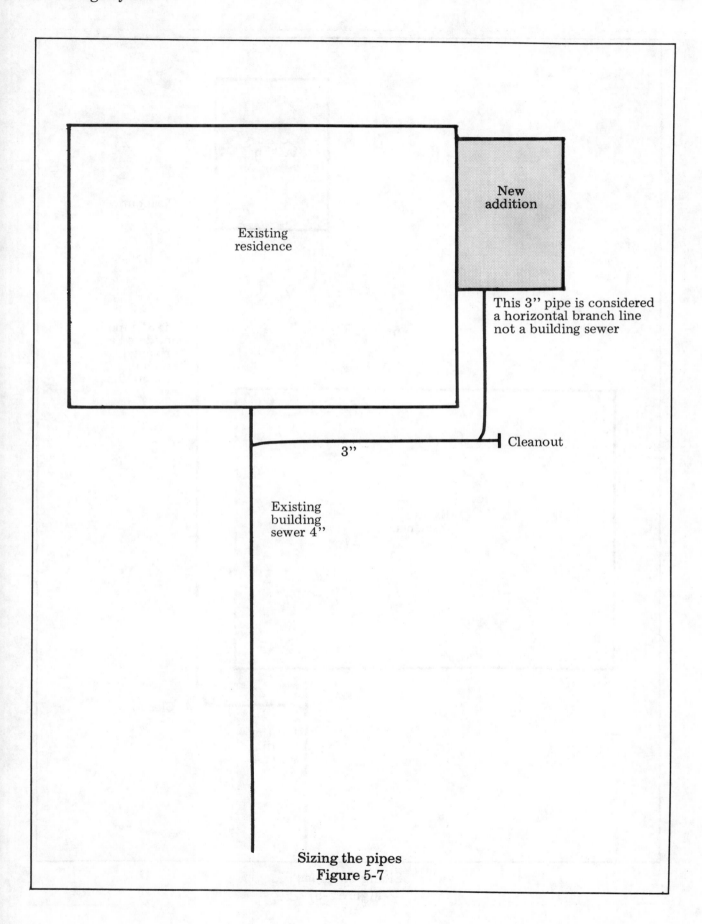

Sizing the pipes
Figure 5-7

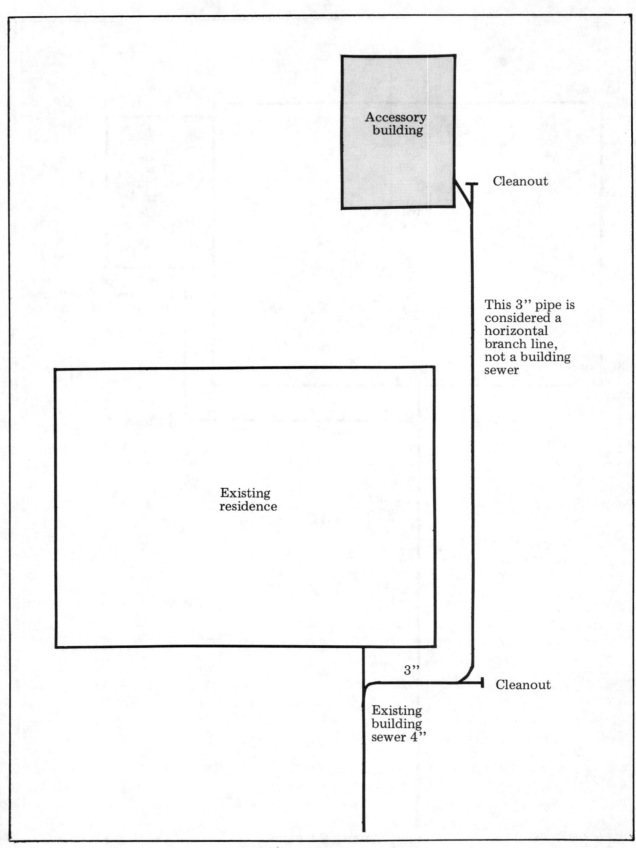

Sizing the pipes
Figure 5-8

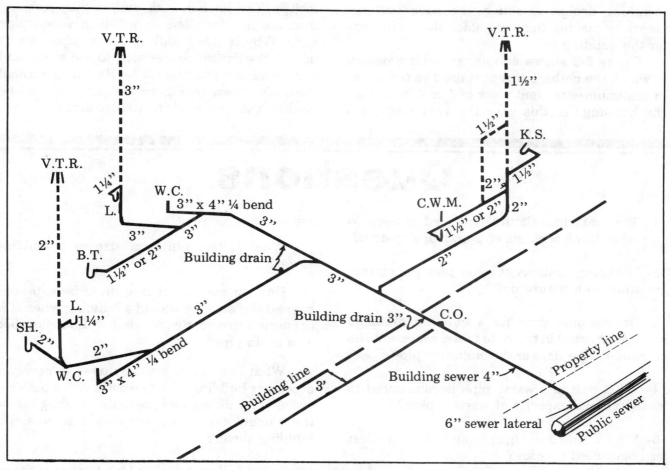

Sizing the pipes
Figure 5-9

For additions to residential buildings when soil and vent lines are inaccessible and it is necessary to install the sewer line outside and around an existing structure, such line shall be considered as a horizontal branch. This horizontal branch must connect into the existing building drain, in this case the building sewer.

A similar exception is also made by the code for accessory residential buildings (not commercial buildings) that are located on the same lot with an existing building having a single building sewer. See Figure 5-8.

The building drain in Figure 5-5 is sized at 3 inches. The minimum size building drain may be 3 inches for all buildings with one or two water closets. If a third bathroom were to be added to this or any building, the building drain would have to be increased to 4 inches at the junction of flow from all three water closets.

Look again at Figure 5-5. The fixture drain

(also known as a wet vent) is sized at 2 inches. Why? The code states that the minimum size vent (wet or dry) to serve a water closet is 2 inches.

The sizing of the waste pipe leading to the kitchen sink and clothes washing machine in Figure 5-5 also needs clarification. Table 5-3 shows that at 1/8-inch fall per foot, a 2-inch pipe may be used to convey up to 10 fixture units. The other two model codes permit 6 fixture units on this same pipe. Why is the 3-inch pipe used? Again, the code states that no kitchen sink receiving greasy waste can be installed on a cross installation with a pump discharge fixture (a clothes washing machine, for instance) of less than 2½ inches. The code also requires that each building have at least one minimum-size vent stack of not less than 3 or 4 inches extending through the roof.

Since both bathrooms have a 2-inch wet vent and since the code requires a 3-inch pipe up and through the sanitary tap cross, it is good plumbing design to satisfy the code requirements by making this vent stack the main vent for the building.

Figure 5-9 shows another plumbing design in which the bathroom vent is used as the main, or minimum-size, vent stack of 3 or 4 inches for the building. In this case the vent stack is 3 inches. The kitchen sink and clothes washing machine are installed on a 2-inch waste pipe here. Why is this legal? A 2-inch pipe can be used if the fixture connections to the waste and vent stack are at different levels and if a relief vent, as shown in Figure 5-9, is used on the clothes washing machine fixture drain.

Questions

1- The maximum fixture unit load is used to size what three sections of a drainage system?

2- How many gallons of water flow per minute are equal to a fixture unit?

3- Do the pipe sizes for a building drainage system, as listed in the code table, represent the *maximum* pipe sizes or the *minimum* pipe sizes?

4- What part of a waste pipe is considered to be the normal capacity of waste piping?

5- Why should drainage piping be the smallest size permitted by code?

6- What are the two primary differences in model plumbing codes in determining waste pipe sizes?

7- What is the minimum size of a building sewer?

8- How far may a building drain be extended beyond the exterior wall of a building when it is connected to a septic tank and still be classified as a drain pipe?

9- What size sewer must be used to serve an accessory building with one bath, when both the accessory building and the main building are on the same lot and are connected to a single building sewer?

10- When is it permissible for a kitchen sink and a clothes washing machine to be installed at different levels on a 2-inch waste stack?

The Vent System

Sizing The Vent System

Vents relieve the pressure that builds up as water from plumbing fixtures is discharged into the sanitary drainage system. These pipes are sized and arranged to provide the best possible relief for each fixture and for the system. The free flow of air within the vent system keeps the back pressure or siphoning action from destroying the fixture trap seal.

In rural areas where inspections are not required, or in municipalities where inspections are lax, inadequate sizing and arrangements of vent pipes make the following problems common:

- Plumbing fixtures drain slowly, as if a partial stoppage exists.

- Water closets need several flushes to remove contents from the bowl.

- Back pressure (called *positive pressure*) within the drainage pipes may, if strong enough, force sewer gases up and through the liquid trap seals and into the building.

- Plumbing fixtures located farther from a vent pipe than is permitted by code may, when the contents are released, siphon the liquid trap seal. (This action is known as *negative pressure*.)

Because of the importance of the free flow of air within the drainage system, in very cold climates where the danger of frost forming on the inside of the vent pipes exists, "increasers", or other devices, may be required. Frost is formed as the warm moist air flows up and out the vent pipes and makes contact with the frigid atmosphere. See Figure 10-31.

All the drain-waste-vent illustrations in this book, and in particular Figures 4-6, 4-7, and 4-8, show the close relationship between the various types of vent pipes and the roughing-in of the whole drainage system

The size of the vent pipe — like the size of its cousin, the soil and waste pipe — is determined by the maximum fixture unit load, the developed length of the vent pipe, type of plumbing fixtures to be vented, and the diameter of the soil or waste stack the vent pipe serves.

Here are the basic principles of venting as summarized from most codes:

For each building with a single building sewer receiving the discharge of a water closet, there must be at least one minimum size vent stack of not less than 3 or 4 inches extending through and above the building roof.

Where there is an accessory building or other building located on the same lot and sharing one common building sewer, the minimum size vent stack or stacks to serve each accessory building can be sized from Table 6-1. Note the exception that if a water closet is located in the accessory building the vent stack must be no smaller than 2 inches.

No wet or dry vent serving a water closet can be less than 2 inches in diameter.

The diameter of the vent stack cannot exceed the diameter of the soil or waste stack to which it connects. (For example, if the waste pipe for a kitchen sink is 2 inches, the vent stack cannot be 2½ inches or larger. It can be 2 inches or smaller so long as it complies with the fixture units listed in Table 6-1).

The diameter of an individual vent stack cannot be less than 1¼ inches or less than one-half the diameter of the drain to which it connects. Here are two examples: (1) If a waste pipe is 2 inches and serves one lavatory, the minimum size vent is 1¼ inches. (2) If a waste pipe is 3 inches, the minimum size vent is 1½ inches.

Table 6-1 compares two model codes and lists fixture units and vent pipe sizes for a single-family residence with simple plumbing installations. In Figure 5-5 the two bathrooms are served by a horizontal 2-inch wet vent. The kitchen sink and clothes washing machine are served by a single 3-inch stack vent. Note in Table 6-1 that there are differences between the two model codes in the number of fixture units permitted. These differences will not affect the sizing of the vent pipes shown in Figure 5-5.

The code book contains complete tables for fixture units, vent pipe sizes and vent lengths. You should use these as references to expand your knowledge for more complicated installations. But the tables in this book are adequate for most single-family residential plumbing.

Maximum Fixture Units Permitted on	South Florida Building Code	Standard Plumbing Code
a horizontal wet vent sized 2 inches	6	4
a horizontal wet vent sized 3 inches	16	18
a vertical dry vent sized 1¼ inches	1	2
a vertical dry vent sized 1½ inches	6	10
a vertical dry vent sized 2 inches	24	20
a vertical dry vent sized 3 inches	72	60

Table 6-1

Stack Venting

Stack venting is commonly used where plumbing fixtures are located on the same wall or where plumbing fixtures are located back to back. Each fixture drain must connect independently to the stack. This type of plumbing installation is very simple and can be used either in a single floor house or on the second floor of a two-story house. The single main pipe receives the liquid waste and also vents the group of fixtures as well.

The stack arrangements shown in Figures 6-2 and 6-3 are acceptable under most plumbing codes. Figure 6-2 illustrates a typical group arrangement: a water closet, lavatory, bathtub and kitchen sink. Figure 6-3 shows two bathrooms back to back, consisting of two water closets, two lavatories and two bathtubs.

Vertical Combination Waste and Vent Stack

The vertical combination waste and vent stack is a unique one-pipe system which both vents and receives waste from certain plumbing fixtures. Although it is used frequently in highrise or commercial buildings, it can be used to advantage in single-family residences having more than one floor.

Low-unit-rated fixtures (like lavatories, bidets, bathtubs, showers and sinks) can use a combination waste and vent stack. If this arrangement is used, remember the following restrictions:

• The stack must be vertical and must extend full-size through the roof. A 3-inch combination waste and vent stack cannot be reduced above the topmost fixture.

• The stack cannot be connected to another existing vent pipe below the roof.

• Check the maximum number of fixture units and pipe sizes carefully. Plumbing codes vary considerably in this area.

• A kitchen sink can not be placed on a two-inch combined waste and vent stack. To illustrate: (1) A kitchen sink connected to a two-inch stack on the first floor cannot be extended to a second floor bathroom and receive the waste from, say, a lavatory. (2) A lavatory on the first floor and a lavatory on the second floor located directly over one another could both be served by a combination waste and vent pipe. (Since they are isolated from the rest of the bathroom fixtures, it would be difficult and expensive to connect each lavatory waste pipe to the existing waste pipes across the room.)

Figure 6-4 shows a vertical combination waste and vent system installation that is acceptable under most codes.

Wet Venting

This special method of venting is generally used in dwellings. Wet venting provides adequate trap protection for certain plumbing fixtures. A wet vent can be vertical or horizontal, as illustrated in Figures 5-5 and 5-9. It can be used on a stack or a flat plumbing installation. This unique venting system is probably used more often than any other type and is generally accepted by all model plumbing codes.

Since it is a single piping system, a wet vent is economical and can serve to vent several adjoining plumbing fixtures located on the same floor level. See Figure 6-5.

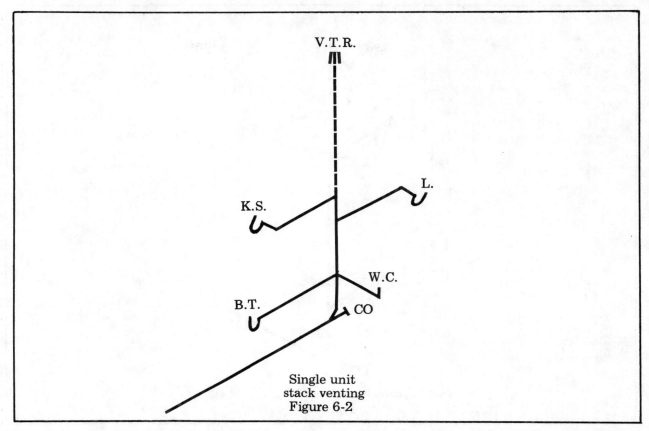

Single unit
stack venting
Figure 6-2

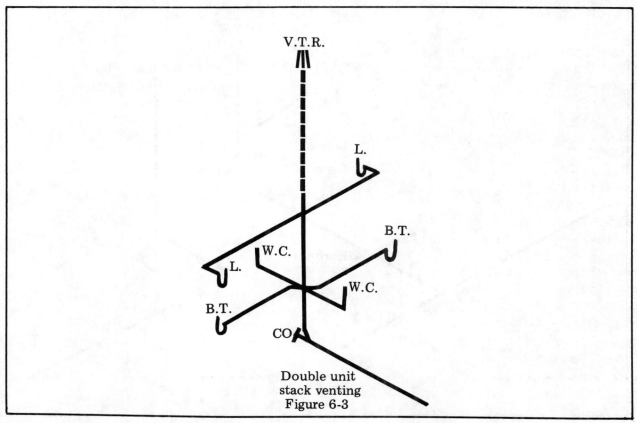

Double unit
stack venting
Figure 6-3

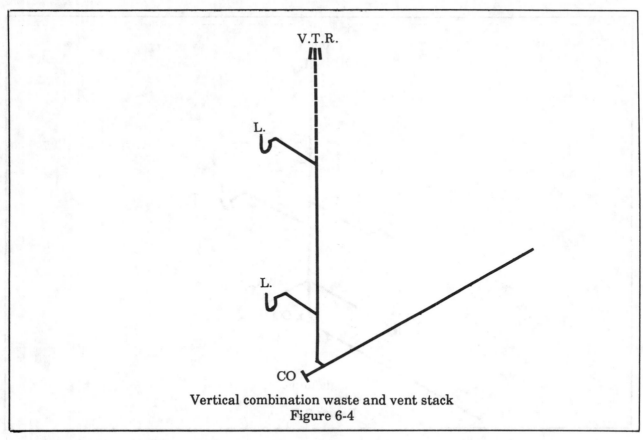

Vertical combination waste and vent stack
Figure 6-4

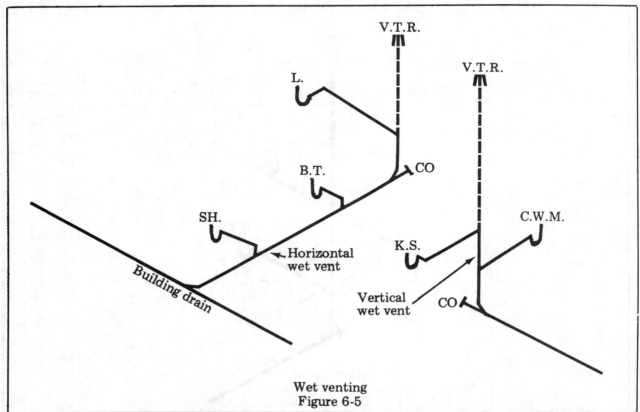

Wet venting
Figure 6-5

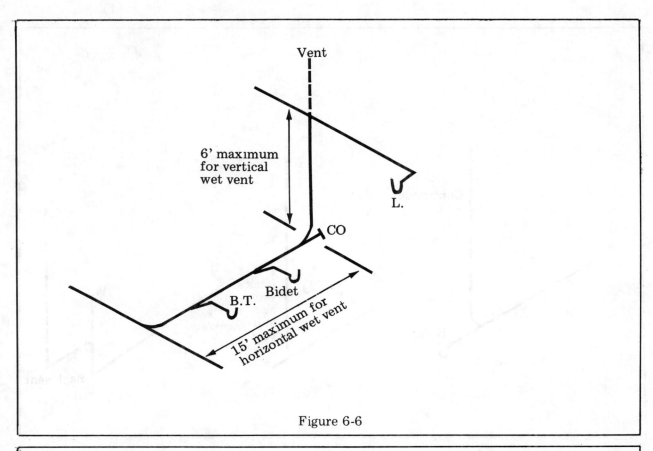

Figure 6-6

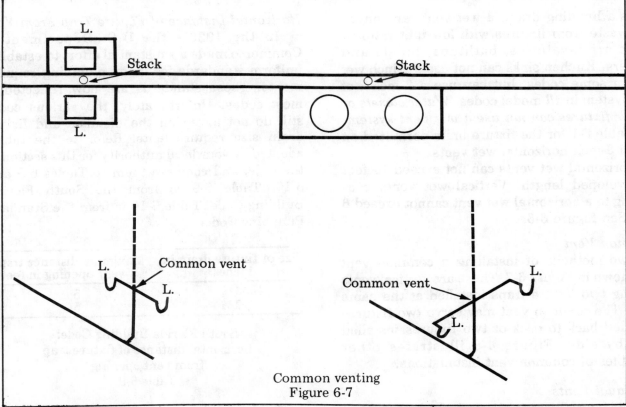

Common venting
Figure 6-7

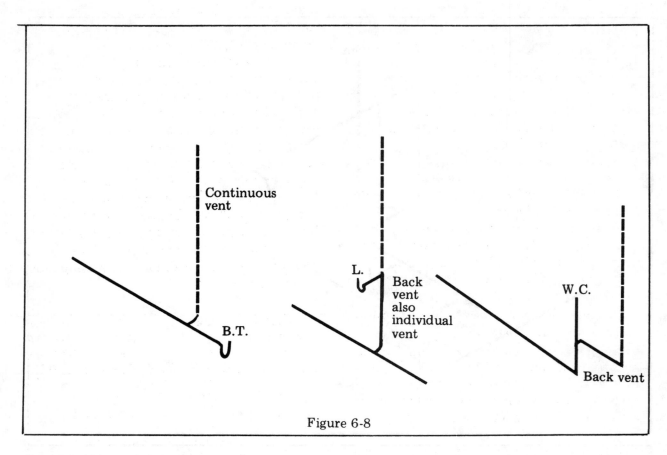

Figure 6-8

As a building drain, a wet vent can convey only waste from fixtures with low unit ratings. These are lavatories, bathtubs, bidets and showers. Kitchen sinks can not use a 2-inch wet vent in some codes, but can use a 3-inch wet vent system in *all* model codes. *Water closets or similar fixtures can not use a wet vent system.* See Table 6-1 for the fixture units permitted on 2- and 3-inch horizontal wet vents.

Horizontal wet vents can not exceed 15 feet in developed length. Vertical wet vents connecting to a horizontal wet vent cannot exceed 6 feet. See Figure 6-6.

Common Vent

Two methods of installing a common vent are shown in Figure 6-7. These are vertical vents serving two fixture traps installed at the same level. The common vent may serve two fixtures installed back to back or two fixtures installed side by side. Figure 4-8 illustrates other examples of common vent installations.

Individual Vents

This type of vent is also called a *back vent* or a *continuous vent*. Figure 6-8 illustrates these common installations.

Horizontal Distance of Fixture Trap From Vent

In the 1920's the U.S. Department of Commerce made a substantial effort to establish uniform standards for many plumbing requirements. Those standards are now reflected in most codes. Unfortunately, the various codes still do not agree on the distance and fixture drain size requirements. Refer to the tables adopted by your local authority for this section of the code, and compare them to Tables 6-9 and 6-10. Table 6-9 is from the South Florida Building Code. Table 6-10 is from the Standard Plumbing Code.

Size of fixture drain in inches	Maximum distance trap to vent opening in feet
1¼	5
1½	5

South Florida Building Code:
horizontal distance of fixture trap
from vent opening
Table 6-9

Table 6-9 is comparatively simple. It has been adopted by local authorities in many states and has proved to be economical and satisfactory. The maximum distance from trap to the vent is 5 feet regardless of the trap size. The slope per foot of the fixture drain and the type of fitting used to connect the fixture drain to the vent are not considered.

Note carefully that the maximum distance from the trap to the vent in Table 6-10 depends on three factors: 1) the size of the fixture drain 2) the size of the trap 3) the fall per foot of the fixture drain.

Remember, *where there is an apparent conflict between two or more codes, it is always best to follow the tables provided in the code adopted by your local authority.*

Some principles for horizontal runs between the fixture trap and the vent are recommended and acceptable by all codes. They are summarized below and discussed in the final section of this chapter.

- The closer the trap to the vent on a minimum slope, the better.
- Every fixture trap must be protected against siphonage and back pressure and must be provided with a vent piping system that permits free admission or emission of air under normal intended use.

Measure the developed length of a fixture drain from the crown weir of a fixture trap to the vent pipe, as shown in Figure 6-11. The measurement must include offsets and turns and must be within the limits prescribed in the code. Table 6-9 and Figure 6-11 show how the developed length is figured.

The fixture drain may be in some instances exceed the limits in Tables 6-9 and 6-10 because of the location of a fixture within a bathroom, kitchen, or utility room. When this occurs, install a relief vent to maintain the limits in the code you are following. For example, a lavatory must be installed with an 8-foot fixture drain from the nearest vent. The drain pipe is 1¼ inches and is installed with ¼ inch per foot slope. The installation methods in Figure 6-12 are taken from Tables 6-9 and 6-10 of the two model codes. One of these methods must be used for this arrangement to be acceptable.

Size of fixture drain (inches)	Size of trap (inches)	Fall per foot (inches)	Maximum distance trap from vent
1¼	1¼	¼	3'6"
1½	1¼	¼	5'
1½	1½	¼	5'

Standard Plumbing Code:
horizontal distance of fixture trap
from vent opening
Table 6-10

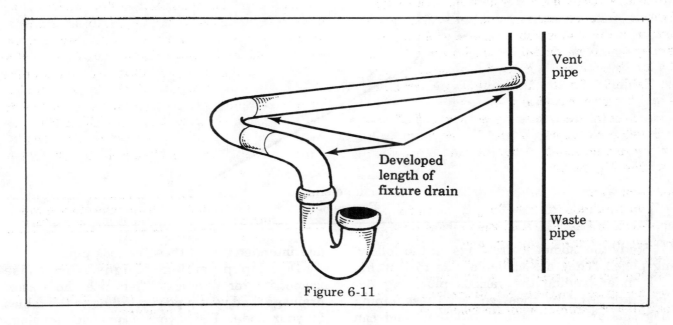

Figure 6-11

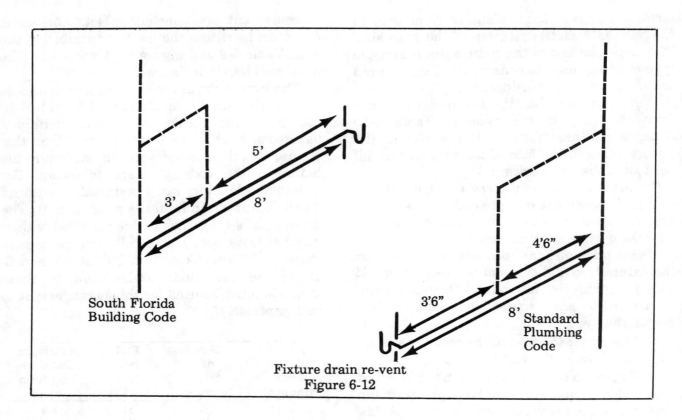

South Florida
Building Code

5'

3'

8'

4'6"

3'6"

8' Standard
Plumbing
Code

Fixture drain re-vent
Figure 6-12

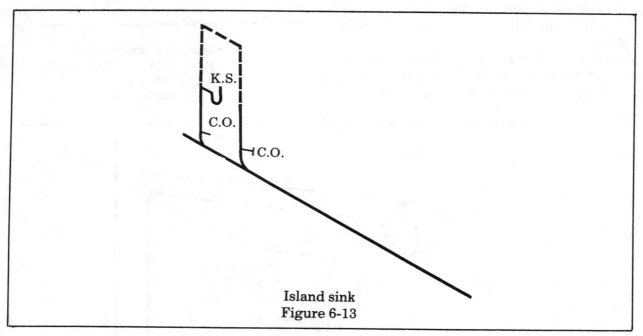

K.S.

C.O.

C.O.

Island sink
Figure 6-13

Study the isometric drawings on the following pages. They are valuable aids to learning and understanding the various plumbing designs approved by most codes. These piping diagrams are for simple installations and can help you see how waste and vent pipes are related. The pipes in these drawings are not sized. The illustrations are to clarify the location and intended use of these various pipes.

Develop plumbing designs from these isometrics for practice. Then size the waste, drainage, and vent pipes according to the tables in your code. Have your isometric drawings checked for accuracy by some qualified person such as your employer, job foreman, or the plumbing inspector who visits the job.

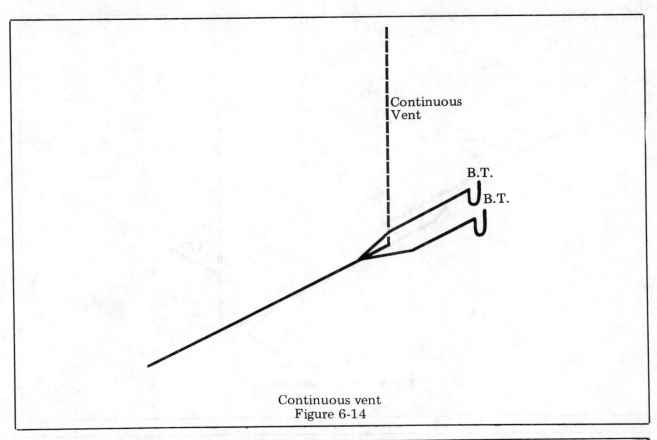

Continuous vent
Figure 6-14

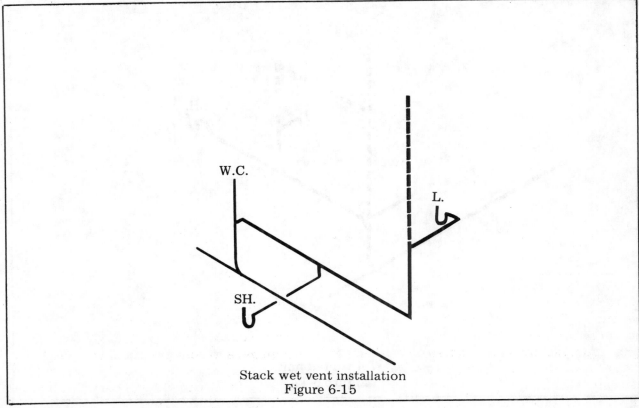

Stack wet vent installation
Figure 6-15

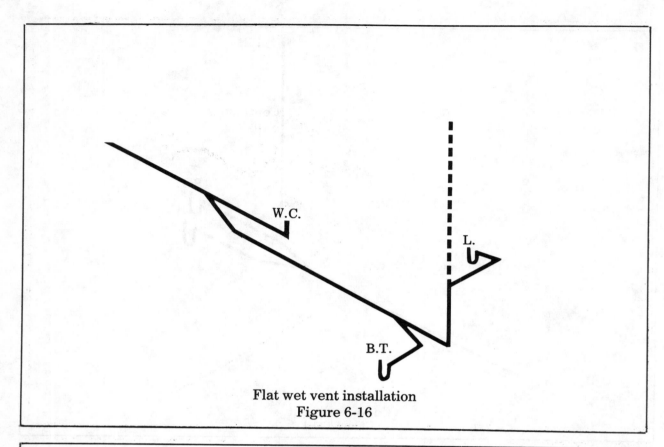

Flat wet vent installation
Figure 6-16

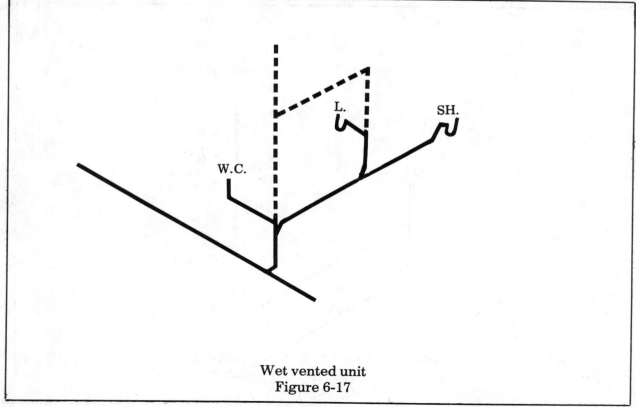

Wet vented unit
Figure 6-17

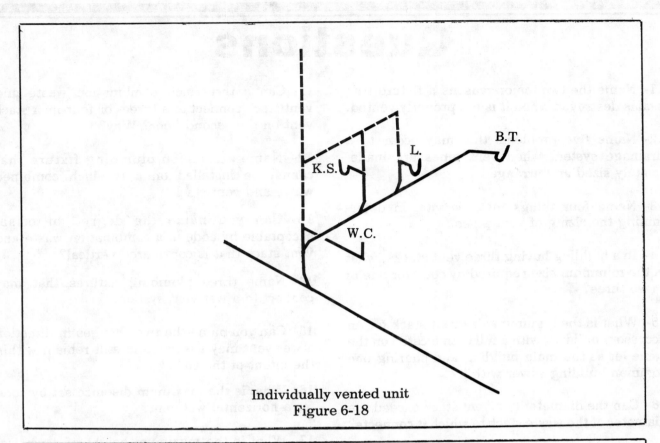

Individually vented unit
Figure 6-18

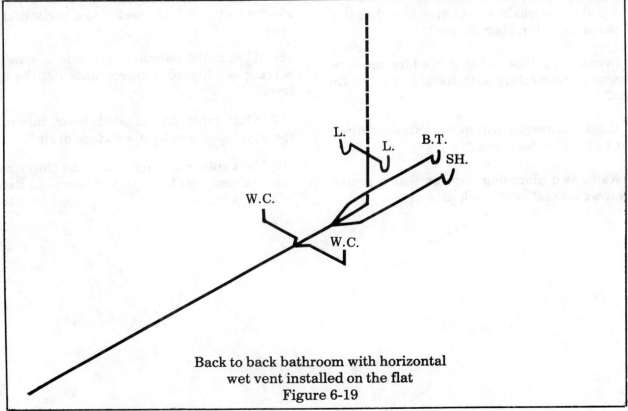

Back to back bathroom with horizontal
wet vent installed on the flat
Figure 6-19

Questions

1- Name the two major reasons a fixture trap seal is destroyed when it is not properly vented.

2- Name five problems that may occur to a drainage system when vent pipes are inadequately sized and arranged.

3- Name four things that are vital in determining the sizing of vent pipes.

4- In a building having three vent stacks, what is the minimum size required by code for one of these three?

5- What is the minimum size vent stack for an accessory building with a full bath located on the same lot as the main building and sharing one common building sewer with it?

6- Can the diameter of a vent stack exceed the diameter of the waste stack to which it connects?

7- What is the smallest vent pipe in inches that can serve any plumbing fixture?

8- In stack venting, what procedure must be followed in connecting each fixture drain to the stack?

9- Stack venting a group of fixtures fulfills what two major functions?

10- Name two plumbing fixtures that a combination waste and vent stack may serve.

11- Can a three-inch combination waste and vent stack connect to a three- or four-inch stack vent on the second floor? Why?

12- Name a common plumbing fixture that cannot be installed on a two-inch combined waste and vent stack.

13- Can you name the degree of offset acceptable by code in a combination waste and vent stack that is considered vertical?

14- Name three plumbing fixtures that may connect to a wet vent system.

15- Can you name the two changes in direction a wet vent may assume and still remain within the intent of the code?

16- What is the maximum distance set by code for a horizontal wet vent?

17- What is the maximum height of a vertical wet vent when it is connected to a horizontal wet vent?

18- What is the definition given to a vent pipe serving two fixture traps connected at the same level?

19- What must be included when measuring the developed length of a fixture drain?

20- Most authorities agree that the closer a trap is to the vent, the better it will serve the fixture. Why is this?

Traps

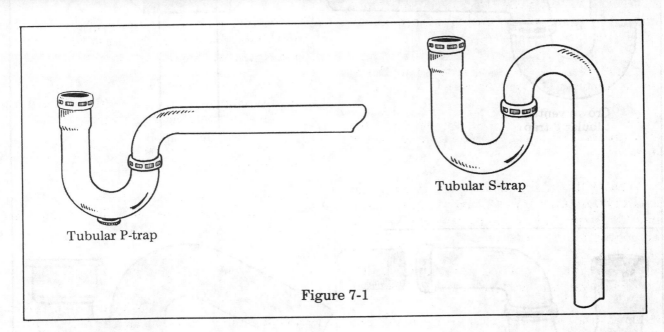

Tubular S-trap

Tubular P-trap

Figure 7-1

Fixture Traps

Plumbing fixtures connected directly to the sanitary drainage system must be equipped with a water seal trap. Each fixture must have its own trap. Fixtures such as water closets have integral traps fashioned within the fixture body.

The trap is designed to provide a liquid seal which prevents drainage system odors, gases and even vermin from entering the building at fixture locations. The trap must provide this liquid seal protection without restricting the flow of sewage or other waste.

The fixture trap design most generally used for lavatories, kitchen sinks, laundry trays, and the like is illustrated in Figure 7-1. It is called a *P-trap* because its shape resembles the capital letter *P*. The *S-trap* (also shown in Figure 7-1) looks like the capital letter *S*.

The code has placed many restrictions and limitations on the P-trap's use because of its unique importance in protecting human health. These restrictions appear in various parts of the code and are listed below.

- Fixture traps must be self-cleaning. This means that the interior of the trap can not

have anything that will retain hair, lint, or other foreign substances.

- No trap outlet can be larger than the fixture drain to which it is connected. In other words, a 1½-inch trap cannot be connected to a 1¼-inch drain with a reducer.

- No trap can be used which depends on the action of movable parts to retain its seal.

- Traps prohibited by most model codes are bell traps, running traps, crown vented traps, pot traps, ¾ S-traps, full S-traps, and traps with slip-joint nuts and washers on the discharge side of the trap above the water seal. See Figure 7-2. You may wonder how full S-traps may still be purchased from most plumbing supply houses when they are prohibited. Many older homes used full S-traps when they were acceptable. Today, they may be used for replacement purposes only, never for new construction.

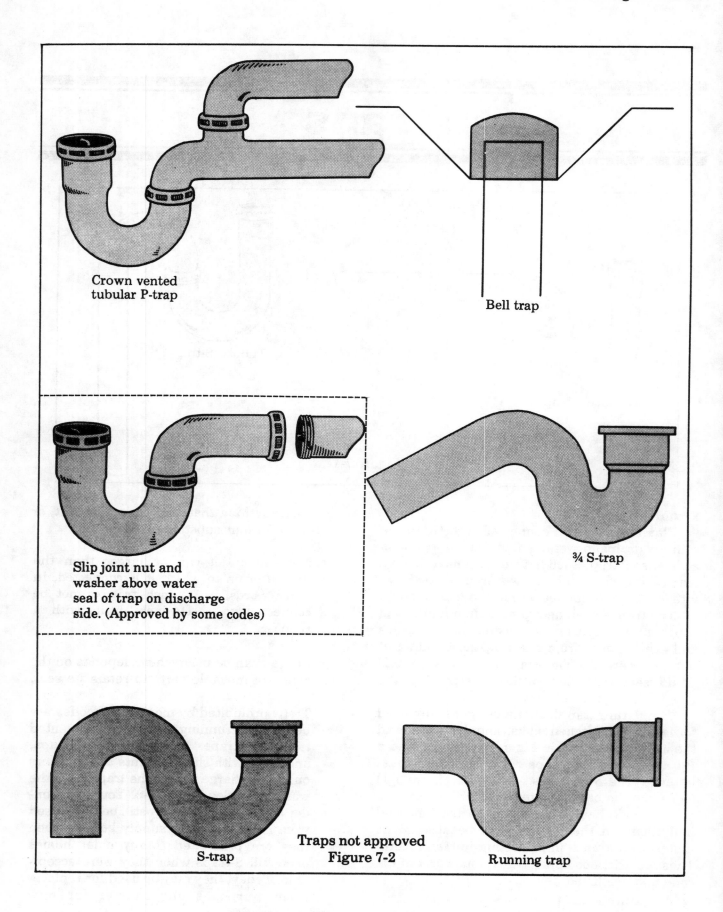

Crown vented
tubular P-trap

Bell trap

Slip joint nut and
washer above water
seal of trap on discharge
side. (Approved by some codes)

¾ S-trap

S-trap

Traps not approved
Figure 7-2

Running trap

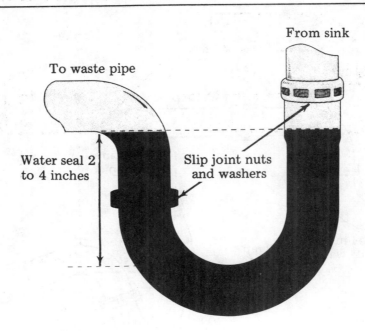

Chrome traps may be of one-piece cast brass, but more often they are of a tubular construction and have slip joints and washers. A slip joint is permitted on the fixture inlet part of the trap and also on the outlet side, provided it is within the trap water seal. An old slip joint and washer located *above* the water seal on the discharge side of a trap can permit sewer gases to enter a building. Therefore, this type of trap is prohibited. Plastic traps may be used for lavatories and sinks where the drainage system is plastic. Plastic pipe can resist heat damage at temperatures up to 180°F. But use plastic traps and tailpieces for kitchen sinks with caution. Many plastic traps and tailpieces in this installation have deteriorated short of their intended life because residents pour boiling water directly into the sink. Codes may prohibit plastic traps and tailpieces in kitchens in the future.

Figure 7-3

- Each fixture trap must have a water seal of not less than two inches or more than four inches. See Figure 7-3.

- All traps must be installed level in relation to their water seals. See Figure 7-4. This is necessary to prevent negative action or self-siphonage from taking place.

- Each plumbing fixture must be separately trapped by a water seal trap with the following exception: water closets or similar fixtures having integral traps cannot be separately trapped. See Figure 7-5.

- Sinks, laundry trays or similar fixtures having two or three compartments may be connected to a single trap with a continuous waste (providing the compartments are adjacent to one another and one compartment is not more than 6 inches deeper than the other). See Figure 7-6.

- No fixture can be double trapped. This means that the liquid waste discharged from a fixture may not go through one trap and then a second trap before discharging into the building stack or drain. The water closet in Figure 7-5 is double trapped as shown and, therefore, would not be acceptable.

- Some codes permit two or three single compartment sinks or lavatories to use a

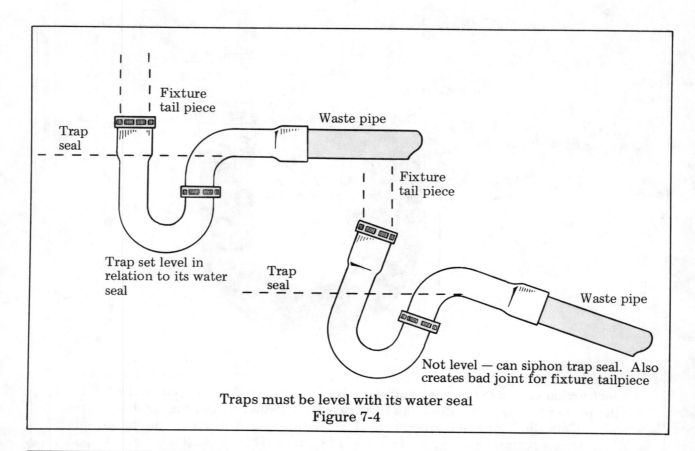

Fixture
tail piece

Trap
seal

Waste pipe

Trap set level in
relation to its water
seal

Fixture
tail piece

Trap
seal

Waste pipe

Not level — can siphon trap seal. Also
creates bad joint for fixture tailpiece

Traps must be level with its water seal
Figure 7-4

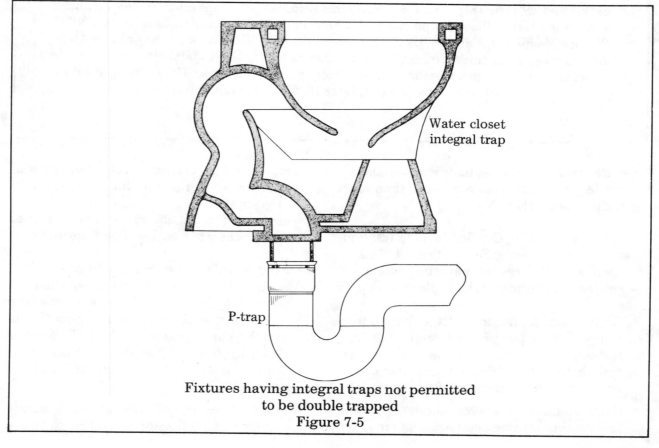

Water closet
integral trap

P-trap

Fixtures having integral traps not permitted
to be double trapped
Figure 7-5

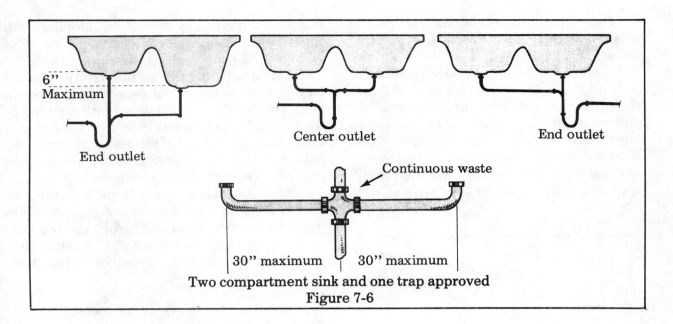

Two compartment sink and one trap approved
Figure 7-6

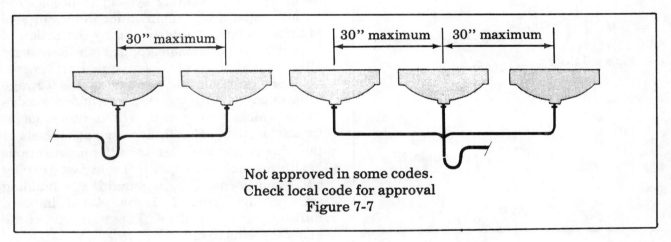

Not approved in some codes.
Check local code for approval
Figure 7-7

single trap, provided that (a) where three fixtures are installed the trap must be located on the center fixture, (b) the fixtures must be set at the same level, and (c) the fixtures must not be spaced more than 30 inches center to center. See Figure 7-7.

- There is a maximum allowable vertical drop from a fixture waste outlet to the trap water seal. This maximum drop further prevents self-siphonage of the fixture trap water seal. The shorter the distance between these two points, the more efficient the fixture trap will be. The longer the fixture tailpiece, the greater the velocity of the waste water as it rushes out a fixture drain. Excess velocity can siphon the trap seal. For sinks, lavatories, showers, bathtubs and all similar fixtures, the vertical

drop (tailpiece) cannot exceed 24 inches (18 inches in some codes) as shown in Figure 7-8. The vertical drop of the pipe

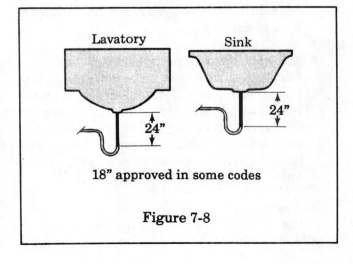

18" approved in some codes

Figure 7-8

serving floor connected fixtures with integral traps, water closets and similar fixtures cannot exceed 24 inches as shown in Figure 7-9. This is generally acceptable by most codes. Of course, a water closet is designed so that the contents can be siphoned with each flush. When a water closet is flushed, the trap seal is lost, and it would remain so if the trap seal were not automatically restored by the refill tube in the flush tank or flush valve.

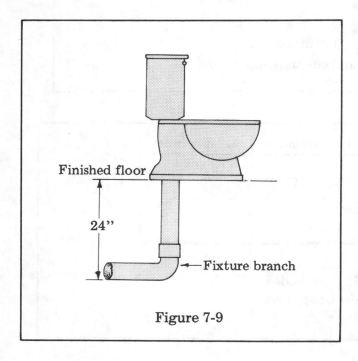

Figure 7-9

- Materials for concealed fixture traps, as for bathtubs, showers and bidets, must be cast-iron, cast brass or lead. In a plastic system, plastic traps can be used. Concealed fixture traps must not be equipped with cleanouts.

- Materials for exposed fixture traps or otherwise accessible traps (except for fixtures with integral traps) must be cast-iron, cast brass, lead, 20 or 17 gauge chrome brass or copper. Plastic traps may be used in a plastic system. Exposed fixture traps may be equipped with cleanouts.

Drum Traps

Drum traps were commonly used in the installation of bathtubs and lavatories not many

years ago. Plumbers should be familiar with drum traps, for they are still used in older buildings. A drum trap consists of a cylindrical metal shell with an inlet for the fixture near the bottom and a waste outlet near the top. The top has a removable screwed cover with a raised or countersunk head. It was installed in the fixture waste line, and the top protruded through the finished floor for easy access. A scatter rug generally concealed it from view.

Today, the use of drum traps is limited by most codes to special fixtures designed for drum traps. Drum traps must be approved by the local authority before installation and must have a minimum diameter of 4 inches. See Figure 7-10.

Building Traps

Before sanitary systems were properly designed with a venting system to protect the fixture traps, it was common for the water seal in these traps to be destroyed by the action of back pressure or siphonage, or both. Rats were able to travel freely from one building to another. Decomposing sewage in the sewage collection system generated gas and offensive odors which were released into buildings at fixture locations. Health department officials of that day recognized this condition as a serious health menace to people living in fast-growing cities and towns. They required that a building trap, as in Figure 7-11, be placed in each building drainage line. These proved to be generally effective.

Building traps at that time provided a secondary safeguard to keep rats, vermin, sewer gases and odors out of a building. The individual fixture traps provided the primary safeguard. The building trap was deemed a necessity until fairly recently. Today most model codes do not require — or actually prohibit — the installation of a building trap in a building drainage line. Refer to your local code for its building trap requirements.

Broken Trap Seals

Trap seals may be broken in the following five ways.

Wind Effect - This is one of the least likely ways a trap seal may be broken. It can happen when the vent pipes are subject to strong upward or downward air currents. The pressure or suction created in the stack may cause the

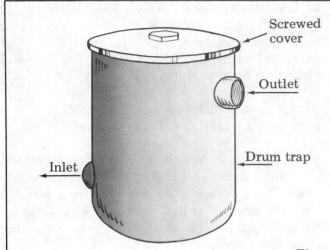

Screwed
cover

Outlet

Drum trap

Inlet

Today the use of drum traps is limited by most codes to special fixtures designed for drum traps. Drum traps must be approved by the local authority before installation and must have a minimum diameter of 4 inches.

Figure 7-10

Formerly a building trap had to be placed in each building drain line. Building traps were considered necessary until recently. But today most codes do not require, or actually prohibit, building traps in building drainage lines.

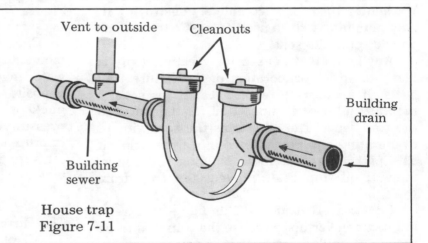

Vent to outside   Cleanouts

Building
drain

Building
sewer

House trap
Figure 7-11

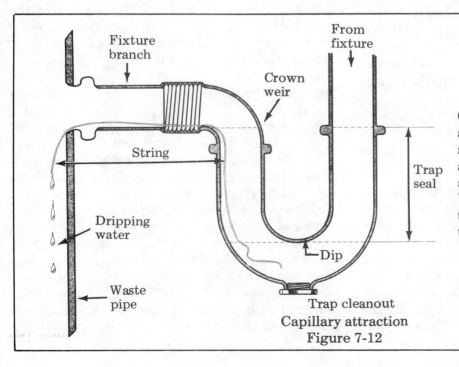

Fixture
branch

From
fixture

Crown
weir

String

Trap
seal

Dripping
water

Dip

Waste
pipe

Trap cleanout
Capillary attraction
Figure 7-12

Capillary attraction can break a trap seal when a length of string enters the fixture drain and reaches through the trap seal into the waste stack. Water follows the string out of the trap into the stack, destroying the trap seal.

water to rise or fall within the trap. Each time it rises within the trap, a small amount can spill into the waste pipes and be lost, thus weakening the trap seal. The trap seal may then be vulnerable to lesser back pressure in a drainage system, allowing sewer gases to penetrate the weakened seal and enter the room at the fixture location.

Evaporation - Evaporation may reduce the depth of the water in a trap seal when a fixture is not used for a long time. When the trap seal is thus weakened, back pressure can break through and allow sewer gases to enter the room at the fixture location. Evaporation in the fixtures can occur when one or more bathrooms or fixtures is not used at least occasionally. Low humidity will tend to evaporate the water from the unused trap or traps faster than usual, weakening the seal or completely destroying it. Any plumbing fixture used at least once a week should retain its seal.

Any home that is used for vacations only is likely to suffer evaporation of the fixture trap seals. If a house is to be vacated for several months, substitute mineral oil for the water in the trap seals. The oil retains the seal without evaporating and prevents vermin from using it as a pathway to enter the building. In winter, mineral oil trap seals prevent the trap from freezing.

Capillary Attraction - This happens rarely, but it can nevertheless cause the trap seal to be broken. Capillary attraction occurs when, for instance, a length of string enters the fixture drain and reaches down into the trap seal and on into the waste stack. The water from the trap seal will follow the string into the waste pipe and leak away, thus destroying the trap seal. See Figure 7-12.

Trap Siphonage - If negative pressure develops within the fixture drain, the fixture trap seal can be siphoned. This may be caused by any of the following 3 conditions:

1) A fixture with a high unit rating is installed over a fixture with a low unit rating. For example, a water closet is installed on the second floor of a residence and a lavatory on a lower floor; both use the same waste pipe. When the water closet is flushed, a large amount of water is sent past the waste opening of the lavatory. A vacuum is formed, pulling the water from the lavatory trap. See Figure 7-13.

2) A fixture is installed creating a ¾ S-trap in the fixture drain. When the fixture is full and the waste water is released, the water rushing through the trap and drain pipe may carry the trap water with it. Thus, the trap will not retain enough water to properly reseal. This is a true siphon action, since the siphonage is caused by the differences in the weights of two columns of water. See Figure 7-14.

3) A fixture installed on a long unvented horizontal drain line can also siphon the water from the trap seal. This is not considered a true siphon, but it has a similar effect on the fixture trap. In this installation, water rushing from the fixture into the trap and drain pipe builds up enough velocity to carry the trap water with it. Thus, the seal is broken. See Figure 7-15.

Back Pressure - Siphoning and water momentum tend to empty a fixture trap. But water may also be forced out of the trap in the opposite direction. This is known as positive or back pressure. This occurs when a higher pressure develops in a drainage system that is improperly designed and vented. This forces air into the fixture drain, through the trap water seal, and into the building. See Figure 7-16.

The Parts of a Trap

All plumbers should know and be able to identify the various parts and materials of the two most commonly used traps in the plumbing trade. Figure 7-17 illustrates the one-piece trap. This trap is generally constructed of the same material as the pipe and is installed in the building's drainage system. A cast-iron system would use a cast-iron trap, a copper system a copper trap, a plastic system a plastic trap, and so on. Usually these traps are concealed and have no cleanout. They are known as P-traps and usually serve bathtubs, showers and bidets.

The most used trap in the plumbing trade is a P-trap that serves lavatories, sinks, laundry trays and the like. See Figure 7-18. P-traps are installed above the finished floor and may or may not have individual cleanouts in their base. Chrome traps may be of one-piece cast brass, but more often they are of a tubular construction and have slip joints and washers. A slip joint is permitted on the fixture inlet part of the trap

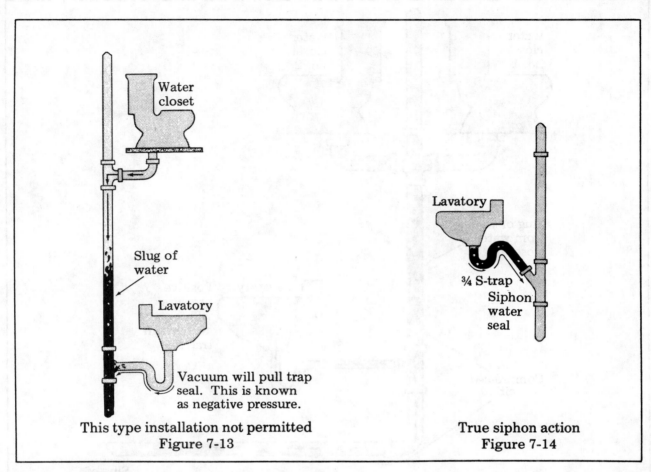

Water closet

Slug of water

Lavatory

Vacuum will pull trap seal. This is known as negative pressure.

This type installation not permitted
Figure 7-13

Lavatory

¾ S-trap
Siphon water seal

True siphon action
Figure 7-14

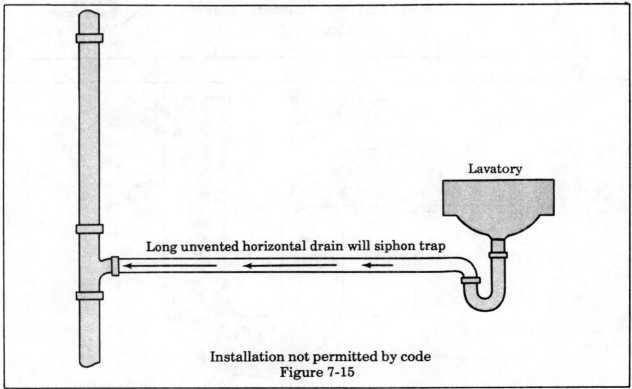

Lavatory

Long unvented horizontal drain will siphon trap

Installation not permitted by code
Figure 7-15

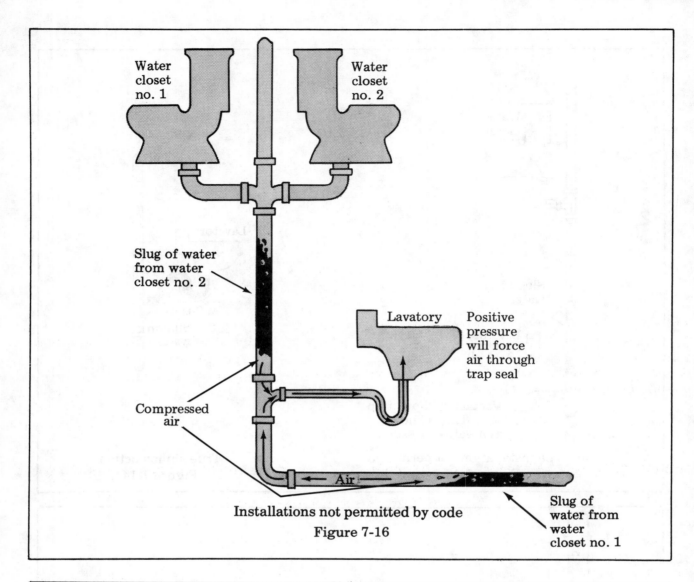

Water closet no. 1

Water closet no. 2

Slug of water from water closet no. 2

Lavatory

Positive pressure will force air through trap seal

Compressed air

Air

Installations not permitted by code

Figure 7-16

Slug of water from water closet no. 1

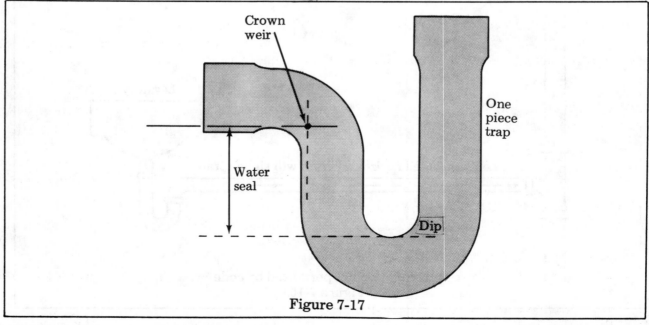

Crown weir

One piece trap

Water seal

Dip

Figure 7-17

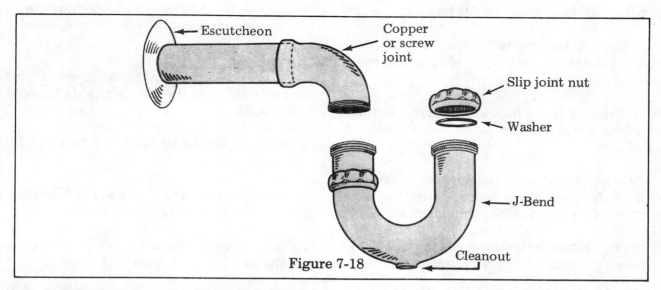

Figure 7-18

and also on the outlet side, provided it is within the trap water seal. See Figure 7-3. An old slip joint and washer located *above* the water seal on the discharge side of a trap can permit sewer gases to enter a building. Therefore, this type of trap is prohibited. Plastic traps may be used for lavatories and sinks where the drainage system is plastic. Plastic pipe can resist heat damage at temperatures up to 180°F. But use plastic traps and tailpieces for kitchen sinks with caution. Many plastic traps and tailpieces in this installation have deteriorated short of their intended life because residents pour boiling water directly into the sink. Codes may prohibit plastic traps and tailpieces in kitchens in the future.

Questions

1- As shown in this chapter, all fixtures must be provided with what to meet code requirements?

2- How does a trap protect human health?

3- What is the most common trap used in the trade?

4- What is the term for a trap that does not collect and retain foreign substances?

5- Why does the code prohibit a 1½-inch trap connected to a 1¼-inch fixture drain?

6- Name three traps that are prohibited by code for use in new construction.

7- What is the minimum depth permitted by code of the water seal for a trap?

8- What is the maximum depth permitted by code of the water seal for a trap other than an interceptor type?

9- Why should all traps be installed level?

10- When two or more fixtures are permitted to use a single trap, what is the waste pipe for this installation called?

11- When two or more fixtures use a single trap, what is the maximum center-to-center measurement allowed?

12- When three individual fixtures are permitted by code on a single trap, on which fixture must the trap be located?

13- What size is the trap for a domestic lavatory?

14- What is double-trapping and when is it permitted?

15- What is the maximum length of the vertical

drop from a lavatory outlet to its trap water seal?

16- When a fixture tailpiece exceeds the maximum length set by code, what may happen?

17- What is the maximum length of the vertical drop from a water closet outlet to the building drain?

18- Water closets are designed so a certain trapping function takes place when they are operated. Can you describe this function?

19- In a tank-type water closet the trap seals are automatically restored by what action?

20- What is the standard trap size for a kitchen sink?

21- What is accomplished by using the shortest fixture tailpiece possible to connect it to its trap?

22- What is prohibited on concealed fixture traps?

23- Where can drum traps be used in today's plumbing?

24- All permitted drum traps must be equipped with what?

25- Why are building traps no longer used in plumbing systems?

26- Name five ways a trap seal may be broken.

27- What is the standard trap size for a domestic shower?

28- Explain the difference between negative and positive pressures within a drainage system.

29- Why is it illegal for a trap to have a slip joint nut and washer on the discharge side above the water seal?

30- What is the standard size trap for a bathtub?

31- Why should caution be used in installing a plastic trap on a kitchen sink waste?

32- S-traps are available today for certain types of fixtures. How may S-traps be used?

33- What happens to a fixture trap when the fixture is installed on an unvented horizontal line exceeding the critical distance set by code?

34- Where is the crown weir located on a trap?

35- What portion of a trap is known as a J-bend?

36- What forms the dip of a trap?

37- What is the minimum size waste pipe to serve a water closet trap?

38- What does a trap seal accomplish at fixture locations?

39- In a two-compartment sink with two different depths, what is the maximum depth of one compartment using a single trap?

Cleanouts

Before cleanouts were required on drainage piping, the plumber had to cut a hole in the blocked drainage pipe to clean out the blockage. He had to insert a cleaning cable to remove the obstruction; then he had to patch the hole with a cement mixture or a similarly impervious material. These patch jobs often deteriorated and allowed raw sewage to seep out and into the ground. This caused a health hazard for the building occupants and their neighbors. Cutting or drilling holes into drainage or vent pipe is now prohibited by code.

Current codes recognize the importance of properly located and accessible cleanouts. They are now an essential part of the drainage system. Today's model codes specify the location, size, distance between cleanouts, and many other requirements. All drainage pipe is subject to stoppage. Making cleanouts accessible saves the serviceman valuable time and the owner unnecessary expense.

Cleanouts or cleanout tees are required where a building sewer connects to the public sewer lateral at the property line. See Figure

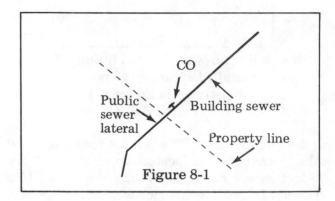

Figure 8-1

8-1. A cleanout at this location serves a dual purpose. This is the point at which a test plug is inserted for performing a water test on a building sewer. See Figure 8-2. The test tee also serves as a cleanout for cleaning any future stoppages that might occur in the public sewer lateral or building sewer. Some codes require that this cleanout be extended up to the finish grade, while others do not.

A full size cleanout *may* be located outside the building at the junction of the building drain and the building sewer, usually within five feet

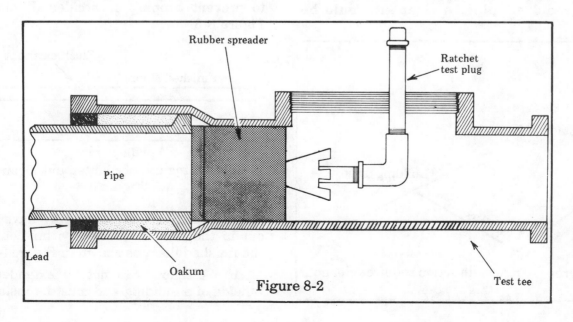

Figure 8-2

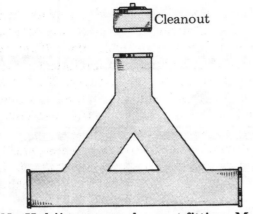

"No-Hub" two-way cleanout fitting. May be
installed at property line as
illustrated in Figure 8-1
Figure 8-3

of the building line. This is not a requirement if
other cleanouts are located upstream and if a
75-foot sewer cable will reach this area. If a
cleanout is used at this location, it must permit
upstream as well as downstream rodding. This
means that a fitting known as a two-way
cleanout must be used. See Figure 8-3. This
fitting may or may not have to be brought to
finish grade, depending on local code requirements.

Accessible cleanouts are required on all
horizontal drainage piping, and separation
distances cannot exceed 75 feet (more or less in
some codes). A cleanout should also be provided
for each change of direction in a building drain
greater than 45 degrees. For example, Figure
8-4 illustrates the change in direction of a
90-degree drain pipe. A cleanout would be

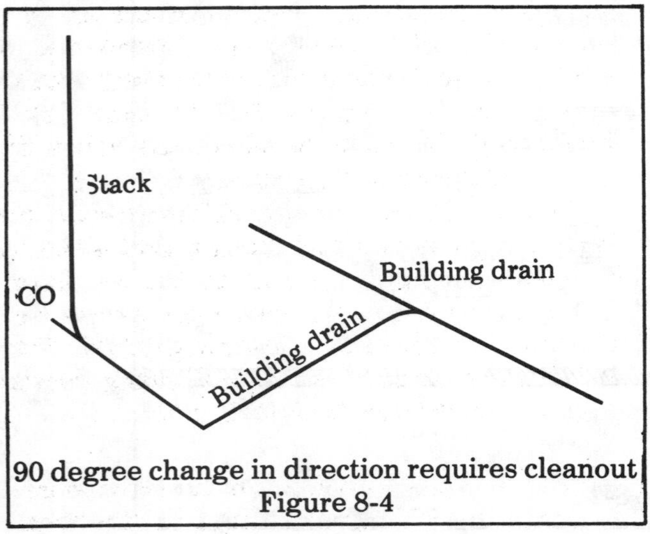

required at the base of the stack or at the end of
the drain pipe. Sometimes the extension of a
cleanout to make it accessible to the outside of a
building creates what is known as a "dead
end." See Figure 8-5. The dead end is created
when the extension is terminated with a
cleanout that has a developed length of two feet
or more. Dead ends are permitted only where
other accessible locations are lacking.

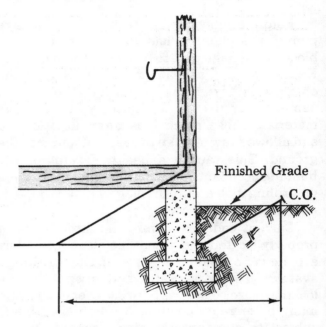

Dead end created with cleanout extension
Figure 8-5

An underground piping cleanout that is
brought to grade and terminates in a walkway,
hallway or room must have a countersunk plug
to prevent tripping or accidental injury. See
Figure 8-6.

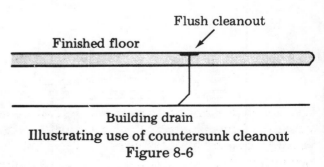

Illustrating use of countersunk cleanout
Figure 8-6

An exposed vertical stack may have a cleanout in the base or located within four feet of
the finished floor, as shown in Figure 8-7.

If a cleanout can not be extended to the
outside of a building and must be installed in a

90 degree change in direction requires cleanout
Figure 8-4

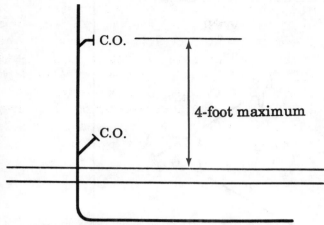

A cleanout must be provided at the base of all required stacks. Either location is acceptable.
Figure 8-7

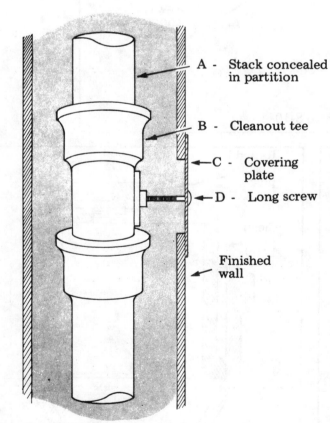

A - Stack concealed in partition

B - Cleanout tee

C - Covering plate

D - Long screw

Finished wall

If a cleanout cannot be extended accessibly to outside, a cleanout tee may be used in the vertical stack. The cleanout plug must be accessible. A covering plate (C) or access door must be provided within four feet of finished floor to permit rodding. The covering plate may be held in place by long screw (D). The raised head portion of cleanout plug may be tapped for this purpose.
Figure 8-8

concealed vertical stack, a cleanout tee may be used. The cleanout plug must be accessible. Study Figure 8-8. A covering plate (C) or access door must be provided within four feet of the finished floor to permit rodding. The covering plate may be held in place by a long screw (D). The raised head portion of the cleanout plug may be tapped for this purpose.

Cleanouts smaller than 3 inches shall have a 12-inch clearance, and cleanouts 3 inches and larger shall have an 18-inch clearance for rodding purposes. See Figure 8-9.

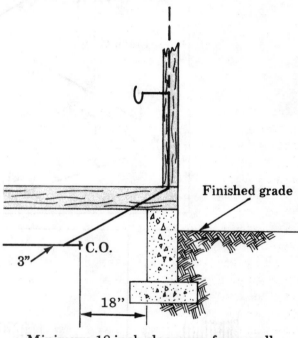

Finished grade

C.O.

3"

18"

Minimum 18 inch clearance from wall
Figure 8-9

Cleanouts in one-story buildings can be omitted if certain code requirements are met. This section of the code is controversial, so check your local code before proceeding with the installation. Use Figure 8-10 to help you to understand the code requirements when this type of installation is permitted.

Vent stacks can sometimes have a dual role. In certain types of installations a vent stack can be used to supply and remove the air from a drainage system and can also serve as a cleanout for inserting a cleaning cable. These installations must meet the following requirements to qualify for this dual role:

The drainage system must not have more than one 90-degree change in direction.

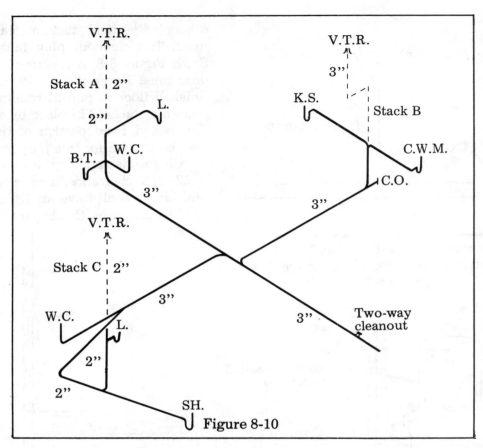

Figure 8-10

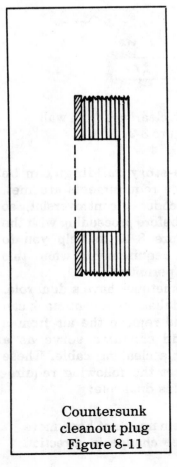

Countersunk
cleanout plug
Figure 8-11

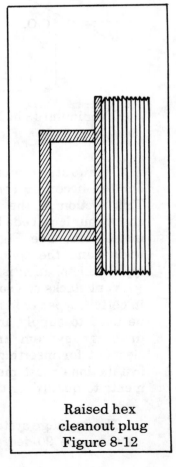

Raised hex
cleanout plug
Figure 8-12

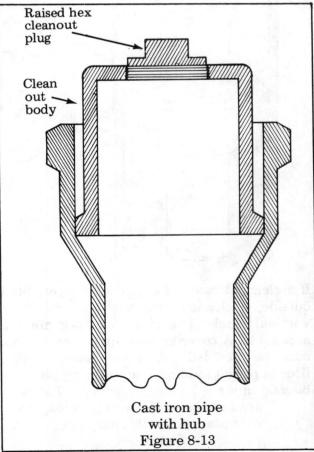

Cast iron pipe
with hub
Figure 8-13

The vent stack must be vertical throughout (without any offsets) and must extend up through the roof.

The vent stack must be of the same size as the waste pipe it serves.

The vent pipe must not be reduced beyond the following minimums:

4'' must not reduce to less than 3''
3'' must not reduce to less than 2''
2'' must not reduce to less than 1½''

In Figure 8-10 a two-way cleanout is located in the building drain and will permit rodding upstream to the base of Stack A. Thus, installing a cleanout in the base of Stack A is not necessary.

Stack B requires a cleanout at its base. The stack is properly sized, but is not vertical throughout as required by the code. Stack C has the same diameter as the waste pipe it serves (2 inches) and is vertical up and through the roof. Thus, installing a cleanout in its base is not necessary.

Figure 8-11 shows a countersunk cleanout plug. Figure 8-12 shows a raised hex cleanout plug. A raised hex cleanout plug with cleanout body installed in the hub of a cast-iron pipe is illustrated in Figure 8-13. The size of these cleanout bodies, up to 4 inches, must be the same nominal size as the pipe to which the cleanout is joined. A waste pipe 2 inches in diameter would require a 2-inch cleanout; a 3-inch waste pipe would require a 3-inch cleanout, and so on. Building sewers up to 8 inches in diameter may be served by 4-inch cleanouts.

Questions

1- Why does the code prohibit cutting a hole in a drainage or vent pipe to remove a blockage?

2- What is the dual role of a cleanout tee when installed at the end of a building sewer?

3- What is the advantage of a two-way cleanout fitting?

4- A cleanout is required in a horizontal drain pipe for each change of direction of how many degrees?

5- What creates a dead end?

6- When is a countersunk plug required for a cleanout?

7- What is the maximum height from a finished floor a cleanout may be installed in a vertical stack?

8- What are the two possible procedures in making a concealed cleanout in a vertical stack accessible?

9- What is the necessary clearance from walls or other obstructions, required by code, for a 3-inch cleanout for rodding purposes?

10- Name three of the four requirements in a one-story building where cleanouts may be omitted.

11- What is the minimum size of a cleanout which serves a 4-inch horizontal drain pipe?

12- What is the distance between cleanouts in a horizontal straight 4-inch drain line?

9

Floor Plans And Layouts

Bathrooms Yesterday and Today

It is estimated that 85 percent of all bathrooms measure 5 feet by 8 feet. Bathrooms with these dimensions are most often found in older homes and in homes having only one bathroom back-to-back with the kitchen sink. The bathrooms of that day were purely utilitarian — clean and antiseptic but completely lacking in style. The plumbing systems were simple installations and maintenance problems were minimal.

But as houses grew in size over the years so did bathrooms. New designs of bathtubs, showers, lavatories and water closets and the increasing popularity of the bidet allow for a personal expression in bathroom design.

Builders currently allot more space for bathrooms in their floor plans, as baths now assume greater importance in the home. Today, the once-neglected bathroom is often a showpiece. Designing the bathroom is almost as important to the overall appeal of the house as designing and decorating the rest of the house combined. Home buyers are looking for more than functionality in their plumbing. This creates a challenge for today's plumber that did not exist for the plumber of a generation ago.

Beginning Your Layout

Before beginning the installation of the rough plumbing for a bathroom addition or a bath in a new residence, draw an isometric layout of the plumbing. The configuration of the layout will be determined by the location and type of fixtures provided by the architect on the floor plan of the construction blueprint. The floor plan will show floor dimensions, walls, doors, windows and other pertinent information. It is of course imperative not to have waste, vent or water pipes installed through door or window openings. Plan the layout carefully to avoid these openings.

Draw the isometric to illustrate the type of installation best suited for the job — a stack system, a flat system, a horizontal wet vent system, or the like. The isometric drawing should reflect the number and type of fittings to be used. These can be listed from the isometric and then ordered from the supplier. The pipe footage can be scaled from the plan, thereby minimizing leftover materials.

Be sure to space the fixtures properly on the isometric so that when you rough-in, you have room. The waste and water pipe inlets must be installed at the right location for the plumbing fixture to fit and function properly. Fixture manufacturers furnish roughing-in booklets for the professional, but it is difficult for the beginner to obtain them. Specially designed fixtures usually come with fixture dimensions and roughing-in measurements. Chapter 18 contains the dimensions and measurements for many of the more commonly used plumbing fixtures.

There are many acceptable ways to lay out the rough plumbing for a particular job and still meet code requirements. Very few professionals lay out the same job twice in exactly the same way.

Seven bathroom floor plans are shown in Figures 9-1 through 9-7. Figures 9-8 through 9-11 show complete floor plans for four residential units. Room dimensions have been omitted as they are not relevant here. The building drain or sewer location is given so that you may have the same starting point for all your isometric drawings. There are two illustrations for each bathroom floor plan in Figures 9-1 through 9-7, one on a flat installation and one on a stacked installation. Figures 9-8 through 9-11 show only one type of layout for each floor plan. These isometric drawings should meet code requirements for anywhere in the country.

Use the selected floor plans and isometric drawings in the figures to help prepare you to make your own rough plumbing layouts for any installation you may be required to make.

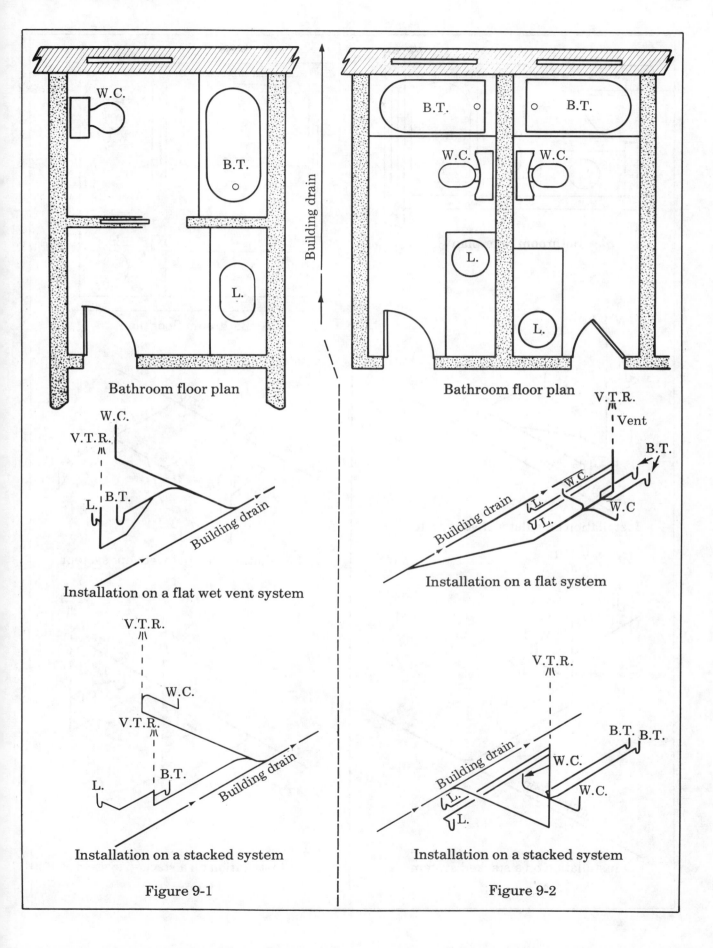

Bathroom floor plan

Installation on a flat wet vent system

Installation on a stacked system

Figure 9-1

Bathroom floor plan

Installation on a flat system

Installation on a stacked system

Figure 9-2

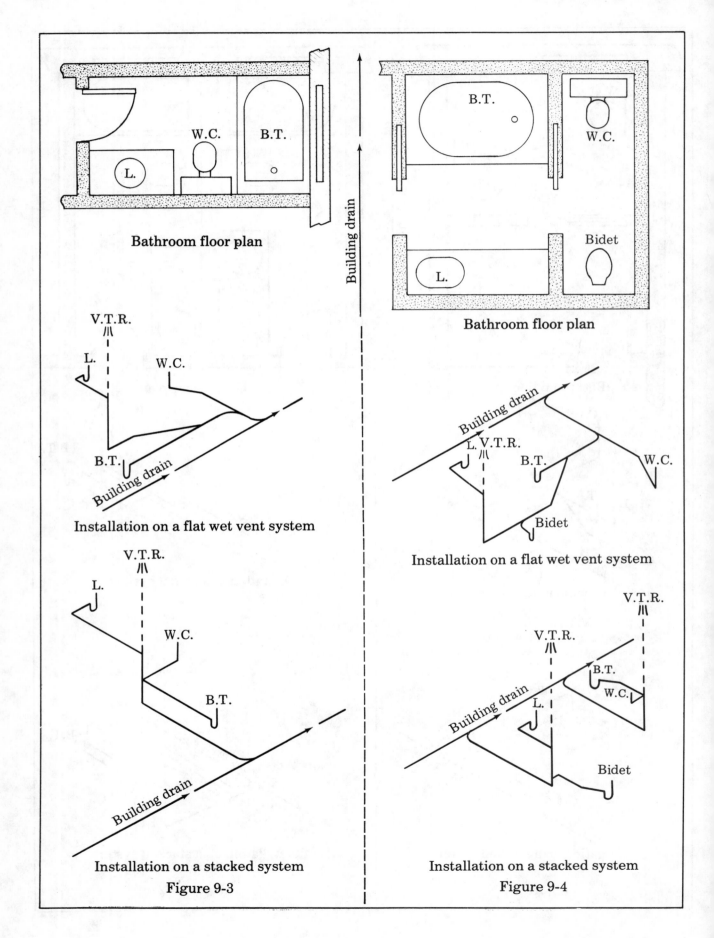

Bathroom floor plan

Bathroom floor plan

Building drain

Installation on a flat wet vent system

Installation on a flat wet vent system

Installation on a stacked system

Figure 9-3

Installation on a stacked system

Figure 9-4

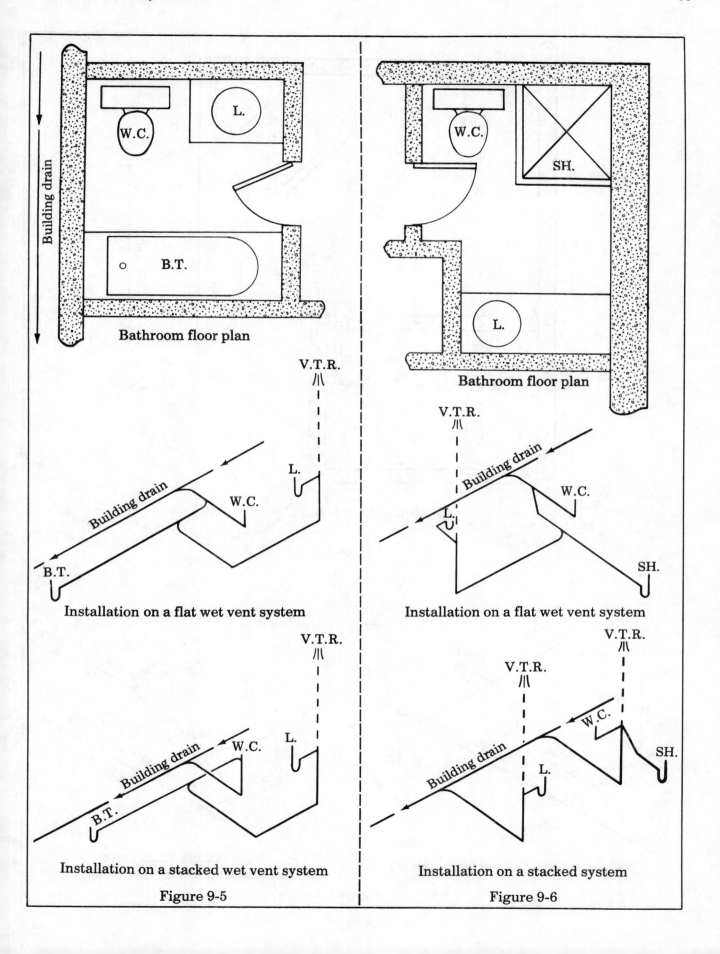

Bathroom floor plan

Bathroom floor plan

Installation on a flat wet vent system

Installation on a flat wet vent system

Installation on a stacked wet vent system

Installation on a stacked system

Figure 9-5

Figure 9-6

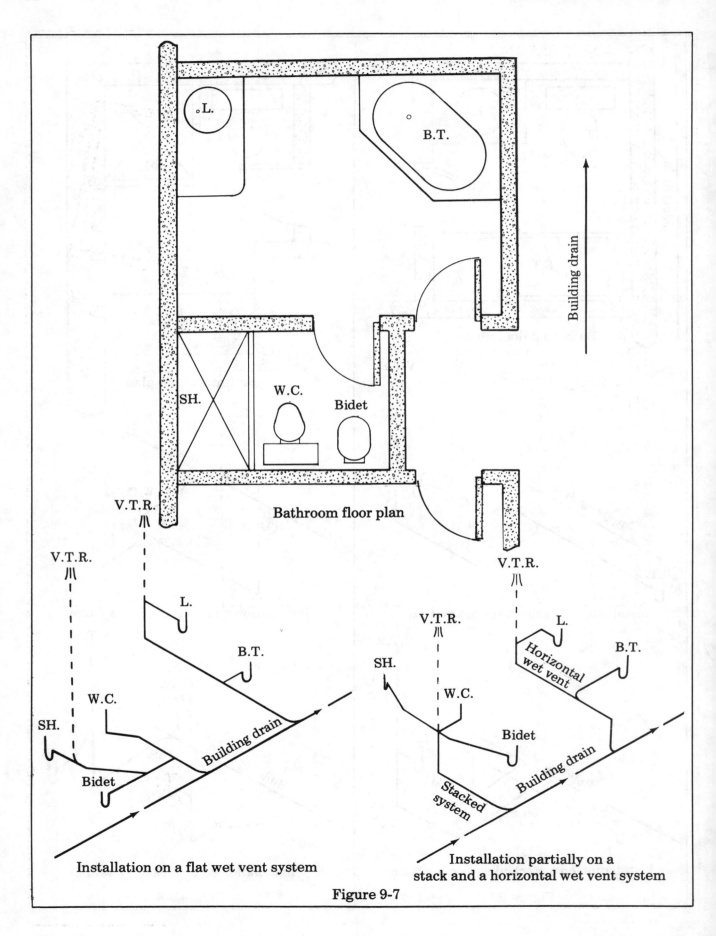

Bathroom floor plan

Installation on a flat wet vent system

Installation partially on a
stack and a horizontal wet vent system

Figure 9-7

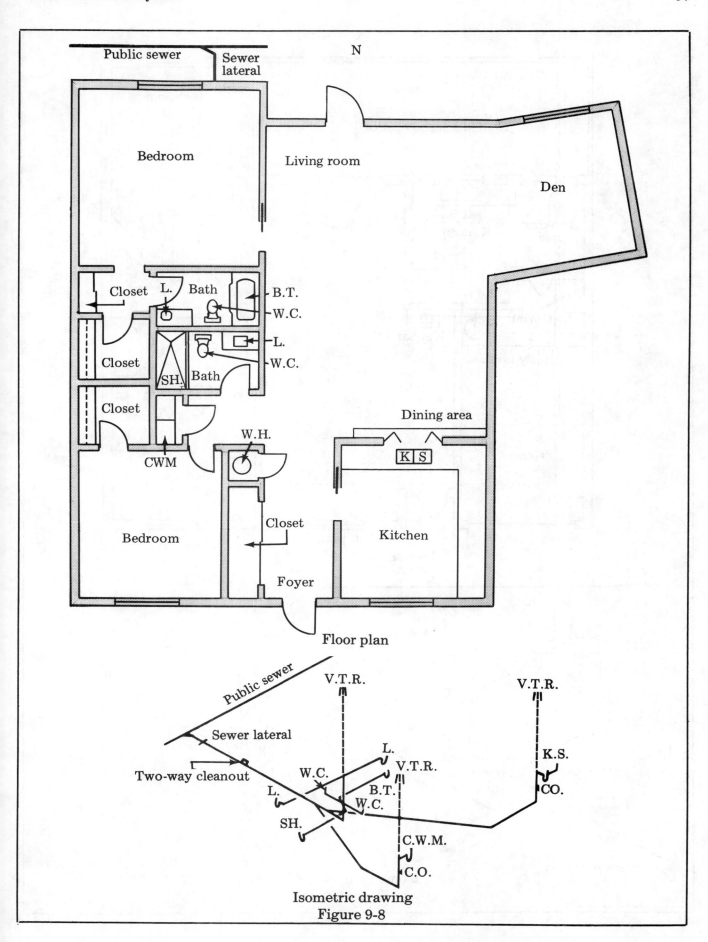

Floor plan

Isometric drawing
Figure 9-8

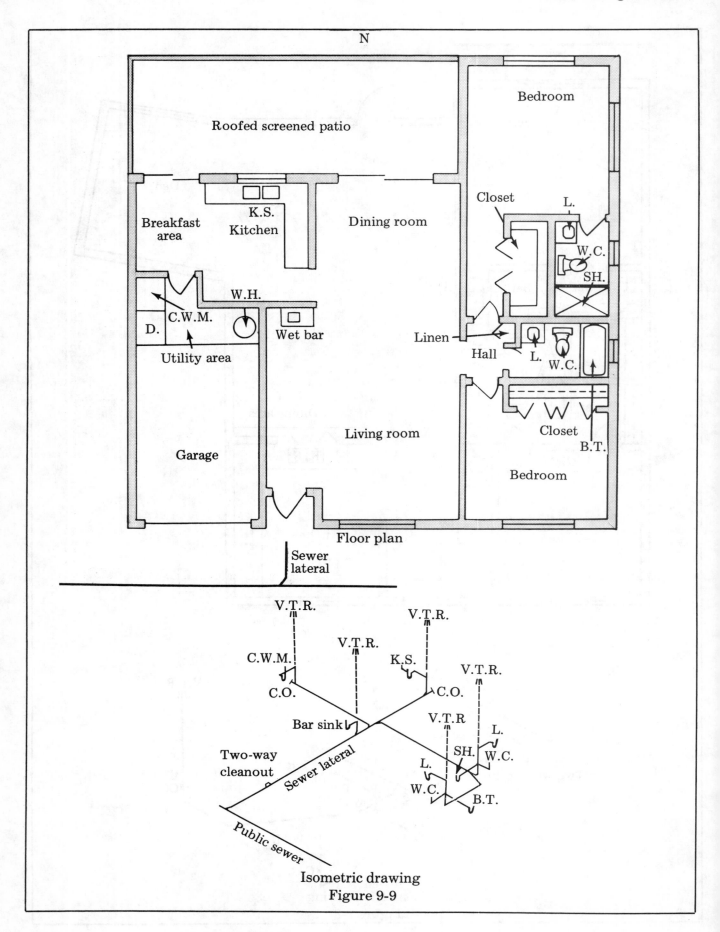

Floor plan

Isometric drawing
Figure 9-9

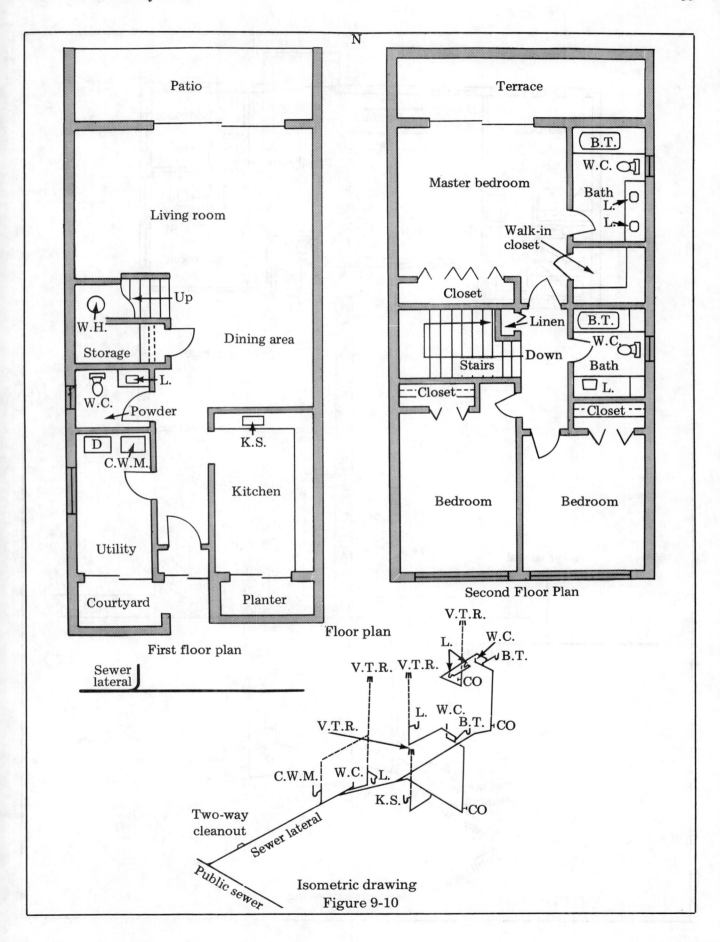

First floor plan

Second Floor Plan

Floor plan

Isometric drawing
Figure 9-10

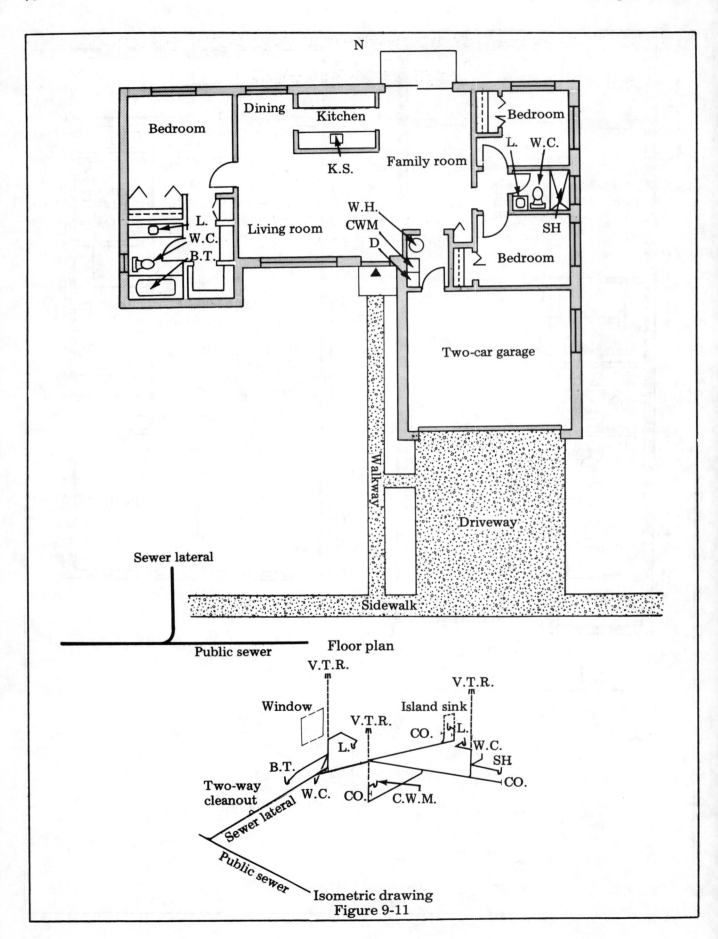

Floor plan

Isometric drawing
Figure 9-11

Questions

1- Why are plumbing maintenance problems easier in older homes?

2- Why have builders allotted more space for bathrooms in today's homes?

3- Why is it necessary to draw an isometric layout of the plumbing before beginning work?

4- Name at least three items that must be included in professionally prepared floor plans.

5- What should an isometric drawing show? Be specific.

6- What are some of the other important steps that should be considered before beginning the installation of rough plumbing?

10
DWV Materials And Installation

The plumbing code ensures that the design, installation, and maintenance of plumbing systems protect the health, welfare, and safety of the public. While the details of plumbing installations can vary from code to code, the same basic principles of sanitation and safety are followed by all codes.

This chapter shows the installation requirements of most drain, waste, and vent systems for residential or other simple buildings. The principles outlined here do not cover every plumbing situation, but they should make the intent of the code clear and understandable. These principles are so important that anyone violating them may be certain the plumbing inspector will not pass the work he has done.

Plumbing fixtures are the end of the potable water supply system and the beginning of the sewage system. Therefore, drainage systems should be designed to prevent fouling or the deposit of solids along the pipe walls. Always use the minimum pipe sizes established by the code. The drainage system must be properly vented to provide free circulation of air. These vent pipes must be sized and arranged to relieve pressure that builds up in any drainage system. Correct sizing prevents back pressure or siphoning action from destroying the fixture trap seal. But stoppages can occur in even the best-designed system. Adequate cleanouts must be installed so that all portions of the drainage system are accessible by cleaning equipment. (See Chapter 8.)

Another important design consideration is direction changes. The code regulates use of 45 degree wyes, long or short sweeps, long or short quarter bends, sixth, eighth, or sixteenth bends and other direction change fittings. Single and double sanitary tees, quarter bends, and one-fifth bends may be used in vertical sections of drainage lines only where the change of direction in flow is from horizontal to vertical.

Proper use of fittings for change in direction is illustrated in Figures 10-1, 10-2 and 10-3. Review these carefully.

Double hub fittings (as in Figure 10-4) are prohibited on drainage lines. Fittings having a drainage pattern in the direction opposite to flow in Figure 10-5 are prohibited. An inexperienced plumber is likely to make this mistake when using the newer no-hub or plastic drainage fittings. When you install these fittings, be sure that the drainage pattern of the fitting is in the same direction of flow as the rest of the drainage system.

The code does not permit drilling or tapping any drainage or vent piping. Drilling or tapping can allow raw sewage to seep on to or into the ground. In the case of a vent pipe, sewer gases could escape into a building.

Plumbing drainage pipes must be installed in open trenches and must remain open until the piping has been inspected, tested, and accepted by the plumbing inspector. This means that you are not allowed to tunnel under driveways or other permanent structures to lay drainage piping. There is no way to properly support drainage piping installed in this manner. This piping would also be likely to sag and cause stoppages.

The code requires that all drainage system piping be water tested with at least a five-foot head. With all other openings plugged, at least one vent pipe in the system must be extended a minimum of five feet higher than the horizontal drainage piping and filled with water. This is to make sure that the joints of the system do not leak. The plumbing inspector may require the removal of cleanout plugs or caps to check whether the water has reached all parts of the system.

Materials

Types of materials and installation methods

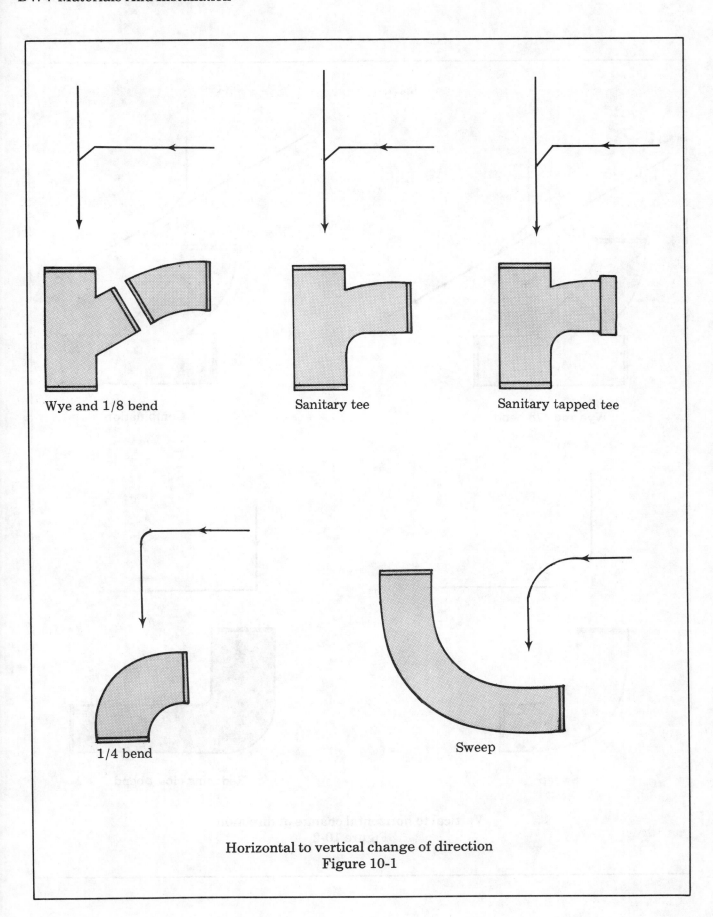

Wye and 1/8 bend   Sanitary tee   Sanitary tapped tee

1/4 bend   Sweep

Horizontal to vertical change of direction
Figure 10-1

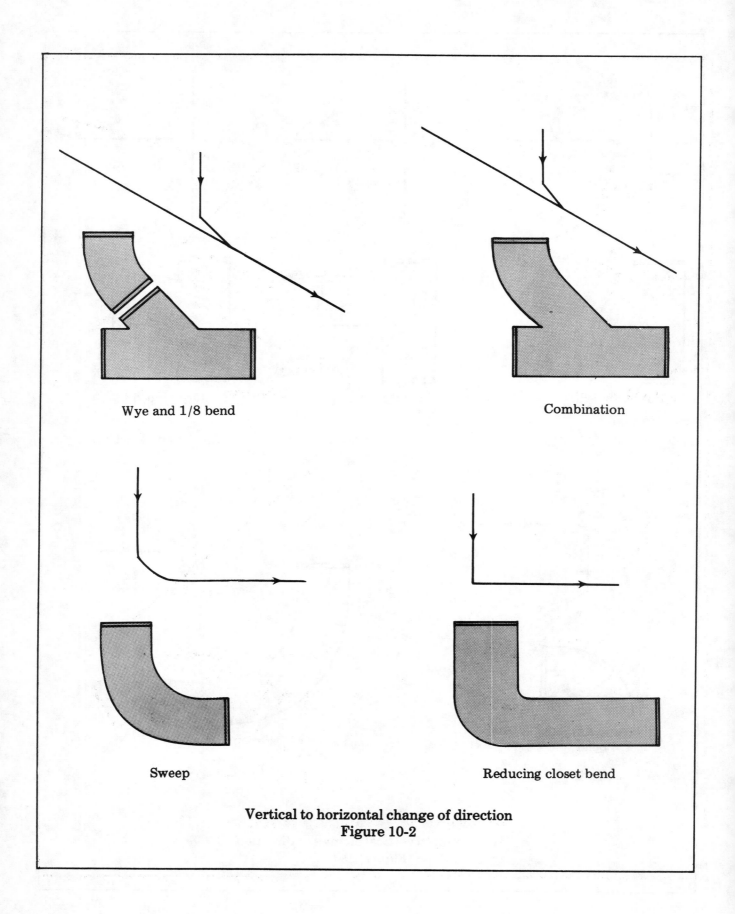

Wye and 1/8 bend

Combination

Sweep

Reducing closet bend

Vertical to horizontal change of direction
Figure 10-2

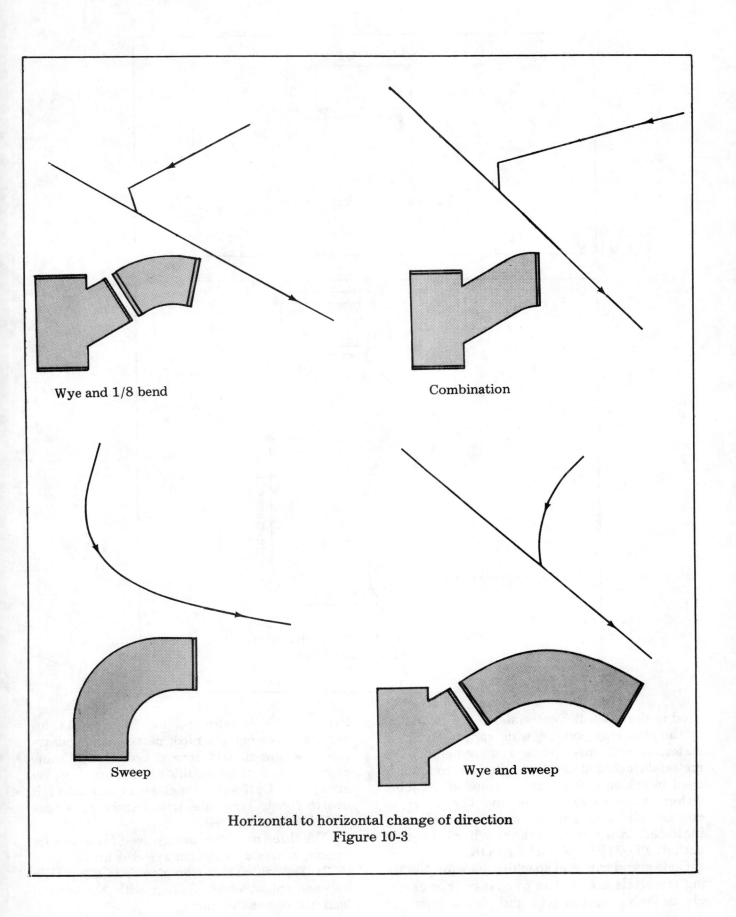

Wye and 1/8 bend

Combination

Sweep

Wye and sweep

Horizontal to horizontal change of direction
Figure 10-3

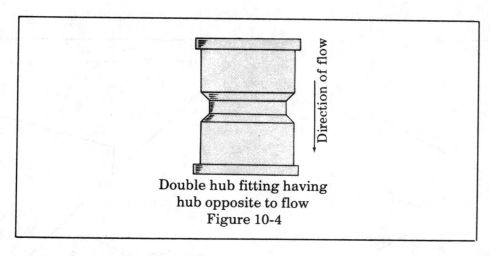

Double hub fitting having
hub opposite to flow
Figure 10-4

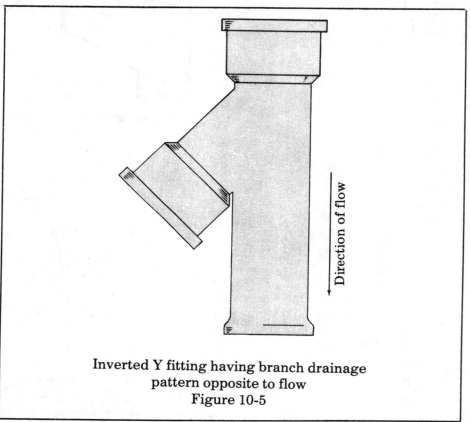

Inverted Y fitting having branch drainage
pattern opposite to flow
Figure 10-5

used in drain-waste-vent systems are regulated by the plumbing code. A wide variety of piping materials will meet code requirements. The materials included in this chapter are commonly used in residential or simple commercial jobs. Other types of materials and their use in commercial and special waste systems are included in a more advanced book by this author, *PLUMBER'S HANDBOOK*.

Code standards and specifications for plumbing materials are subject to change. For example, no-hub cast-iron pipe and plastic pipe and

fittings are considered new materials for plumbing systems. In most parts of the country they are now in widespread use. But in some counties and municipalities they are not yet accepted. Call your local authority or visit nearby construction sites to learn whether or not their use is permitted.

The three most frequently used materials for drainage, waste, and vent systems are:

1) Tar-coated cast-iron pipe and fittings with hub-and-spigot ends. Oakum with hot poured lead completes the joint.

2) Tar-coated cast-iron pipe and fittings of the no-hub type. Approved elastomeric sealing sleeves and stainless steel clamps complete the joint. The cast-iron pipe and fittings may be extra heavy or centrifugally spun service weight. Extra heavy and service weight pipe and fittings should not be mixed within the same system. Where it is permitted by code, use service weight pipe and fittings as it is easier to work with and cut.

3) Schedule 40 ABS or PVC plastic pipe and fittings with cemented joints. The code prohibits mixing ABS and PVC plastic pipe and fittings in the same installation.

Copper pipe type DWV or M and galvanized steel pipe with recessed drainage fittings are commonly used in cast-iron systems for fixture drains. Some codes now permit using plastic pipe and fittings for fixture drains for cast-iron installations.

Vent extensions through the building roof from a cast-iron system are frequently DWV or M copper, galvanized steel or plastic pipe with proper conversion adapters. These materials are lighter in weight than cast-iron and come in long sections so few joints are required. This saves labor and material.

Fittings used in all portions of the drainage system must conform to the material and type of piping used. In other words, a drainage system installed with PVC plastic pipe cannot use ABS plastic fittings, or vice versa. A drainage system installed with cast-iron pipe cannot use plastic fittings, even with approved conversion adapters. Fittings on screwed, copper or plastic pipe used within a sanitary drainage system should be of the recessed drainage pattern type. *Fittings designed for use in water piping installations should not be used in drainage systems.*

Cast-Iron Soil Pipe and Fittings

Cast-iron pipe was imported from England and Scotland in 1813 to replace deteriorated wooden water mains in Bethlehem, Pennsylvania. As the demand for cast-iron pressure pipe increased, foundries of New Jersey and Pennsylvania supplied most of the nation's needs for pressure pipe used in water mains. In the 1880's cast-iron pressure and soil pipe for building construction was manufactured to meet the requirements of the first plumbing code, published in Washington, D.C., in 1881.

Cities installed water works and sewage systems at a rapid pace. By 1894 there were 64 cast-iron foundries producing pressure pipe and soil pipe.

The center of the soil pipe industry shifted from the Northeast to the South by 1915. By 1940 approximately 70% of total soil pipe production in the United States was in southern foundries, most of them in Alabama.

Cast-iron soil pipe and fittings are used primarily in buildings for sewers, drains, soil and vent piping. It is available in two weights: centrifugally spun service weight and extra-heavy cast-iron. The service weight, unless prohibited by code, is recommended for use in all residential construction.

Cast-iron soil pipe can have a hub and spigot joint using lead and oakum to make a watertight, gastight joint. The waterproofing characteristics of oakum fiber have long been recognized by the plumbing trades. To get an idea of how well oakum can seal a joint, try the following. When a job has been completed and the lead joints are caulked, fill the system with water. Disregard any leaks. Let the water stand overnight before calling for inspection. The wet oakum will expand overnight and fill each crevice, making joints water and gas tight. This technique will save the time it would take to recaulk each seeping joint.

Cast-iron soil pipe is available with no-hub joints. This plumbing concept supplements the lead and oakum hub and spigot joint. The no-hub joint uses a one-piece neoprene gasket, a stainless steel shield and retaining clamps. In limited-access areas (such as close to ceilings or in corners) the great advantage of using a no-hub system is that the joints are easy to make up. Since the pipe and fittings do not have hubs, they can be used in thinner partitions. Installation is fast and efficient. No-hub joint pipe can be used in combination with a lead and oakum joint pipe to meet the needs and specific requirements of any particular job. Properly installed, lead and oakum joint pipe and no-hub pipe lasts as long as cast-iron soil pipe. Generally, lead and oakum joint pipe and no-hub pipe can be installed and forgotten.

Certain characteristics of cast-iron soil pipe make it the preferred material for sewers. Its strength, durability and resistance to trench loads are generally considered superior.

When installing cast-iron soil pipe, be sure to keep the pipe barrel in firm contact with solid ground. Always excavate for the hub (bell) or for the steel shield on no-hub pipe. The weight should be evenly distributed along the full length of the pipe barrel. See Figure 10-6.

5-foot and 10-foot lengths. When a piece of pipe shorter than 5 feet is needed, it is usually cut from a double hub pipe to avoid waste. See Figure 10-7. This procedure will leave two usable lengths of pipe, each with a hub.

Installations using no-hub pipe would not require this technique. Standardized cuts from any piece of no-hub pipe would leave pipe sections long enough for use elsewhere. This procedure eliminates waste.

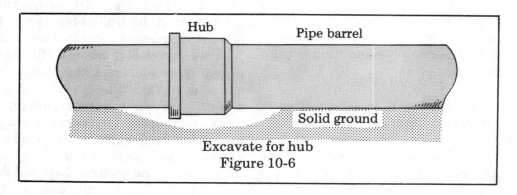

Excavate for hub
Figure 10-6

Because of cast-iron pipe's high resistance to trench loads, depth is not too important. However, don't backfill with large boulders, rocks, cinder-fill or other materials that could damage or corrode the pipe.

Cast-iron soil pipe is manufactured in sizes of 2 inches (inside diameter) and up. It comes in

Soil Pipe Fittings

Figures 10-8, 10-9, 10-10, and 10-11 illustrate some of the more commonly used hub and spigot soil pipe fittings. No-hub fittings for the same uses will have the same designs (see Figures 4-3, 4-4, and 4-5), but without the hub and spigot ends.

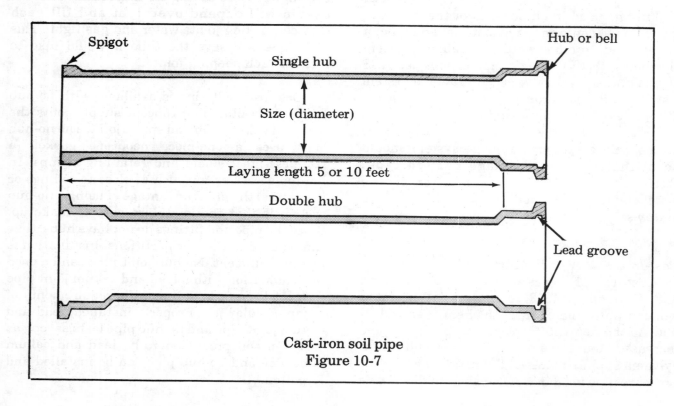

Cast-iron soil pipe
Figure 10-7

Tees

Tees (sometimes called T's or T-branches) are used to make 90-degree angles to the main pipe serving this fitting. See Figure 10-8. The main pipe must be vertical and may be either a soil, waste or vent stack. This and other types of horizontal to vertical directional change fittings are illustrated in Figure 10-1.

Many types of cast-iron tees are in general use. Sanitary tees are designed to carry waste substances. Straight tees are designed to carry air only and are used in a vent system. Tees may be tapped or may have a hub and can be used to connect threaded or unthreaded branch drains or vent lines to the stack vent.

The side takeoff of a sanitary tee enters the straight through section of the fitting on a downward curve. This is known as a *drainage pattern*. This changes the direction of flow smoothly from the horizontal to the vertical, thus helping to prevent stoppages at this location. Tees are available as singles, as double horizontals, as crosses and as angles.

The *test tee* is a fitting with a screw plug. It is generally used for testing a system for leaks and can be used as a permanent cleanout after the test. See Figures 8-2 and 8-8.

Wyes (*See Figure 10-9*)

Wyes, or Y-branches, are used to make a 45-degree change in direction. They are used mainly to connect horizontal waste pipes and branch drains to the building main drain. Their design allows a smoother change of flow direction than either straight or sanitary tees, which are prohibited in horizontal installations. Some of the more commonly used fittings and combinations of fittings for horizontal to horizontal and vertical to horizontal changes of direction are shown in Figures 10-2 and 10-3.

A 45-degree Y-branch, also called simply a "Y-branch," has the side takeoff entering the straight through section of the fitting at a 45-degree angle. If the side takeoff is the same size as the through section (3 inches, for example), it is called simply a "3-inch wye," as all three openings are the same size. The side takeoff may be smaller (never larger) than the through section. Then the fitting is called a "reducing Y-branch" (3 x 2, for example). Wyes are available in single as well as double design.

A 90-degree Y-branch is also known as a "combination Y and 1/8 bend." (It may be called a "combination" in the trade.) These are manufactured as one-piece units. Where this particular fitting is required but is not available, a Y and a 1/8 bend of the proper size can be quickly assembled into a combination wye using only an extra joint.

Combinations are made in various shapes and sizes. They may be single, double or reducing. As in the 45-degree wye, the side takeoff may be the same size as the through section. If the side takeoff is 3 inches, for example, the fitting would simply be called a "3-inch combination," as all three openings are the same size. The side takeoff may be smaller (2 inches, for example), but never larger than the through section (say 3 inches). In such a case, the fitting is called a "reducing combination," known correctly as a 3 x 3 x 2 reducing combination or simply a 3 x 2 combination. The through section is always sized first, reading from the spigot end to the hub end, and then the branch is sized. This is the correct method for sizing all plumbing fittings, whatever their use.

Bends

Bends are classified by the degree of turn and the radius of the curve. Bends are used to change the direction of either horizontal or vertical drainage or vent lines. Figure 10-10 shows a 1/8 bend, a 1/5 bend, two 1/4 bends, and a short sweep. The 1/16 bend turns at 22½ degrees, the 1/8 bend turns at 45 degrees, the 1/6 bend turns at 60 degrees, the 1/5 bend turns at 72 degrees, the 1/4 bend turns at 90 degrees. The sweep also turns at 90 degrees. A 4-inch 1/4 bend has a turning radius of 4 inches, while a short sweep has a turning radius of 6 inches. A long sweep has a turning radius of 9 inches.

Bends are classified according to the radius of the curve as regular, short or long. As the radius increases, the change in flow direction becomes smoother.

The *closet bend* illustrated in Figures 4-3, 4-4 and 10-2 is a special fitting to receive the waste from water closets. Closet bends are made in different styles to fit different types of construction and local code requirements. One type has a scored end (marked with lines) which fits into the closet floor flange. See Figure 4-4. The scoring makes it easier to cut the bend to the desired length for a given connection.

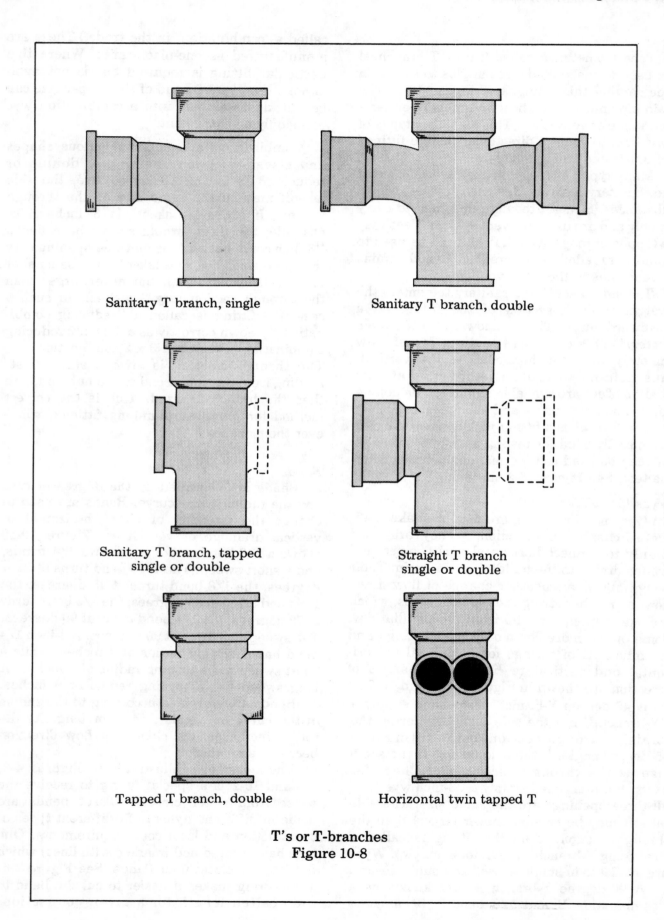

Sanitary T branch, single

Sanitary T branch, double

Sanitary T branch, tapped
single or double

Straight T branch
single or double

Tapped T branch, double

Horizontal twin tapped T

**T's or T-branches
Figure 10-8**

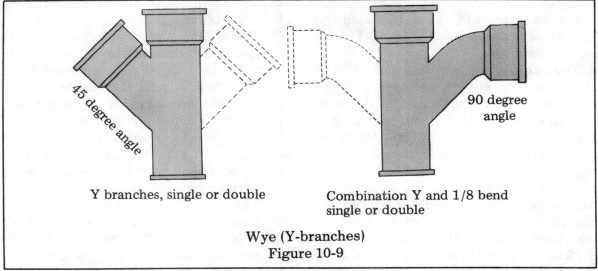

Y branches, single or double

Combination Y and 1/8 bend
single or double

Wye (Y-branches)
Figure 10-9

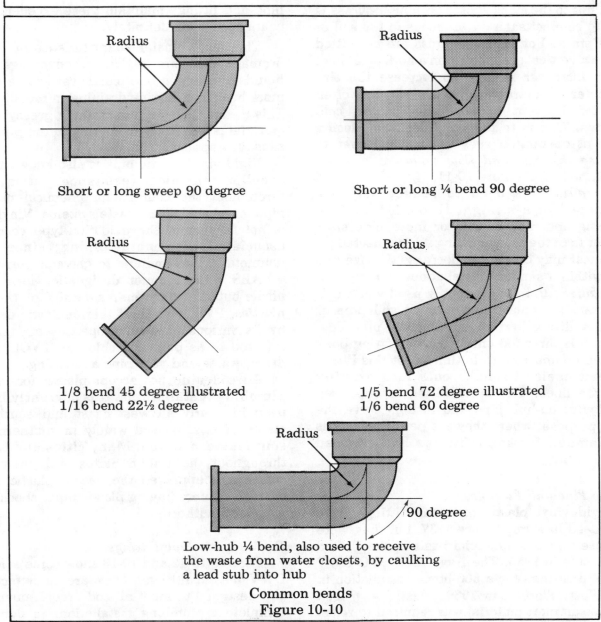

Radius

Radius

Short or long sweep 90 degree

Short or long ¼ bend 90 degree

Radius

Radius

1/8 bend 45 degree illustrated
1/16 bend 22½ degree

1/5 bend 72 degree illustrated
1/6 bend 60 degree

Radius

90 degree

Low-hub ¼ bend, also used to receive
the waste from water closets, by caulking
a lead stub into hub

Common bends
Figure 10-10

Other Fittings

Other types of common and seldom used soil pipe fittings are shown in Figure 10-11 and explained below.

A *1/8 bend offset* is used to carry the soil or waste lines past an obstruction such as a tie beam. This bend is manufactured in different sizes and lengths. Two 1/8 bends will do the same job with an extra joint, but the one-piece 1/8 bend offset gives a smoother transition than the two-piece unit.

A *cross* joins two lines that are in the same plane and perpendicular to each other. The two types of cross, the straight cross and the sanitary cross, are identical except for the takeoffs. The straight cross is used to connect vent lines to the stack vent. The sanitary cross is used to connect sanitary branches to the soil or waste stack. Both types of cross are permitted only in a vertical pipe, never on a horizontal line.

An *increaser* is used to increase the size (diameter) of a straight through line. It is often used for the vent stack terminal in very cold climates. This is to prevent frost from closing the vent opening. *Never use an increaser to increase the size and flow within a sanitary installation.* See Figure 10-11.

A *reducer* is used to reduce the size (diameter) of a straight through line. The reducing end may be one or more pipe sizes smaller than the pipe or fitting it is connected to. A reducer may be used to decrease the size and flow within a sanitary installation.

A *kafer fitting* is commonly used when it is necessary to replace a section of soil pipe or install a fitting into an existing soil pipe line. The hub is threaded onto the kafer fitting body and can be unscrewed. In times past this fitting made it easier to make cuts into existing cast-iron lines. The kafer fitting is still used today. But no-hub pipe and fittings serve the same purpose, where they are permitted. Thus the demand for kafer fittings has dropped considerably.

Plastic Pipe and Fittings

Rigid vinyl plastic pipe and fittings have been used in Europe since 1937, but they were first used on a commercial basis in the United States only in 1955. The Navy used PVC DWV pipe and fittings for a 500-home installation in Key West, Florida, in 1960. Plastic was used because an inert material was required to resist the corrosive effects of the salty soil dredged from the ocean.

The use of plastic DWV pipe and fittings for residential construction has grown rapidly since then. By 1968 there were over 650,000 plastic DWV installations. It is estimated that 95% of currently manufactured mobile homes use plastic DWV pipe and fittings.

The smooth inner surface and superior resistance to deposit formation makes plastic drain, waste and vent piping ideal for residential sanitary systems.

The manufacturers of plastic pipe and fittings claim that their products allow a consistently higher flow rate than pipe of brass, copper, cast-iron, or steel. Moreover, plastic pipe and fittings maintain these excellent flow characteristics indefinitely.

The heat transfer characteristics of plastic drainage pipe are a distinct advantage. Hot liquids discharged into plastic sewer line retain more heat than liquids draining in metal pipes. This means that water containing greasy waste material is less likely to solidify in the pipe and cause a stoppage.

Rigid vinyl plastic pipe is also known for its excellent chemical resistance. It is used throughout the country for chemical process piping and acid drain waste systems. Vinyl pipe is not subject to the oxidation-type corrosion associated with metal piping; since it is nonmetallic it is immune to galvanic corrosion.

ABS is the common designation for acrylonitrile-butadiene-styrene, a family of *thermoplastics*. PVC is an inert thermoplastic defined by its various physical properties and simply referred to as polyvinyl chloride (PVC) plastic drain, waste and vent pipe and fittings.

All codes do not accept plastic for use in plumbing installations. It is currently being used in nearly all west coast and southwest states. It is also used widely in southern and southeastern states. Many cities and towns throughout the United States with their own codes have approved the use of plastic DWV systems. Before buying plastic pipe, check with your local authority.

DWV Plastic Pipe Fittings

Figures 10-12 and 10-13 show some selected DWV plastic fittings. They are manufactured and designed to meet all code requirements to complete a plumbing installation. In domestic

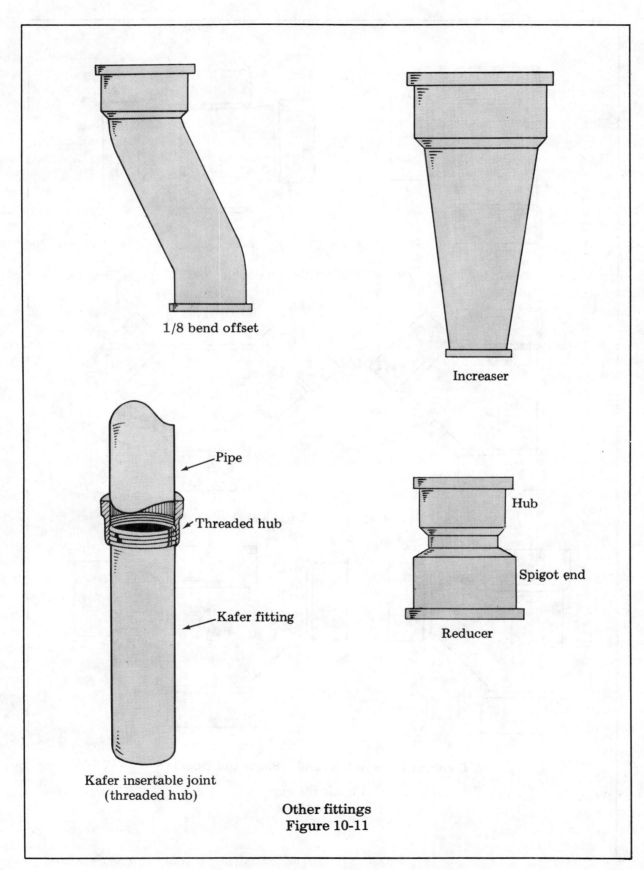

1/8 bend offset

Increaser

Pipe

Threaded hub

Hub

Spigot end

Kafer fitting

Reducer

Kafer insertable joint
(threaded hub)

**Other fittings
Figure 10-11**

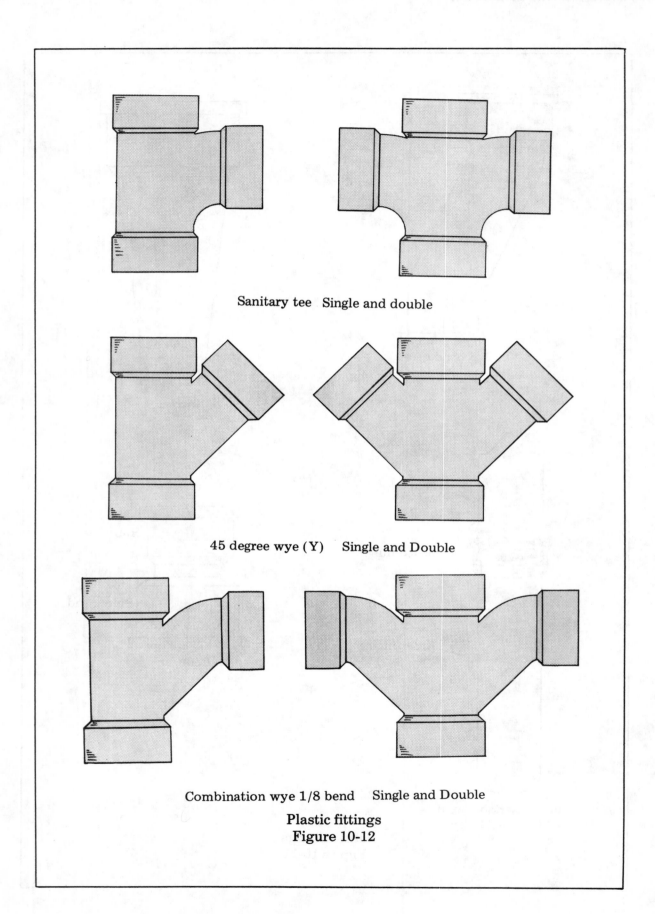

Sanitary tee   Single and double

45 degree wye (Y)   Single and Double

Combination wye 1/8 bend   Single and Double

Plastic fittings
Figure 10-12

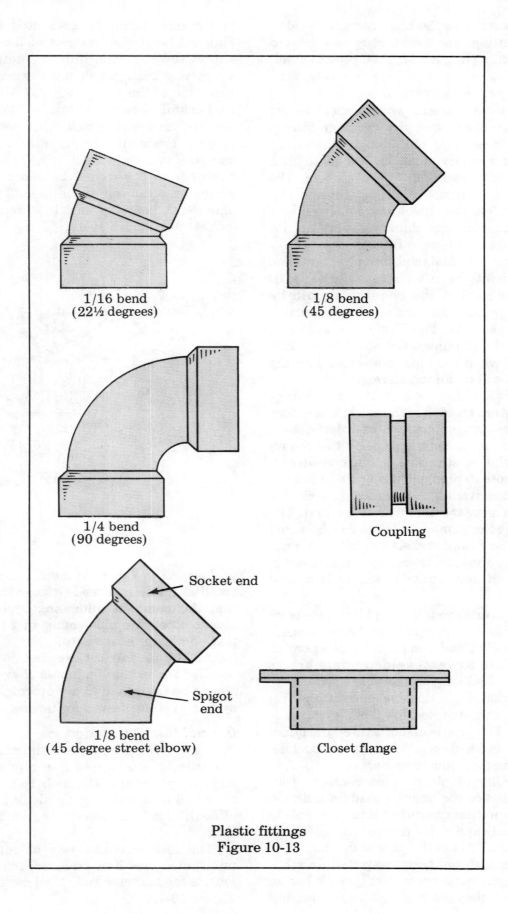

1/16 bend
(22½ degrees)

1/8 bend
(45 degrees)

1/4 bend
(90 degrees)

Coupling

Socket end

Spigot
end

1/8 bend
(45 degree street elbow)

Closet flange

Plastic fittings
Figure 10-13

sanitary system installations, solvent welded (cemented) fittings are used throughout. Where necessary, connecting ABS or PVC pipe to steel, cast-iron or other types of pipe is easily done with correctly sized conversion adapters.

Plastic fitting patterns are identical to the familiar cast-iron screwed pipe drainage fittings except for a socket-type hub. Rough dimensions and sweeps are nearly identical. Adequate pitch per foot is provided in all branch sockets. The roughing-in dimensions of plastic fittings (except in hub lengths) meet the specifications given in American Standards Association for cast-iron screwed drainage fittings.

Molded fittings designed for solvent-welded (cemented) joints have a slight taper. The ID (inside diameter) at the opening should be slightly larger than the maximum OD (outside diameter) of the pipe. The ID at the seat should be less than the minimum OD of the pipe. This ensures that when the pipe is inserted into the fitting there will be an interference fit.

The user should make sure that the fittings he purchases are molded to the correct tolerance to ensure the proper amount of interference. Use only fittings recommended by the manufacturer of the plastic pipe being considered.

As with no-hub pipe, plastic pipe and fittings can be used to advantage where access areas are limited (such as near ceilings or in corners). The solvent-welded or cemented joints can be made with ease. The fitting sockets are much thinner than cast-iron hubs and therefore can be used in thinner partitions. Installation is fast and efficient.

Another advantage to using plastic pipe is its ease of handling. A complete bathroom assembly can be easily lifted into place by one person. Properly made solvent-welded joints are as strong and last as long as the pipe itself.

The weight of plastic pipe and fittings recommended and accepted for DWV systems is *schedule 40*. The pipe is manufactured in 20-foot lengths and is ideal as a labor saver for long runs such as sewer installations.

The fragility of plastic pipe requires that special installation methods be used for building sewers. The bottom quarter of this pipe should be continuously and uniformly supported by the trench bottom. Support the pipe with 4 inches of fine, uniform material that will pass through a ¼-inch screen. Some codes now permit use of up to ¾-inch screen material. Hub and coupling

projections should be excavated as shown in Figure 10-6 so that no part of the pipe load is supported by the hub or coupling. The supporting material should extend 4 inches on each side of the pipe.

Backfill should be firmly compacted with selected material which will pass through a ¼-inch screen. Where permitted, this material may be ¾-inch screen. The backfill should extend from the trench bottom to a point six inches over the top of the pipe. The minimum pipe depth below ground level must be twelve inches. See Figure 10-14.

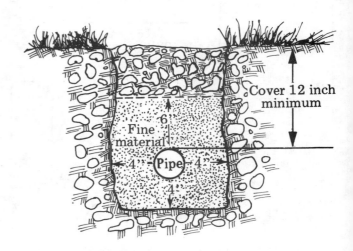

Figure 10-14

Plastic DWV pipe is manufactured in sizes as small as 1¼ inches inside diameter. This means that the complete plumbing system can use plastic pipe. The pipe comes in 20-foot lengths that eliminate most waste.

Although solvent welded joints can be handled two or three minutes after joining, they should be permitted to dry overnight before the system is water tested for inspection.

General Installation Methods

Cutting cast-iron pipe - Uninstalled cast-iron pipe can be easily rolled to any angle and is fairly simple to cut. The tools required are easy to use. But pipe that is rigidly in place is more difficult to cut and generally requires a special tool.

The most common way of cutting free or uninstalled cast-iron pipe, especially on small jobs, is the hammer and chisel method shown in Figure 10-15.

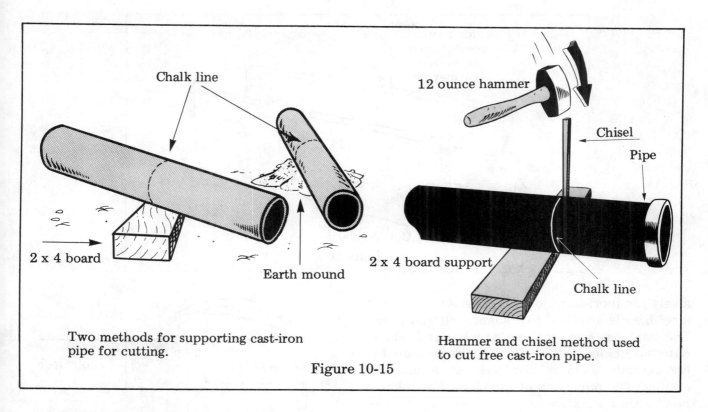

Two methods for supporting cast-iron pipe for cutting.

Hammer and chisel method used to cut free cast-iron pipe.

Figure 10-15

1) **Make the cut square.** Unless you have a very good sense of direction, draw a chalk mark around the entire circumference of the pipe where the cut is to be made.

2) **Lay the pipe over a 2 x 4 board,** or use a mound of earth to support the pipe while it is turned.

3) **Score the pipe barrel** around the chalk mark with a ¾-inch cold chisel which is not too sharp. Use a 12 ounce ball pein hammer. Tap the chisel lightly, turning the pipe until it has been evenly scored all around.

4) **Continue to turn the pipe** and strike the cold chisel with increasingly heavy blows. Most often the pipe breaks evenly at the line of the cut on about the third turn.

For small installations. cut free cast-iron pipe with a handheld skill saw equipped with a metal cutting blade. Use a chalk mark to produce a square cut. A second person should turn the pipe while the first makes the cut. Be sure to wear protective eye shields. This method often causes bits of metal to fly off the blade.

The best tool to cut cast-iron pipe that is rigidly in place is a *soil pipe cutter*. See Figure 10-16. This simple tool can be rented inexpen-

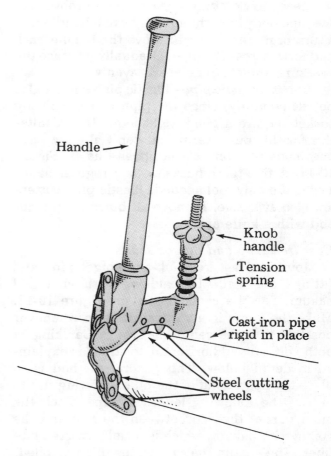

Soil pipe cutter, cutting a sewer pipe in trench
Figure 10-16

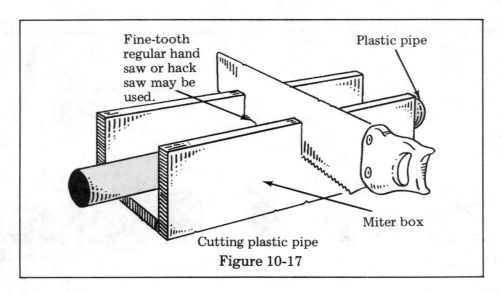

Cutting plastic pipe
Figure 10-17

sively for short-term jobs. It consists of a short steel handle attached to a slightly curved base. The base has two steel cutter wheels. A chain with additional steel cutter wheels can be hooked into slots at the base to assist the cutting. The other end of the chain has a knob handle that works against a spring to provide the necessary cutting tension. Cut a groove into the pipe body by tightening the knob handle half a turn or more at a time. Move the handle back and forth across the pipe. Gradually increase the pressure and the pipe snaps evenly.

Cutting plastic pipe - Plastic pipe should also be cut squarely, since the pipe and the fitting socket require a good, tight joint. Use a miter box for this purpose; if you don't already have one, construct one at the jobsite as in Figure 10-17. A fine-tooth hacksaw or a regular hand-saw is the only tool needed. Plastic pipe cutters are also available. Remove all burrs on the cut end with a knife or file.

Cast-Iron Hub and Spigot Joints

Joints in cast-iron hub-and-spigot pipe and fittings are made by caulking with lead and oakum. See the caulked joint in Figure 10-18. Make the joints in a vertical position whenever possible to make centering easier. Caulking in both vertical and horizontal positions is explained in detail below. The general method is to place the spigot end of the pipe or fitting in the hub of the following length of pipe. Pack the inner part of the gap between the hub and the spigot with oakum, which is usually twisted jute fiber. The oakum may be either plain or oiled. Pack the oakum well into the gap, leaving one inch at the outer end of the hub to receive the

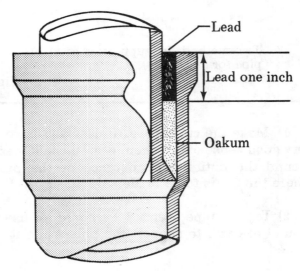

Lead and oakum joint
Figure 10-18

molten lead. A joint should take an estimated one pound of lead per inch of pipe circumference. For example, a two-inch joint should take two pounds of lead, a three-inch joint should take three pounds, and so on. Figure 10-19 shows the tools required to make a lead joint.

Plumber's ladles for holding and pouring the hot lead come in different sizes. Use a ladle large enough to make the joint in one pour. Allow a minute or two for the lead to harden before attempting to caulk the joint as explained below. CAUTION: *Never dip a cold or damp ladle into a pot of hot lead. The lead will explode.* Heat the ladle over the edge of the melting furnace before dipping it into the lead. Take the same precaution when adding new lead to the pot.

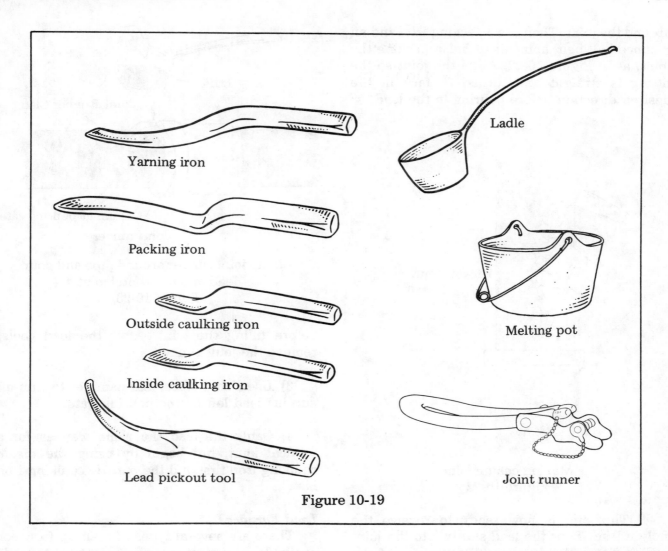

Figure 10-19

Caulking a Vertical Joint

1) If the weather is damp, wipe the hub and spigot ends of the pipes to remove moisture and foreign matter. Water or dampness causes the melted lead to spatter and can result in serious burns.

2) Place the spigot end of the fitting into the hub of the pipe and align the joint so the spigot end is in the center of the hub as in Figure 10-20.

3) Using a yarning iron (Figure 10-21) firmly pack a strand of oakum into the hub completely

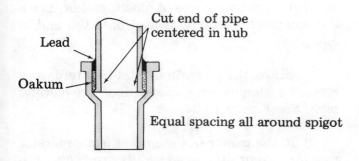

Figure 10-20

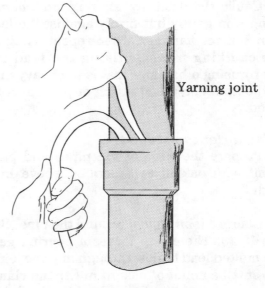

Making oakum joint
Figure 10-21

around the joint. Then use a packing iron and an 8-ounce ball pein hammer to finish packing the joint, as in Figure 10-22. Align the joint so the fitting is straight and turned to face in the desired direction before pouring in the lead.

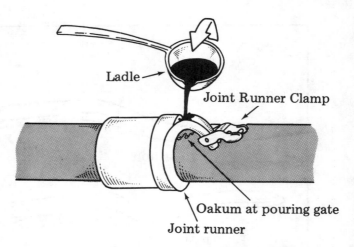

Clamp joint runner around pipe and pour
lead into horizontal joint
Figure 10-23

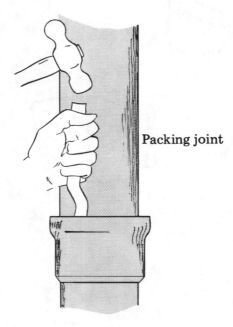

Making oakum joint
Figure 10-22

4) The joint is now ready to receive the molten lead. Pour the lead slowly into the joint until it rises slightly above the top rim of the hub.

5) Caulk the lead by striking an *outside* caulking iron gently but firmly against the lead with an 8-ounce hammer. Then repeat using an *inside* caulking iron. Caulking the lead too tightly by using a hammer that is too heavy may crack the hub. A cracked fitting or pipe must be replaced.

Caulking a Horizontal Joint

1) Prepare the ends of the pipes and pack the joint with oakum as described above in 1 through 3.

2) Clamp a joint runner around the pipe. See Figure 10-23. The clamp leaves a pouring gate for the melted lead to flow through into the joint. Pack a small amount of oakum under the clamp to prevent wasting the lead. Pour the lead fast (faster than in a vertical joint) so it will not cool

before filling the joint. After the lead cools, remove the joint runner.

3) Use a cold chisel and hammer to trim off surplus lead left by the pouring gate.

4) Caulk the lead the same way as for a vertical joint, but begin by using the *inside* caulking iron first and the *outside* caulking iron last.

Lead Furnaces

There are several types of melting furnaces available. The most common type used today is a tank burning propane gas as a fuel, known as an *insto tank*. It has a hood that receives the lead pot. The propane burner can be controlled easily. Regardless of the type of furnace used, always follow the manufacturer's instructions.

Cast-Iron No-Hub Joints

The no-hub joint is one of the easiest joints to make in a plumbing installation.

1) To join two pieces of pipe together, place the one-piece neoprene gasket onto the end of one pipe.

2) Slide the stainless steel shield and retaining clamps over the end of the second pipe. See Figure 10-24.

3) In the center of the gasket is a separator ring (See Figure 10-25) which fits firmly over the end of the pipe. Position the heads of the

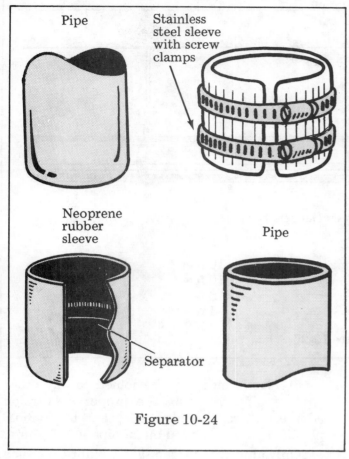

Figure 10-24

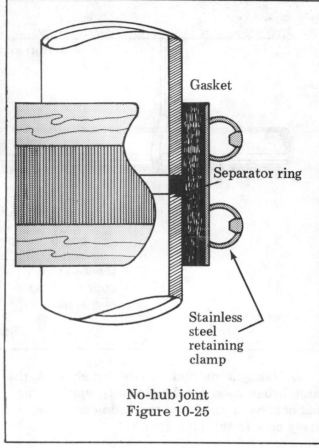

No-hub joint
Figure 10-25

retaining clamps where they are accessible for tightening. Push the end of the second pipe into the neoprene gasket tight against the separator ring. Slide the stainless steel shield over the neoprene gasket until it is covered completely. Push both pipes against the separator ring and tighten with a special torque wrench.

The torque wrench is the only tool needed to assemble no-hub pipe and fittings. It is preset to give 60 inch-pounds of torque. This eliminates guesswork and prevents undertightening or overtightening the retaining clamp.

Plastic Pipe Joints

Plastic pipe has become popular because it is light in weight and easy to assemble. But remember that rigid plastic is best worked with when the air temperature is above 40 degrees F. When the air is too cold, it slows the action of the solvent-cement, thus interfering with the bonding process.

As with any material, care in installation is essential for proper performance. Plastic pipe should be installed aligned and on the correct grade so no bending is necessary to work the pipe into position. Always install plastic pipe and fittings so that identifying marks are visible to the inspector.

Assembly - Follow these steps to obtain welded plastic joints with maximum strength.

1) Wipe or brush off all dirt from the end of the pipe, both inside and outside. Wipe the fitting socket clean. Then use *one* of the following methods to prepare the pipe and the inside of the fitting socket for the application of solvent-cement:

- Apply a cleaner to the spigot and socket contact areas using a natural bristle brush about half the diameter of the pipe in width. Immediately wipe these surfaces dry with a clean cotton cloth to ensure that these surfaces are clean and that surface gloss has been removed.

- Remove the gloss with emery cloth and wipe away the dust.

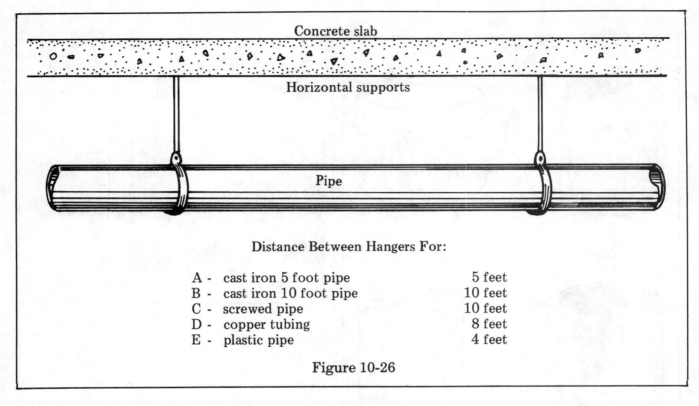

Distance Between Hangers For:

A -	cast iron 5 foot pipe	5 feet
B -	cast iron 10 foot pipe	10 feet
C -	screwed pipe	10 feet
D -	copper tubing	8 feet
E -	plastic pipe	4 feet

Figure 10-26

2) **Using a natural bristle brush (not the same brush used for the cleaner), apply a light coat of solvent cement to the fitting socket and a heavy coat to the pipe spigot.**

• Immediately insert the pipe to the full socket depth while rotating it one quarter turn in the fitting to ensure complete and even distribution of the cement. A bead of cement around the fitting socket shows that the proper amount of cement was used in making the joint. No bead means not enough cement was used.

• Wipe away any excess cement on the pipe or fitting.

3) **Within two minutes after joining. the assembly has begun to set and may be handled with care.** Setting time is longer at temperatures below 75° F. Check the fitting for correct angle and direction immediately; after the joint is assembled and begins to set, any attempt to adjust the joint might result in joint failure. The only correction possible for joint failure is to replace the fitting.

4) **Let all the joints dry overnight before filling the system with water for inspection.**

5) **Plastic pipe may be joined to any other type of pipe by using the proper conversion adapters.** Use threaded adapters to connect plastic pipe to threaded metal pipe and fittings.

Only approved thread tape or lubricant such as Teflon tape or petroleum jelly are recommended as joint sealers. *Do not use conventional pipe compounds.* Threaded joints should not be over-tightened. After hand tightening the joint, it is sufficient to take one-half to one full turn with a strap wrench.

6) **Any conventional pipe clamps, hangers or brackets with a bearing width of ¾-inch or more can be used to support plastic pipe.** Supports for horizontal runs of pipe 1½ inches or less in diameter should be on a maximum of three-foot centers. Supports for pipe of larger diameters should have a maximum spacing of four feet. See Figure 10-26. Trap arms should be supported at the trap discharge. Vertical pipes should be supported to maintain alignment. Where expansion and contraction of the piping will occur, supports should permit the pipe to move without binding. Where pipe expansion is to be absorbed at an elbow or tee, do not place clamps or brackets next to the fitting in a way that restricts the movement of the fitting during expansion or contraction.

7) Plastic traps in summer and vacation homes can be protected from freezing by filling them for the winter with one of the following solutions or mixtures:

- Four quarts of water mixed with five quarts of glycerol.

- Two and one-half pounds of magnesium chloride dissolved in one gallon of water.

- Three pounds of table salt dissolved in one gallon of water.

The salt solutions are effective to approximately 10° F. If lower temperatures are anticipated, the traps should be filled with the glycerol solution.

Be aware of the following restrictions placed by the code on plastic systems.

- Plastic pipe or fittings cannot be installed to support the weight of any plumbing fixture. Be certain the weight of the fixture is supported by approved brackets or other means.

- It is never acceptable to mix ABS and PVC pipe or fittings within the same system.

Drainage, Waste, and Vent Piping Within a Building

The code has certain requirements for underground or above ground installations within a building that you should be aware of before starting to install a DWV system.

- Underground or horizontal drainage, waste and vent piping must be adequately supported. Approved hangers or masonry supports may be needed to keep the pipe in alignment and prevent sagging.

- Building drains passing beneath foundations must have a clearance from the top of the pipe to the bottom of the footings of at least two inches. See Figure 10-27.

- Drainage pipe passing through cast-in-place concrete should be sleeved to provide a ½-inch annular space around

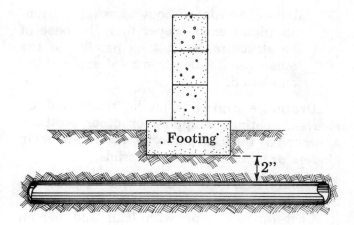

Figure 10-27

the entire circumference of the pipe. This prevents stress on the pipe in case of building settlement or shift. It also protects the pipe from corrosive effects of concrete. This annular space between the sleeve and the pipe must be tightly caulked with coal tar or asphaltum compound, lead, or other approved materials. This provides sufficient flexibility for movement and also prevents insects from using this opening as a passageway to and from the interior of the building.

- Occasionally it is necessary to excavate a trench parallel to and deeper than the foundation of the building. No excavation for drainage piping can be placed within a 45° angle of pressure from the base of an existing structure to the sides of the trench, unless a special design is approved by the building official. See Figure 10-28.

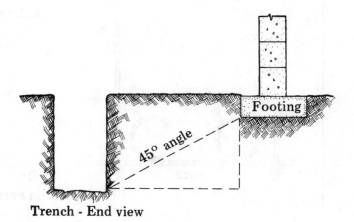

Trench - End view

Figure 10-28

In other words, without approval a drainage pipe trench deeper than the base of the structure can not be parallel to the foundation if it is within a 45° angle of the foundation.

Drainage piping must be protected by sleeves, coating, wrapping or other methods approved by the local authority when installed in cinders or other corrosive materials.

Horizontal Drainage, Waste, and Vent Piping Supports

Cast-iron soil pipe with lead and oakum joints must be supported with hangers or other approved means at not more than 5-foot intervals. Pipe lengths over 5 feet may be supported at intervals of not more than 10 feet. See Figure 10-26. Supports must be placed within 18 inches of the hub or joint. No-hub cast-iron joints must have supports immediately adjacent to each joint or coupling if the developed length exceeds 4 feet. This is necessary because the joints are not rigid and will sag at these points.

Screwed pipe is stronger at the joint but still must be supported at approximately 10-foot intervals when it is used in a horizontal drainage and vent system. Copper pipe used as horizontal drainage or vent piping must be supported at approximately 8-foot intervals. Plastic horizontal drainage or vent piping must be supported at intervals not exceeding 4 feet. Some of the more commonly used horizontal pipe supports are illustrated in Figure 10-29.

Vertical Drainage, Waste, and Vent Piping Supports

In one- or two-story residential construction, vertical supports are needed more to keep the pipe in alignment than to carry any of the weight of the installation. In two-story structures the floor clamp shown in Figure 10-29 could be placed around the stack where it penetrates the second floor level.

Vent Terminals

Extensions of vent pipes should terminate at least 6 inches above the roof. This ensures that gases and odors in all parts of the drainage system discharge well above the roof surface.

Where a portion of a two-story house is used as a sun deck, the vent should extend a minimum of 7 feet above the roof deck. Since

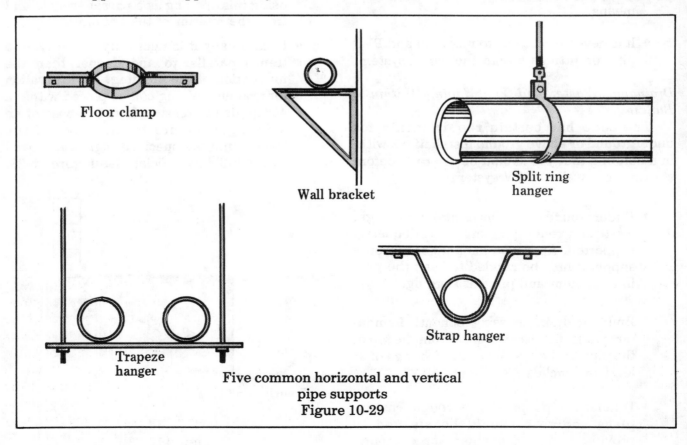

Floor clamp

Wall bracket

Split ring hanger

Trapeze hanger

Strap hanger

Five common horizontal and vertical pipe supports
Figure 10-29

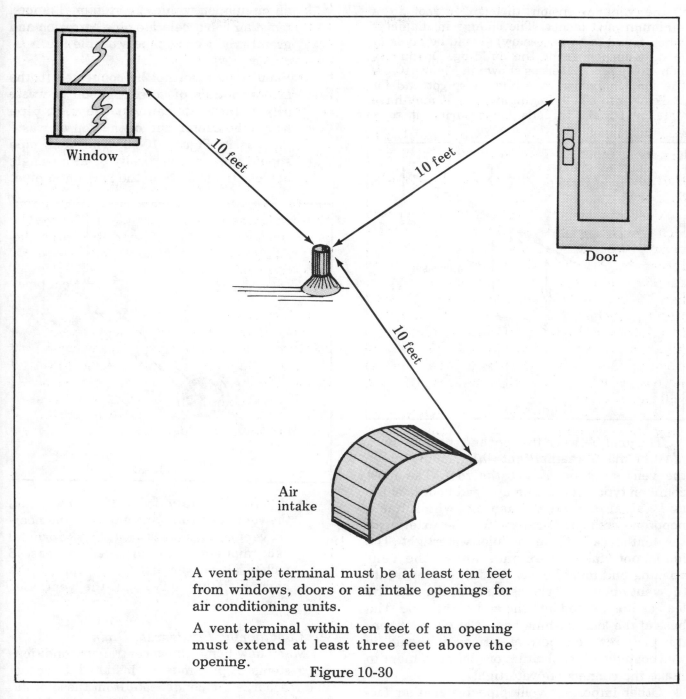

Window

Door

Air
intake

A vent pipe terminal must be at least ten feet
from windows, doors or air intake openings for
air conditioning units.

A vent terminal within ten feet of an opening
must extend at least three feet above the
opening.

Figure 10-30

cast-iron pipe can be extended above a roof deck
only 3 feet, the material for such an extension
should be galvanized steel or copper. Plastic
pipe can be used, but it must be well protected
and supported because of its lack of strength.

In multi-level houses, the vent terminal of a
sanitary system projecting from a lower level
should not terminate within 10 feet of any door,
window or ventilating opening on an upper
level. See Figure 10-30. Where this situation
cannot be avoided, the vent opening should
extend a minimum of 3 feet above the height of

that door, window, or opening.

In very cold climates, the terminal opening
of the vent stack may become closed with frost.
This is caused by the flow of warm, moist air
rising through the vent stack. When this rising
air comes in contact with the frigid outside air, it
forms frost on the interior of the vent stack. If
frost should completely close the opening of the
vent stack, the fixture traps could be siphoned.
If siphoning does not take place, fixtures may
drain very slowly. The most effective way to
reduce frost danger is to increase the diameter

of the vent extension through a roof to a minimum of 3 inches. The change in diameter (when found to be necessary) should be made at least 12 inches inside the building. Do this by using a long increaser, as shown in Figure 10-31. The long increaser may have a spigot end for caulking into a hub in the attic, or it may have male or female threads for securing it to a galvanized threaded vent stack.

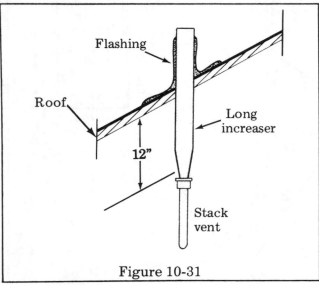

Figure 10-31

The joint between the roof and the vent stack must be made weathertight with flashing where the vent stack penetrates the roof. The most common types of flashing are lead boot flashing and galvanized steel flashing, which has a neoprene collar that snugly fits the exterior of the vent stack to make the joint watertight. The lead boot flashing extends above the vent opening and must be cut and turned down into the vent opening. Take care not to damage the lead as it is tapped into the end of the pipe. The base of the lead flashing has a flat plate. Where this plate rests on the roof surface, it should be well coated with asphaltum or roofing cement to make the connection watertight.

Other important vent pipe installation factors you should know about are outlined below:

- Horizontal vent piping should be installed and sloped to drain to the vertical pipe section. Improper grading or sags in the pipe allow condensation to collect in low places, thus restricting air circulation and reducing the venting capacity. Vertical vent piping connected to a stack vent should be installed in an upward slope. This prevents the entrapment of warm, moist air in the vent pipes and allows free

air circulation within the system. Trapped moist air can accelerate pipe corrosion and greatly reduce the pipe's usable life.

- Vent pipes must not be connected to the side or bottom of horizontal soil or waste pipes. Instead, connect the vent pipe above the center line of the soil or waste pipe as in Figure 10-32. The vent pipe should then rise vertically or at an angle not less than 45° from the horizontal pipe.

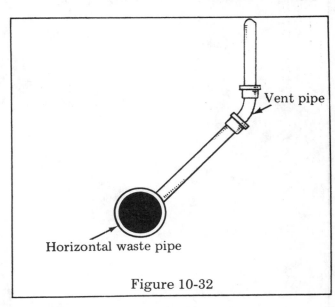

Figure 10-32

- Where a fixture is located in a room and the vent must run horizontally to the nearest vent stack or stack vent, this horizontal vent pipe must be installed at least 6 inches above the flood-level rim of the highest fixture served by the vent. See Figure 10-33.

Air Conditioning Condensate Drains

Many new homes have central air conditioning systems. A pipe must be installed to receive and discharge the condensate from these units. The following drainage method is acceptable by most authorities.

Where it is permitted by your local code, use only plastic pipe and fittings as indirect waste pipe for air conditioning drains. (Of course other piping materials may be used.) The waste piping in a home constructed with a slab on fill should be installed 2 inches below the bottom of the floor slab. Install this waste piping after fill and compaction are completed to prevent sags in the pipe. Lay the piping on a firm base for its entire length and backfill with 2 inches of sand.

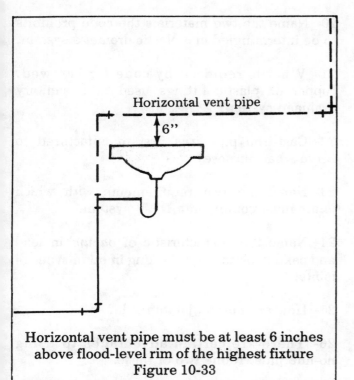

Horizontal vent pipe must be at least 6 inches
above flood-level rim of the highest fixture
Figure 10-33

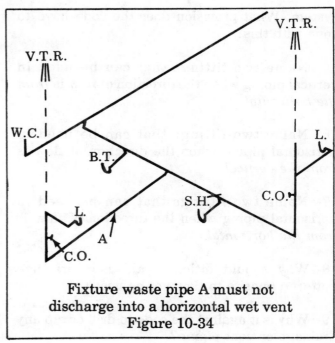

Fixture waste pipe A must not
discharge into a horizontal wet vent
Figure 10-34

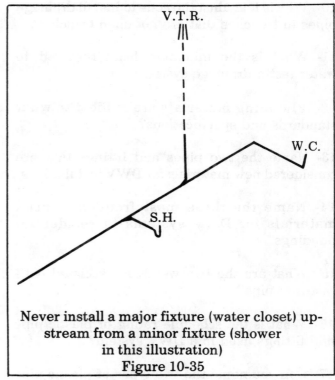

Never install a major fixture (water closet) upstream from a minor fixture (shower
in this illustration)
Figure 10-35

All risers passing through the slab must be sleeved. Waste piping installed in a house with hardwood floors should be strapped securely to the floor joists to prevent sagging.

The minimum size for air conditioning condensate drain piping under a slab is 1¼ inches. It is not necessary to vent a condensate drainage system.

Air conditioning units not exceeding 5-ton capacity (most residential units have less) may discharge their waste upon a pervious area such as grass or bare soil. A dry well may also be used. It should be constructed of a buried pipe 10 inches in diameter by 24 inches long and open at ground level. It is filled with ¾-inch rock. No cover is required for a dry well.

The two faulty installations shown in Figures 10-34 and 10-35 are commonly done by inexperienced plumbers. The inspector will reject these installations. Study the figures carefully to avoid these mistakes.

Questions

1- Why do we need a plumbing code?

2-Why should the drainage system be designed carefully?

3- For what purpose should vent pipes be sized and arranged?

4- Stoppages occur in the best-designed

system. What provision does the code have to cope with this?

5- Name two fittings that can be used in vertical piping when the direction of flow is *from the horizontal.*

6- Name two fittings that can be used in horizontal piping when the direction of flow is *from the vertical.*

7- Name two fittings that can be used in horizontal piping when the direction of flow is *from the horizontal.*

8- Why should fittings with their drainage pattern opposite to flow *not* be used?

9- Why is it against the code to drill or tap any drainage or vent piping?

10- Why is it against the code to install drainage pipes in trenches that are not open trenches?

11- What is the minimum head required to water test a drainage system?

12- Plumbing materials are subject to what standards and specifications?

13- Name the two pipes and fittings that are considered new materials for DWV installations.

14- Name the three most frequently used materials for DWV systems in residential buildings.

15- What are the two weights associated with cast-iron pipe?

16- What is the schedule rating of plastic pipe and fittings used in a DWV system?

17- Can you name the two types of plastic pipe and fittings used in DWV systems?

18- Name two piping materials commonly used for fixture drains (branches) in a cast-iron system.

19- Plastic or copper pipe is frequently used for vent extensions in a cast-iron system with what fitting?

20- Name the two materials the code prohibits to be intermingled in a plastic drainage system.

21- What is required by code for screwed, copper or plastic fittings used in a sanitary drainage system?

22- Cast-iron pipe was first manufactured to serve what purpose?

23- Name the two requirements with which joints must comply in a DWV system.

24- Name the characteristic of *oakum* in lead and oakum joints that is lacking in other types of joints.

25- How are no-hub joints made?

26- There are certain advantages to using no-hub pipe. Name two of these.

27- How long will a well-made lead and oakum joint last?

28- What characteristics of cast-iron soil pipe make it superior as a building material for sewers?

29- Why is it important to keep the pipe barrel in firm contact with solid ground?

30- What is the minimum size cast-iron soil pipe manufactured for use in a DWV system?

31- When a cut is necessary in a hub and spigot system, from what type of pipe should the piece be cut in order to avoid waste?

32- Why is there less waste in a no-hub or plastic system?

33- Tees or tee branches must be installed only in soil, waste, or vent_____.

34- What is the term used for a tee that enters the straight through section of the fitting on a downward curve?

35- Straight tees are used in what section of a DWV system?

36- What are sanitary tees designed to carry?

37- What two purposes does a test tee commonly serve?

38- Y-branches are used to make changes of direction of how many degrees?

39- Combination Y and 1/8 bends are used to make changes of direction of how many degrees?

40- Wyes and combinations are used in the building's main drain to serve what purpose?

41- What is saved when using a combination instead of a Y and 1/8 bend?

42- The side takeoff from a wye or combination may never be larger than what part of the fitting?

43- What method is used for sizing plumbing fittings?

44- How are bends classified?

45- Give the degree of turn for a 1/16, 1/8 and 1/6 bend.

46- When are 1/8 bend offsets generally used?

47- How do a straight cross and a sanitary cross differ?

48- How many vent lines may connect to a straight cross installed in a stack vent?

49- What two types of stacks may a sanitary cross use?

50- Where may an increaser be used?

51- An increaser is generally used for what purpose?

52- Where and for what purpose may a reducer fitting be used?

53- When and for what reason are kafer fittings used?

54- What makes a kafer fitting unique?

55- In the United States, what city first used plastic pipe and fittings for sanitary installations?

56- What are two advantages to using DWV plastic pipe and fittings?

57- Describe the advantageous heat transfer characteristics of plastic drainage pipe over those of *metal* pipe.

58- Rigid vinyl plastic pipe is noted for its resistance to what substances?

59- What must one do before buying plastic DWV material in the United States?

60- What type of joint is used in a DWV system?

61- Plastic fitting patterns are identical to what familiar type of fitting?

62- Plastic pipe and fittings can be used advantageously where access areas are limited. Why?

63- Why is it possible for only one person to work with plastic installations?

64- Why is it impossible to remove a plastic fitting once it is cemented to the pipe?

65- Plastic pipe usually comes in lengths of how many feet?

66- Because of the fragility of plastic pipe when it is used for building sewers, what precautions must be taken?

67- What is the minimum depth below ground a plastic building sewer can be installed?

68- What is the minimum size for plastic pipe and fittings used in a DWV system?

69- What is the most common method used on small jobs for cutting cast-iron pipe?

70- To prevent cast-iron pipe from breaking unevenly, what supports must be considered before the pipe is cut?

71- Why is a chalk mark necessary for the

inexperienced plumber cutting pipe?

72- Why must a soil pipe cutter be used to cut cast-iron pipe that is rigidly in place?

73- What device is used so that plastic pipe can be cut square?

74- What are the most common tools used to cut plastic pipe?

75- What is the depth of lead required by code in a lead and oakum joint?

76- Approximately how much lead is required to pour a 4-inch joint?

77- What may happen if a cold or damp ladle is plunged into a pot of hot lead?

78- What tool is used first in making a lead and oakum joint?

79- In caulking a vertical lead and oakum joint, which caulking iron should be used first?

80- What tool is necessary to retain hot lead poured into a horizontal joint?

81- In no-hub installations where access areas are limited, why is it important to position the heads of the retaining clamps in a certain direction?

82- What is the name of the tool used to assemble no-hub pipe and fittings?

83- Why is it best to work with plastic joints when the air temperature is above 40° F?

84- What must be visible when installing plastic pipe and fittings?

85- What must be removed in preparing plastic pipe and fittings for the application of solvent-cement?

86- What has been accomplished when a bead of cement shows around the plastic fitting?

87- What does it mean when no bead shows around the fitting?

88- What type of joint sealer should *not* be used on plastic threaded joints because of the injurious elements it contains?

89- What must be adequately provided for when you install *horizontal* drainage, waste and vent piping?

90- What is the minimum clearance from the top of a horizontal pipe to the bottom of the footing of a building?

91- How much annular space must be provided for around the entire circumference of the pipe when the pipe must pass through cast-in-place concrete?

92- What procedure must be followed when excavating a trench parallel to and deeper than the foundation of the building?

93- What is the maximum distance between hangers for horizontal cast-iron pipe in 10-foot lengths?

94- What is the maximum distance from a no-hub joint that a hanger should be placed?

95- What is the maximum distance between hangers for 3-inch plastic pipe installed in 20-foot lengths?

96- What common support is used to keep vertical pipe in alignment?

97- How high above the roof should vent pipes terminate?

98- What is the minimum distance from a window that a vent pipe opening may be installed?

99- What is used to weathertight a vent stack where it penetrates the roof?

100- What can happen if vent pipes are improperly installed?

101- What happens when moist air is trapped in a horizontal vent pipe?

102- Where must a vent pipe be connected to a horizontal soil or waste pipe?

11

Septic Tanks
And Drainfields

Cesspools and outhouses were once the common means of disposal of human waste in both urban and rural areas. The inadequacy of cesspools and outhouses became clear when population densities increased. Now, local authorities who enforce model codes no longer accept outhouses or cesspools for human waste disposal. Some codes still permit cesspools as a means of sewage disposal for limited, minor, or temporary use, but only when they are first approved by the local authority.

Contamination from sewage is a real possibility in rural areas where drinking water is taken from open or closed wells. Cross-connections allow untreated sewage to enter the drinking water, hastening the spread of such diseases as cholera and typhoid.

Where public sewers are not available, the most acceptable method for sewage disposal is the septic tank system. There is still some controversy over whether the septic tank is truly danger-free. The Septic Tank Association contends that there are no proven cases of septic tanks contaminating drinking water. But the Department of Environmental Resource Management (DERM) has questioned septic tank safety. The controversy and field testing continue. In many areas, DERM has been granted the authority to regulate the installation of septic tanks, and DERM has established strict guidelines that have been adopted by most model code organizations.

Septic tanks and drainfields may vary in design, construction and installation methods. But the basic function of all septic tanks and the principles of sanitation and safety remain the same regardless of the code or geographical location.

Septic tank and drainfield installations are an important part of plumbing work, but few plumbing contractors and fewer homeowners attempt this type of work. A licensed septic tank contractor usually does the installation and maintenance for these systems. The plumbing professional is usually equipped to handle only the connection of the building drainage system to the inlet tee of the septic tank. But whether or not you intend to do this work as a plumber, be aware of the proper way to install the entire treatment system.

A properly designed and installed septic tank and drainfield needs no more maintenance than the removal of the accumulated solids (sludge on the bottom and scum on the top) every few years. The drainfield requires no attention until it ceases to function after about fifteen or twenty years. When this occurs, the drainfield must be dug up and the old drain pipe and gravel removed and disposed of. Replace the drainfield with new drain pipe and gravel to meet present code requirements.

How The Septic Tank Functions

A septic tank is a watertight receptacle which receives the sewage discharge of a drainage system through an inlet tee. See Figure 11-1. It is designed to separate solid from liquid wastes. The solids are usually about three quarters of a pound in each 100 gallons of water. The heavier portions settle to the bottom of the tank while the lighter particles and grease rise to the top of the liquid. *Anaerobic* bacteria, which feed in the absence of air or free oxygen, decompose the solid matter, transforming it into gases and harmless liquids. The resulting gases agitate the tank contents and hasten the action of the bacteria. As new sewage is discharged from the plumbing system and enters the septic tank, the gases are forced up and through the drainage vent pipes and into the atmosphere above the building roof. The tank is sized to have a capacity equal to approximately 24 hours

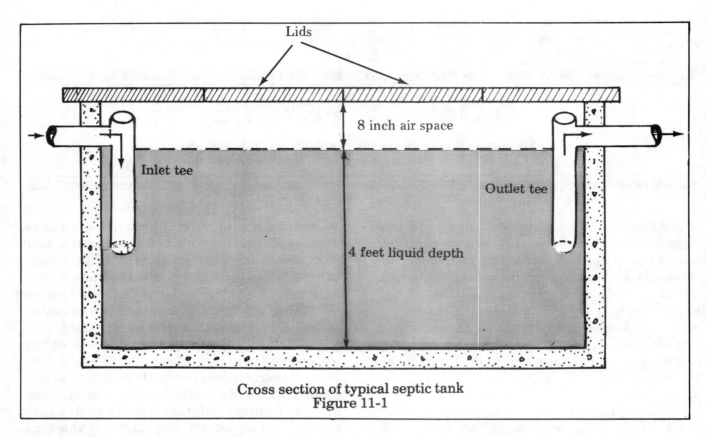

Cross section of typical septic tank
Figure 11-1

of anticipated flow. This is necessary so that the bacteria have adequate time to fully digest the solids.

The resulting clear liquid, called *effluent,* is forced through the outlet tee (see Figure 11-1) and enters a subsurface system of open-joint or perforated piping installed on a bed of washed rock, gravel, slag, coarse cinders or other approved materials. The best designed drainfields allow air to penetrate to the rock bed so that final *aerobic* bacterial action can take place in the presence of free oxygen. The effluent seeps out between the joints or through the holes in the piping, undergoing further nitrification as it oxidizes and evaporates in the rock bed. The treated effluent is thus returned to the soil beneath the drainfield. Make sure that the ground surrounding the septic tank and drainfield can absorb the flow of effluent properly. Poor ground absorption can be the cause of unpleasant odors.

Locate drainfields in full sunlight, if possible, to allow proper oxidation and evaporation of the waste. Aerobic bacteria work more effectively when they are warm. In cold climates mound dirt over the drainfield as insulation.

Avoid planting trees or shrubbery on or near a drainfield. The roots will penetrate the

drainfield and considerably reduce its useful life.

Size and Location of Septic Tanks

The minimum capacity requirements for residential and duplex septic tanks vary slightly between different model codes. For this section of the code, refer to the table adopted by your local authority. Compare Tables 11-2 and 11-3 with your own.

Number of Bedrooms	Minimum Liquid Capacity Required (Gallons)
2 or less	750
3	900
4	1,050
5	1,200

South Florida Building Code
Capacity of Septic Tanks
Table 11-2

All codes base the capacity of septic tanks on the number of bedrooms. The septic tank sizes in Tables 11-2 and 11-3 are designed to provide for sludge storage capacity as well as for domestic food waste disposal units. For each extra bedroom over the number listed in the

Single Family Dwellings Number of Bedrooms	Minimum Septic Tank Capacity in Gallons
1 or 2	750
3	1,000
4	1,200
5 or 6	1,500

Standard Plumbing Code
Capacity of Septic Tanks
Table 11-3

tables add 150 gallons. These extra liquid capacities will provide the necessary 24-hour retention period to permit the bacteria in the tank time to adequately digest the solids.

The code has placed a number of restrictions on the locations of septic tanks. These are grouped and listed below:

• The tank cannot be located under or within 5 feet of any building.

• The tank cannot be located within 5 feet of any water supply line.

• The tank cannot be located within 5 feet of property lines other than public streets, alleys, or sidewalks.

• The tank cannot be located within 50 feet (25 feet in some codes) of the shore lines of open bodies of water.

• The tank cannot be located within 50 feet (100 feet in some codes) of a private water supply well which provides water for human consumption, bathing or swimming.

• The excavation for a septic tank must not be made within a 45° angle of pressure as transferred from the base of an existing structure to the sides of an excavation. See Figure 10-28.

• Circulation of air within the septic tank and drainfield must be through the inlet and outlet tees of the septic tank only. No other circulation is permitted.

Note that when a septic tank is to be abandoned it must be pumped dry by a certified professional, the bottom of the tank broken in, and the tank filled with clean dirt.

The Design and Construction of Septic Tanks

Blocks, brick, or sectional tanks are not permitted by most plumbing codes. Metal tanks are approved by local authorities in some areas. Fiberglass tanks are relatively new and have been approved in some areas as long as they are properly installed and protected. Precast or cast-in-place concrete tanks are the most common types and meet the requirements of practically all plumbing codes. The interior walls of all septic tanks must be finished so that they are smooth and impervious. Voids, pits, or protuberances on these walls are prohibited by code.

Inlet and outlet tees for concrete septic tanks must be of terra cotta or concrete. They must have a wall thickness of at least one inch. The diameter of these tees must be not less than, nor more than two times greater than, the diameter of the building sewer pipe. The outlet invert must be a minimum of one inch and a maximum of three inches lower than the inlet tee. Inlet and outlet tees must be installed at opposite ends of the septic tank and be a maximum of 5 inches above and 18 inches below the liquid level line. See Figure 11-1.

The connection of the building drain pipe to the inlet tee, as well as the tight-jointed pipe used from the outlet tee, must be made watertight with a rich mixture of cement and sand. To protect metal tanks from corrosion, the inside and outside must have a bituminous coating applied. Some codes require that concrete septic tanks also have a bituminous coating applied to the inside only.

Access to septic tanks for cleaning purposes must be provided by *either* manholes at least 20 inches in diameter (one located over each tee) or removable sectional lids. Septic tanks must be installed level.

Sizing Drainfields

It is essential that the soil surrounding a septic tank be able to absorb the effluent properly. Codes vary on this requirement. Differences in soil absorption requirements are generally due to two factors: 1) topography (the lay of the land) and 2) absorption capacities and rates of soils. Check your local code for any

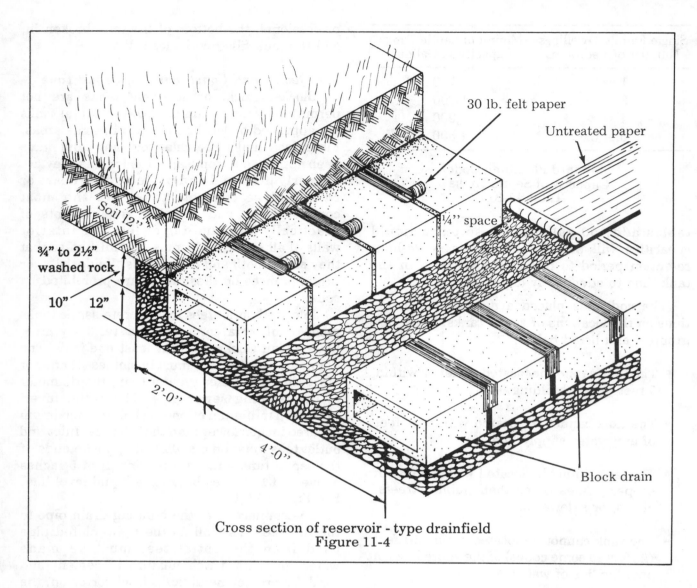

Cross section of reservoir - type drainfield
Figure 11-4

special requirements.

Topography Where the lay of the land is relatively flat and the soil is sand or gravel, no special problem exists and a conventional subsurface drainage system may be used. See Figures 11-4, 11-5, and 11-6.

Sometimes the site is steep or the soil is predominately sugar sand, rock, clay or other impervious formations. In such cases seepage pits or underdrains may be installed, if the original subsurface system cannot absorb all the daily effluent. See Figures 11-7 and 11-8.

Underdrains are open-joint drain tiles buried about twice as deep as the regular subsurface drainage system. See Figure 11-8. The underdrains are not connected to the distribution box or to the regular subsurface drainage system. They form a separate system and are installed preferably midway between the subsurface

drainage lines, as in Figure 11-7. An underdrain filter bed may be installed to help prevent saturation of the drainage area due to poor percolation.

Rated Absorption Where soil porosity appears to be less than usual (such as in heavy clay, rock, sugar sand and the like) conduct a percolation test before construction begins.

Dig or drill at least three test holes to a depth equal to that of the planned drainage bed. Place a 2-inch layer of gravel in the bottom of each hole and fill the hole with water. If the soil is tight or has a heavy clay content, the test hole should stand overnight. If the soil is sandy and the water disappears rapidly, no soaking period is needed. Then pour water into the hole to a depth of 6 inches above the gravel. Measure the depth of the water every 10 minutes over a 30-minute period. The drop in water level

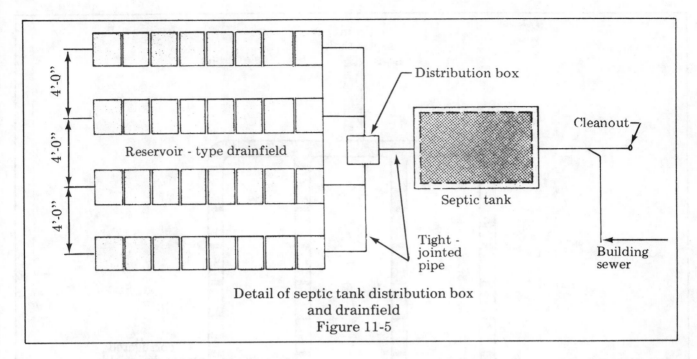

Detail of septic tank distribution box
and drainfield
Figure 11-5

during the *final* 10 minutes determines the percolation rate of the soil. If a test hole needs more than 15 minutes to absorb 1 inch of water, the soil is not suitable for a subsurface drainage system. Take the average percolation rate for the three holes to figure the minimum effective drainfield absorption area. Table 11-9 shows the required absorption area in square feet per bedroom based on the percolation rate in minutes. The most acceptable rate is three minutes per inch, which equals 100 square feet per bedroom, with a minimum drainfield of 200 square feet for one or two bedrooms. (Some codes permit 150 square feet as minimum.) Any number below or above the three-minute rate would constitute a special condition and would require that the drainfield be sized accordingly.

Find the required drainfield absorption area per bedroom for single family and duplex residences on Table 11-9, as follows:

If the percolation rate is 5, you need 125 square feet of percolation area per bedroom. If you are sizing a drainfield for a four-bedroom residence, multiply four times 125. The required drainfield size for bedrooms would be 500 square feet of trench bottom. Either individual trench bottoms or the overall square footage of a single drainfield bed (length times width) may make up the total 500 square feet required for this installation. Substitute

your own percolation rate for your own job and the number of bedrooms in the residence. Then calculate the square footage of drainfield trench bottom required.

Drainfield Construction and Installation

Several materials can be used for drainfield piping. The most common are open-jointed or perforated clay tile, perforated bituminous fiber pipe, block or cradle-type drain units, and the newer corrugated plastic perforated tubing. Each type of piping has its own code requirements and installation methods, and these will be covered later in the chapter. All drainfield piping must be able to carry and distribute the effluent evenly throughout the drainfield bed. The end of each distribution line must be sealed by capping or by cementing a block to it, as in Figures 11-6 and 11-7.

Tile Drainfields

Drainfield tile must have a minimum inside diameter of 4 inches. The open-joint drainage tile should have a gentle downward slope of no more than ½-inch per 10 feet to insure even distribution of the effluent. Lay the tile on a 12-inch deep bed of ¾ to 2½-inch washed rock. Level another 2 inches of rock on top of the tile. This is a total depth of rock of 18 inches for the full width of the trench. See Figure 11-10.

Lay the tile with a space of ¼-inch between the tile ends. Then cover the top and sides of each ¼-inch space with a 4-inch wide strip of 30-

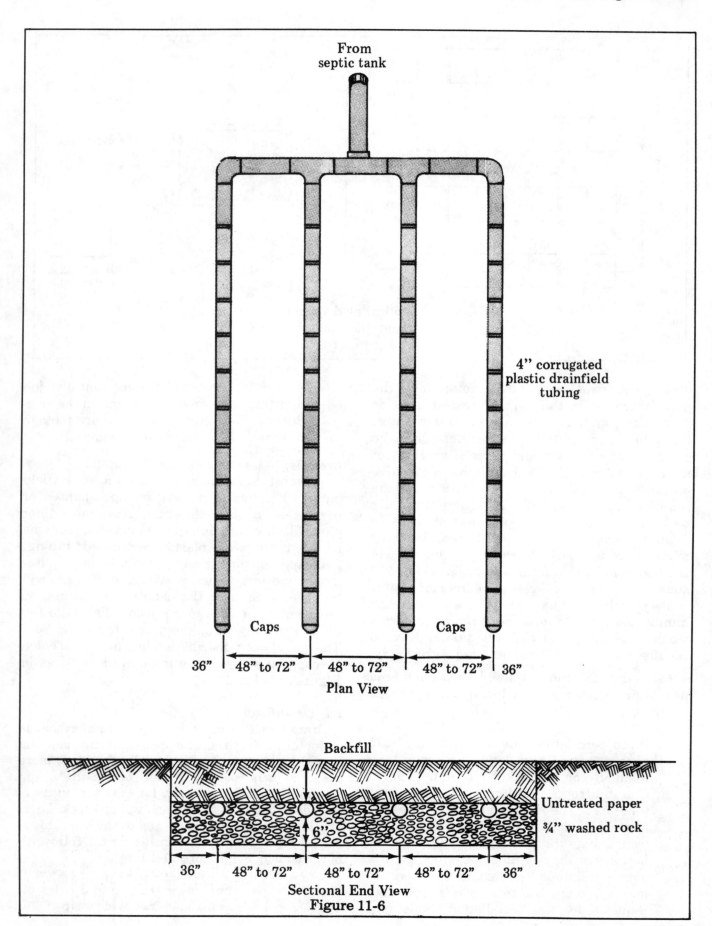

From
septic tank

4" corrugated
plastic drainfield
tubing

Caps   Caps

36"   48" to 72"   48" to 72"   48" to 72"   36"

Plan View

Backfill

Untreated paper

¾" washed rock

6"

36"   48" to 72"   48" to 72"   48" to 72"   36"

Sectional End View
Figure 11-6

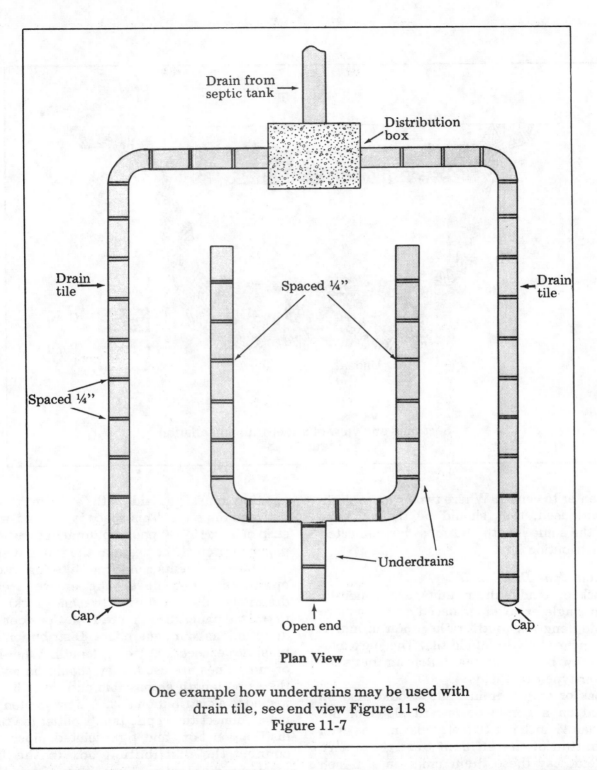

Drain from septic tank

Distribution box

Drain tile

Spaced ¼"

Drain tile

Spaced ¼"

Underdrains

Cap

Open end

Cap

Plan View

One example how underdrains may be used with
drain tile, see end view Figure 11-8

Figure 11-7

pound bituminous saturated paper. This prevents sand or other small particles from filtering into the openings. Then cover the entire length and width of the trench with untreated paper. This prevents sand and other small particles from filtering down and through the washed rock when the trench is backfilled.

Regardless of the width of the trench,
consider each linear foot of drainfield tile to be one square foot. Each trench must have a minimum width of 18 inches with a maximum width of 36 inches.

The maximum length of a single tile drainfield trench is 100 feet. Where more than one tile drainfield trench is required, the trenches must be spaced at least 6 feet apart

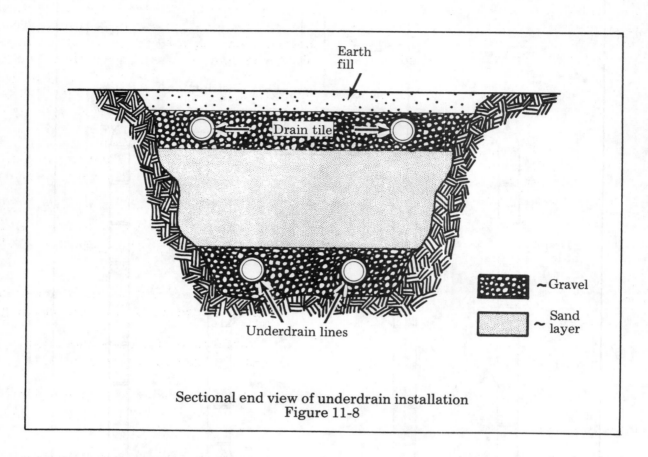

Sectional end view of underdrain installation
Figure 11-8

from center to center. Where two or more drain lines are used, they should be the same or nearly the same length and must be connected by a distribution box.

Reservoir-Type Drainfields

Block or cradle drain units are generally used in single excavations based on the square foot rule (length x width) rather than in individual trenches like drainfield tile. These excavations may be known as leaching beds or reservoir-type drainfields.

Block or cradle drain units should also be installed on a gentle downward slope of not more than ½ inch per 10 feet. This ensures even distribution of the effluent over the entire drainfield. Lay these drain units on a 6-inch deep bed of ¾- to 2½-inch washed rock. Then cover the drain with more of the same rock, leveling the top layer at 10 inches above the bottom of the units. This yields a total of rock of 22 inches for the full width of the drainfield.

Drain units that have a fixed opening to provide seepage may be butted tight against each other. But drain units without a fixed opening must be laid with a ¼-inch space between the ends. This space is covered with a strip of 4'' x 16'' 30-pound bituminous saturated paper to protect the seam at the top and sides. The paper prevents sand from filtering into the openings. Then cover the entire area of drainfield filter material (washed rock) with untreated paper. See Figure 11-4. Consider each drain unit as four square feet. Distribution lines must not exceed 100 feet in length. Where two or more lines are used, they should be as near the same length as possible and must be connected to a distribution box. A tight-jointed pipe must connect the septic tank's outlet tee to the distribution box, and tight-jointed pipes must connect the distribution box to the fixed reservoir distribution lines. See Figure 11-5.

Corrugated Plastic Perforated Tubing Drainfields

Plastic tubing drainfields are most often used in single excavations based on the square foot rule (length x width) rather than in individual trenches. In recent years this material has been approved for drainfields by many codes and seems to work well.

Plastic tubing drainfields must also have a gentle downward slope of not more than ½-inch in each 10 feet. Lay the tubing on a bed of 12-inch deep ¾- to 2½-inch washed rock, then cover the tubing with more rock to a level of 6 inches above the tubing bottom. This is a depth of rock of 18 inches for the full width of the drainfield.

Since the tubing is manufactured with adequate fixed openings, all joints are tight and no extra precautions are necessary to prevent sand and other particles from entering the drain tubing. The plastic is lightweight and easy to install. Lengths of tubing can be cut with a knife or a fine-toothed saw to the desired measurements. Fittings replace the distribution box, as in Figure 11-6. Cover the entire area of drainfield filter material (washed rock) with untreated paper.

The distance between centers of plastic drain lines must not exceed 72 inches. The outside drain lines should be a minimum of 36 inches from the excavated wall, as in the sectional view in Figure 11-6. Corrugated plastic drain lines should not exceed 100 feet in length. Where two or more lines are used, they should be as near the same length as is practical. See the plan view in Figure 11-6.

Percolation Rate Time in Minutes for Water to Fall One Inch	Absorption Area in Square Feet per Bedroom
1 or less	70
2	85
3	100
4	115
5	125
10	165
15	190
Over 15	Unsuitable for absorption field

Table 11-9

Drainfield Replacement

In time, most drainfields fail. The soil can no longer absorb the effluent properly, resulting in offensive odors. "Dead" drainfields cannot be repaired, in most cases, but must be replaced. All drain lines and used filter rock must be removed and replaced with new materials. Until

recently, most codes permitted replacing an existing drainfield with a new drainfield of the same type and size as the original. Now, however, any replacement drainfield must be sized according to current construction requirements. (See Table 11-9.) *Check with the local authority before replacing an existing drainfield and find out what is acceptable in your area.*

Basic Restrictions on Drainfields

- Drainfields must have a minimum separation distance of 8 feet from basement walls or terraced areas.

- Drainfields must have a minimum earth cover of 12 inches and a maximum of 18 inches.

- Drainfields must not be located under any building or within 8 feet of any building.

- Drainfields must not be located within 5 feet of water supply pipe lines.

- Drainfields must not be located within five feet of property lines other than public streets, alleys or sidewalks.

- Drainfields must not be located within 50 feet of shorelines of open bodies of water.

- Drainfields must not be located within 100 feet of any private water supply well which provides water for human consumption, bathing or swimming.

Care of Septic Tanks and Drainfields

As with any plumbing system, septic tanks and drainfields should be treated with care. If not abused, they should give years of trouble-free service. Solid materials such as rags, sanitary napkins, heavy paper, cooking fats, coffee grounds, and food leftovers should not be discharged into the waste system. These items cannot be decomposed by anaerobic bacteria. As a result, sludge and scum form faster than normal, meaning that the tank must be cleaned more frequently.

In normal service, the tank must be cleaned of undigested particles and accumulated sludge and scum. Most codes require that a tank be cleaned by a certified professional who has the equipment to dispose of the waste properly.

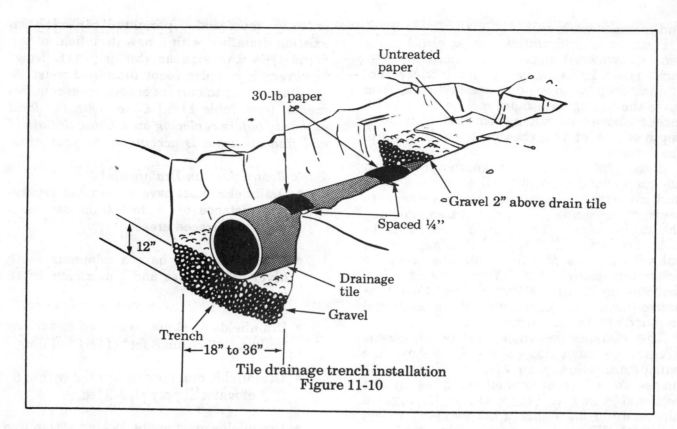

30-lb paper

Untreated paper

Gravel 2" above drain tile

Spaced ¼"

12"

Drainage tile

Gravel

Trench

←18" to 36"→

Tile drainage trench installation
Figure 11-10

This waste is dangerous to health and it should not be buried or used as fertilizer on private property.

Paint thinners, chemical cleaners, or photographic chemicals should never be disposed of through a plumbing system that depends on a septic tank and drainfield for sewage disposal. *Chemical drain cleaner should never be used to unclog drain pipes.* These chemicals will kill the bacteria and the septic tank will cease to function. Solids will no longer be digested. If undigested particles find their way through the outlet tee to the drainfield, it too will fail to function. This could cause a serious plumbing problem and unnecessary expense.

Septic tanks can become sluggish; the building drain fails to drain as rapidly as it should. If you can find no blockage in the line, the trouble might be that there are simply not enough bacteria to do an adequate job. You can increase the population of these microscopic creatures by flushing a yeast cake through the water closet once a week or so.

Odor in the drainfield can be corrected in several ways:

1) You can reduce the flow of water into the septic tank and, thus, into the problem drainfield. Disconnect the clothes washing machine and reconnect its waste to a separate drainfield or drywell, if this is permitted by the code.

2) The percolation rate of the soil may not be as high as it was when the drainfield was originally installed. Dig a seepage pit at the far end of the drain pipes to receive the effluent that has not yet seeped into the soil.

3) Check whether trees or shrubbery have been planted too close to the drainfield. Their roots could be clogging the drain pipes, making them useless.

4) Check for excess shade of the drainfield. Install and keep the drainfield area in full sunlight if possible.

Questions

1- Under what circumstances may a cesspool be used?

2- What is the term used when drinking water becomes contaminated?

3- Name two diseases that can be caused by lack of proper sanitary facilities.

4- What is the most acceptable method for sewage disposal where public sewers are not available?

5- The plumbing professional or homeowner usually connects the building sewer or drain to what portion of a septic tank?

6- What two substances must a septic tank be kept free of in order to maintain it in good working order?

7- What must be done to correct a drainfield that stops working?

8- What is a septic tank designed to do?

9- Anaerobic bacteria live and feed within a septic tank because of the absence of what element?

10- Into what form or substance do bacteria living in a septic tank transform solids?

11- What takes place within a septic tank when new sewage is discharged into it?

12- Approximately how many hours of anticipated sewage flow must a septic tank be sized for?

13- What is the liquid called after it has been processed?

14- The liquid enters a subsurface system containing what type of pipe?

15- Name the substance that is permitted to penetrate to the bedrock of a properly designed drainfield.

16- Why should a drainfield be located in full sunlight?

17- What happens if trees or shrubbery are planted over or near a drainfield?

18- What determines the size of a septic tank?

19- What determines the size of a drainfield?

20- How close can a septic tank be located to a building?

21- How many feet must a septic tank be located from a private (potable) well?

22- An excavation for a septic tank must be made at no less than a _____ degree angle from the base of an existing structure.

23- How is air circulation within a septic tank and drainfield accomplished?

24- What must be done when a septic tank is abandoned?

25- Who is authorized by most codes to clean a septic tank?

26- What types of material are not permitted by most codes in the construction of a septic tank?

27- What is the material that is most often used in septic tank construction because it meets most model code requirements?

28- What is the minimum distance the outlet tee invert must be below the inlet tee?

29- Where must the inlet and outlet tees be placed in a septic tank in relation to each other?

30- What must be done to protect metal septic tanks from corrosion?

31- Give the two ways the lid of a septic tank can be arranged to permit access for cleaning purposes.

32- What is the essential characteristic of soil suitable for a septic tank installation?

33- Name two factors necessary for sizing drainfields.

34- What is the purpose of underdrains?

35- What is required where soil porosity appears to be less than usual?

36- What constitutes a ''special condition'' when sizing a drainfield?

37- What must be done to the end of each distribution line in a drainfield to meet code requirements?

38- What is the minimum inside diameter of drainfield tile?

39- What is the acceptable downward slope for all types of distribution lines?

40- Why should the entire width of filter material be covered with untreated paper?

41- What is the minimum center-to-center spacing for individual trenches?

42- On what rule is the sizing of reservoir type drainfields based?

43- What is the minimum depth of rock required under drain units of a filter bed?

44- What must drain units be provided with to avoid spaces between the ends of each unit?

45- What is the maximum distance between centers of distribution lines for reservoir-type drain units?

46- What type of pipe must be used to connect the septic tank outlet tee to the distribution box?

47- Why is no seepage space needed when plastic tubing is used?

48- What is the maximum distance between the centers of drain lines in a plastic distribution system?

49- What is the maximum permitted length of a drain line?

50- What is the minimum depth of earth required to cover a drainfield?

51- What is the minimum distance of separation between a drainfield and a basement wall?

52- What is the minimum separation space between a drainfield and a water supply line?

53- To help maintain a septic tank in good working condition, what solid waste materials should *not* be disposed of through a drainage system?

54- Name two chemicals that should not be disposed of through a drainage system served by a septic tank.

55- What may be used to help the growth of bacteria in a septic tank?

12 Water Supply And Distribution System

The sources of water for public water supply systems are lakes, rivers, and deep wells. This raw water is generally not acceptable for human consumption. Unpleasant taste, odors and impurities are removed through treatment before the water is distributed through mains.

The water main for a public water distribution system is usually located in the street, alley, or dedicated easement adjacent to each owner's parcel of land. The main carries public water for community use, and is a common pipe installed, maintained and controlled by local public authorities.

The water main provides drinkable (potable) cold water at a pressure that meets certain standards. Pressures in the main usually range from 40 to 55 pounds per square inch (psi). If the pressure in certain instances exceeds 80 psi, a pressure-reducing valve should be installed in the water-service line where it enters the building.

Each property owner requests that the utility company tap into the public water main adjacent to their property and install a water meter located near the property line. (See Figure 12-1 for a typical water service connection detail.) The building water service pipe is then connected at the outlet side of the water meter.

Private Water Supply

Nearly one home in five in the United States is without a public water system. Approximately 18.7% of all households or 47 million people get their water from lakes, streams or private wells (see Chapter 17). These private sources must provide water that is safe to drink. In many cases the homeowner must provide his own treatment equipment.

Pumps are generally used to pump water from the source to a storage tank near the building. The pump also pressurizes the water in the storage tank so the water distributing pipes can supply the plumbing fixtures with enough water to function properly. The building water service pipe is then connected to the storage tank.

Note that local authorities usually require that a connection be made to a public water supply system when one becomes available.

How The Water System is Sized

The maximum rate of flow or the demand in a building's water supply system cannot be determined exactly. You cannot predict how many fixtures will be in use at the same time. But you still have to estimate the maximum demand as accurately as possible to ensure that there will be an adequate water supply to all fixtures when it is needed. The objective is to avoid undersizing but still be as economical as possible. See the typical water piping diagram in Figure 12-2.

Some of the factors that affect the sizing of the water supply piping are the types of flush devices to be used on the fixtures, the pressure of the water supply in pounds per square inch at the source, the length of the pipe in the building, the number and kinds of fixtures installed, and the number of fixtures used at any given time.

A simplified method for sizing building water supply systems is based on the *demand load*. In this method sizing is done in terms of water supply fixture units (1 fixture unit = 7½ gallons per minute). The water supply systems of many buildings have been sized adequately and properly by this method. It works well for all buildings supplied from a source that has enough water pressure to supply the highest and most remote fixtures during peak demand.

You can use the method described in the following section for almost all one and two

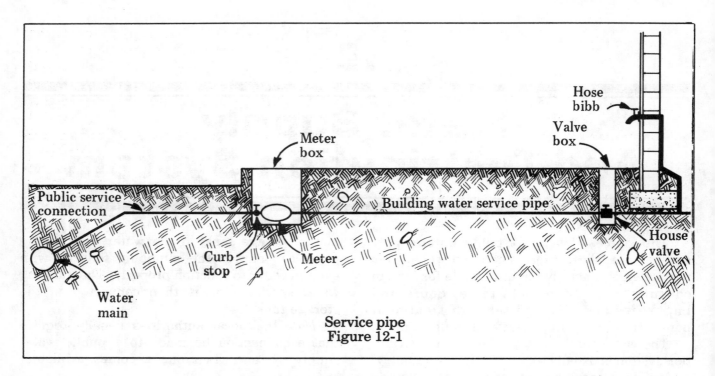

Service pipe
Figure 12-1

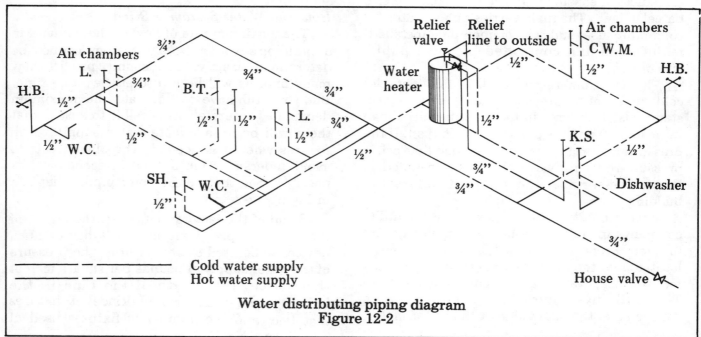

Water distributing piping diagram
Figure 12-2

family dwellings, if the minimum water pressure is at least 50 psi. In these types of buildings 50 psi is usually more than enough to overcome the static head and ordinary pipe friction losses. Pipe friction is *not* an additional factor to consider in sizing.

This method takes into account 1) velocity limitations recognized as good engineering practice and 2) the recommendations of manufacturers of piping materials. The calculations in Table 12-4 are considered adequate by plumbing authorities in most areas. This table takes into account the major physical properties that reduce or restrict the flow of water. Anyone wishing to do simple water pipe installations will find this sizing method easy to use. Both Table 12-3 and Table 12-4 are similar to those you will find in many codes. Examples and figures have been included to explain how the tables are used. More complete tables for larger installa-

Fixture Type	Pipe Size (inches)
Water closet	½
Shower	½
Bathtub	½
Lavatory	½
Kitchen sink	½
Clothes washer	½
Hot water heater	¾
Hose bibb	½ or ¾

Note: For fixtures not listed, the minimum supply branch may be sized as for a comparable fixture.

Size of fixture supply pipe
Table 12-3

tions can be found in *PLUMBER'S HANDBOOK* by the same author.

Sizing The Water Service Pipe

The water service pipe is the heart of the water supply system and must be sized first. This pipe must convey the necessary volume of water at an acceptable velocity to the water distributing pipes within a building if the fixtures are to operate properly. The minimum practical size for the water service line is ¾-inch, and that size should be used even when calculations show a smaller one could be used in certain instances.

The size of the water service pipe must not change when it enters the building to become the building water main, whether the pipe is horizontal only or horizontal and vertical as in a two-story dwelling. After water distributing branch pipes are connected to the building main, the main may be reduced in size with proper fittings. As the main progresses through the building, the demand likely to be placed on the line decreases.

Tables 12-3 and 12-4 show the correct pipe sizes for residential buildings. The tables have certain sizing restrictions which are not difficult to follow when sizing water supply pipes. These tables are adequate if these restrictions are kept in mind:

- Table 12-4 applies only where water main pressure does not fall below 50 psi at any time.

- In single family residential buildings more than two stories in height, use the next larger pipe size.

- Buildings located where the water main pressure falls below 50 psi should use the next larger pipe size.

- Table 12-3 shows the minimum sizes for fixture supply pipe from the main or from the riser to the wall openings.

- For fixtures not listed in Table 12-3, size the minimum supply branch for a similar fixture.

Number of Bathrooms and Kitchens		Diameter of Water Service Pipe	Recommended Meter Size	Approximate Pressure Loss Meter and 100' of Pipe
Tank-Type Closets				
Copper	Galvanized	Inches	Inches	P.S.I.
1-2	--	¾ or 1	5/8	27
--	1-2	¾ or 1	5/8	40
3-4	--	1	1	22
--	3-4	1	1	24
5-9	--	1¼	1	28
--	5-8	1¼	1	32

For residential use not exceeding two stories in height

Minimum water service pipe size
for one and two story buildings
Table 12-4

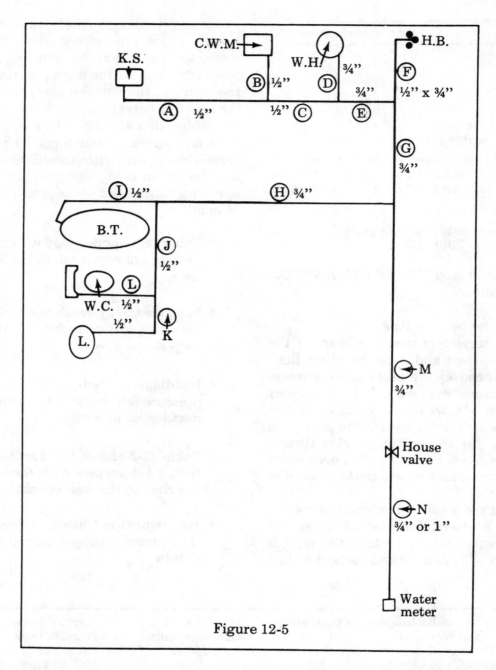

Figure 12-5

- Only two fixtures can connect to a ½-inch cold water supply branch.

Note that more fixtures can be installed on a copper system than on a galvanized steel system of the same size. The load values (fixture units) assigned to fixtures are the same for sizing water supply pipes as for sizing drainage, waste and vent pipes. (Refer again to Table 5-1.)

Test your knowledge of Tables 12-3 and 12-4 with several examples. Assume that you are sizing the cold water service piping in the one-bath dwelling shown in Figure 12-5. Ignore the hot water piping when sizing the water service and distribution lines. The horizontal

and vertical piping in Figure 12-5 serves one kitchen sink, one clothes washing machine, one hot water heater, one hose bibb, one water closet, one lavatory and one bathtub. Study the next few paragraphs carefully to see how the fourteen sections of pipe (A through N in Figure 12-5) are sized using the tables.

The street level water pressure is 53 psi and the water piping material selected for this installation is galvanized. Begin with Table 12-3, Supply Pipes. Without hesitation, you should be able to size pipe sections A, B, F, I, K, and L as ½-inch pipe and section D as ¾-inch pipe.

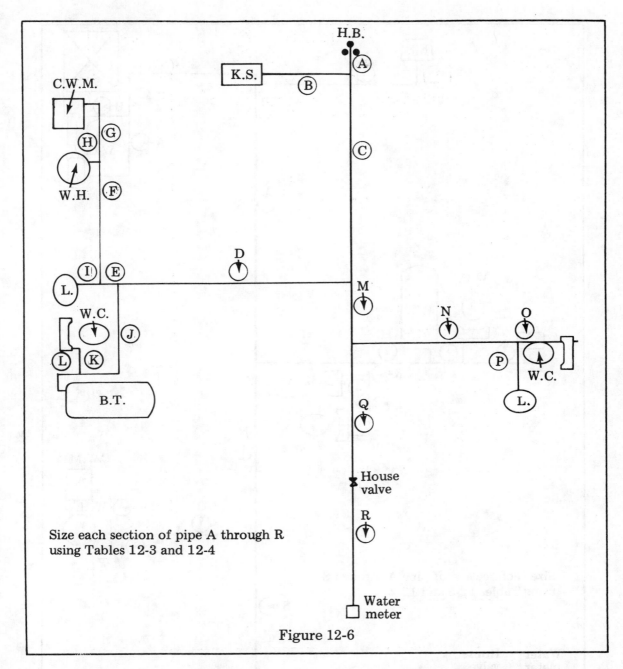

Size each section of pipe A through R
using Tables 12-3 and 12-4

Figure 12-6

The minimum pipe sizes for individual fixtures are shown in Table 12-3. To size pipe section C and J, you must recall that no more than two fixtures can connect to a ½-inch water supply branch. Since only two fixtures are connected to sections C and J, they may also be ½-inch. Pipe sections E and G must be ¾-inch to supply pipe section D, which supplies the hot water heater. Pipe section H would also be sized at ¾-inch, as this water supply branch supplies water to more than two fixtures (a water closet, bathtub and lavatory). Pipe section M would have to be sized at ¾-inch to the house valve, as no water supply pipe can be reduced between the source of supply and demand.

To size pipe section N refer to Table 12-4. Under "Number of Bathrooms and Kitchens," column two, "Galvanized," note that a ¾-inch water service pipe (1-inch in some codes) is sufficient for a 1 or 2 bathroom dwelling.

Two other piping examples follow. See Figures 12-6 and 12-7. Use them with the tables to compute the cumulative demand for two dwelling units. To find the right answer, calculate the sizes of each section of pipe by using Tables 12-3 and 12-4 as illustrated in Figure 12-5. The street level water pressure is

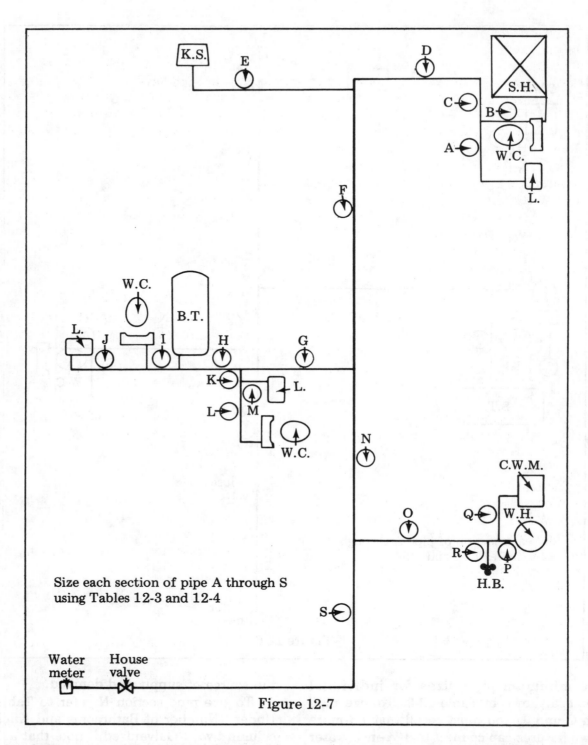

Size each section of pipe A through S
using Tables 12-3 and 12-4

Figure 12-7

50 psi, and the water piping material selected for these two installations is copper.

MAINTENANCE OF WATER SYSTEMS

Some piping materials have limited usefulness in areas where hydrogen sulphide gas or other injurious elements are present. Be aware that in areas which have been recently filled (such as below high tide, as in coastal or tidal regions) the life of galvanized or copper water service pipe is very short. The code requires that some types of pipe be protected by sleeving it in polyethylene plastic tubing, painting it with asphaltum paint, or using other approved protection methods.

Any type of corrosion is a matter of great concern. Preventive measures can greatly prolong the installed life of piping. Corrosion actually breaks down the walls of the pipe, causing a thinning or pitting of the metal.

Types of Corrosion and Their Effects on Piping

Corrosion caused by electrolytic action (chemical breakdown by electric current) is one of the most common types. This is sometimes known as galvanic corrosion. It occurs in a plumbing installation in which two different kinds of metal pipe are used in the same system (such as galvanized steel pipe and copper pipe). The two metals conduct electricity at different rates. When they are covered by an electrolytic solution such as water, a weak electric current is produced. By a process known as ionization, the metal from one pipe passes into the electrolyte and is plated onto the other metal.

Another example of galvanic corrosion occurs when a water heater with a galvanized steel tank is installed in a system with all copper supply piping. In this system, weak electric currents generated by the two metals cause the galvanizing from the tank to pass into the water and plate the copper pipe. This reduces the thickness of the tank walls and, if the process continues, the entire thickness of the metal could be corroded away in some spots.

Water with a high acid content is another cause of corrosion. Over a period of time, acidic water will corrode the inside surface of galvanized steel pipes and tanks, resulting in rust formation. In this type of corrosion the iron in the pipe forms a compound with the oxygen dissolved in the water, and this compound coats the metallic surface. Rust formation increases with any rise in temperature. Thus, hot water piping is affected by rust more readily than cold water piping.

Corrosion also varies with the velocity of water moving through the pipes. The more rapid the flow, the greater the opportunity for corrosion.

Scale is still another type of corrosion. Hard water contains large amounts of calcium and magnesium compounds. These compounds prevent soap from lathering and form a scum which is deposited on the inside of pipes. This slows the flow of water through the pipes. The deposits harden and form scale that can in time completely close the interior of the pipe.

How to Reduce Corrosion

There are several ways of slowing corrosion and prolonging the life of a water supply system:

Galvanic corrosion of water tanks can be controlled by placing dielectric unions or other approved fittings in the cold and hot water takeoffs from the tank. These dielectric unions or special fittings have a fiber washer or special material which insulates the tank from the rest of the water piping installation and stops the weak flow of electric current from the tank to the piping.

Water heaters with galvanized tanks are equipped by the manufacturer with magnesium or other approved rods which protect against rust and corrosion. These rods act as electrolytic cells in which the magnesium particles go into solution, flow through the water, and are deposited on the metal to be protected. The electrolytic action dissolves the rods, which must be replaced periodically.

Water distributing pipe must be electrically isolated from all ferrous pipes, electrical conduit, building steel and reinforcing steel. This means that the pipe can not come into physical contact with any good conductor of electricity. Where contact is unavoidable, a piece of felt or other insulation material should be securely fastened around the pipe to separate the two metals.

Control corrosion caused by excess velocity by installing a reducing valve in the line where the pipe enters the building. Using the next larger pipe size will also reduce water velocity. Small pipes corrode more quickly than large pipes when carrying the same quantity of water under high pressure.

There are three ways of reducing the formation of scale within pipes due to hard water.

• Use copper pipe, or CPVC plastic where permitted. The smooth interior of copper and plastic tends to resist scale formation.

• Regularly flush the hot water heater of sediments (calcium and magnesium deposits) to keep them out of hot water distributing lines.

• Use a water softener. Water softeners usually contain the mineral zeolite. When hard water containing calcium and magnesium sulfates passes through a zeolite water softener, the calcium and magnesium are exchanged for the sodium in the zeolite. The softened water is then piped into the distribution system. Zeolite water softeners must be recharged regularly by adding sodium chloride (table salt).

Protection Against Freezing

In climates where water piping is subject to freezing, lay the water service supply pipe below the frost line. Water service supply pipe entering a building above ground or in areas not protected from the cold must be thoroughly insulated to prevent freezing. Remember that a building's heating system can fail and allow the temperature inside the building to drop below 32° F. Pipe exposed to continued freezing temperatures accumulates ice. Since ice takes up about one-twelfth more space than the same amount of water, the pipes can burst from the expansion of the contents. The ice transmits pressure along the cold water line to the weakest point. The point of rupture is not necessarily the point where the ice formed.

Insulation

Pipe exposed to continued freezing temperatures in crawl spaces, in unheated basements, and at the point where it enters a building above ground should be insulated. The insulation materials in Figure 12-8 and 12-9 are recommended to prevent freezing.

Woolfelt (Figure 12-8) is made of matted

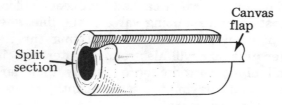

Woolfelt pipe covering
Figure 12-8

wool fibers or wool and fur or hair, pressure-rolled into a compact material. Woolfelt comes in a canvas jacket in thicknesses of ½- to 1-inch and is available in 3-foot lengths. It fits most pipe sizes.

Frostproof insulation (Figure 12-9) is made

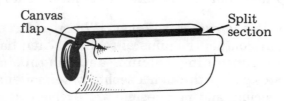

Frostproof insulation
Figure 12-9

for cold water service lines located outside or in unheated areas. Frost proof insulation usually has five layers of felt. Three of these are woolfelt and two are asphalt-soaked asbestos felt. It is sold in 3-foot lengths with a canvas cover and is 1¼ inches thick. It fits most pipe sizes.

Installing Pipe Coverings

The coverings described are easy to install because each section is split in half and the canvas has a flap for quick sealing. Joint collars should be used to cover joint seams on piping exposed to outside temperatures. Cheesecloth can be used instead of canvas, but it must be glued in place. A thick paste of flour and water is an ideal glue for this purpose, but it must be applied in an area protected from the weather. Metal straps are usually supplied with the insulation and should be used to hold the insulation firmly in place. See Figure 12-10.

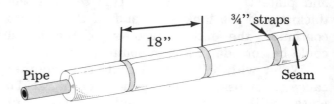

Pipe insulation with metal straps in place
Figure 12-10

These straps should be at least ¾-inch wide and spaced no more than 18 inches apart. If the insulation is exposed to view, a special type of white enamel paint can be used on the insulated pipe.

Properly installed insulation requires little maintenance. Insulation exposed to weather or damage from sharp objects should be inspected frequently. If the canvas cover is torn or punctured, it should be patched with another piece of canvas. Only a waterproof paste should be used when installing or repairing outside insulation.

Occasionally you will find loose straps or loose insulation around valves and fittings. The straps should be re-tightened and the loose insulation replaced or pasted down. Make inspections before cold weather arrives. This could save the inconvenience and expense of frozen water lines.

Procedures for thawing frozen lines are described and illustrated in Chapter 20.

Questions

1- What is the source of water for public water supply systems?

2- What is potable water?

3- What is the approximate number of people in the United States that depend on a private water supply system?

4- Most plumbing fixtures operate satisfactorily when the water main pressure is____psi.

5- When should a pressure reducing valve be installed on a water service line?

6- What does the law require owners of private water systems to do when public water is available?

7- Name three factors affecting the sizing of water supply piping.

8- Each water supply fixture unit represents how many gallons per minute flow?

9- What part of a water supply system must be sized first?

10- What is the minimum practical size for water service line?

11- When may a building main be reduced in size?

12- What rule of thumb is recommended when you are sizing the water piping in a building with a water main pressure of *less* than 50 psi?

13- How many fixtures may connect to a ½-inch cold water supply branch?

14- Why is hot water piping not considered when sizing cold water supply piping?

15- What is the minimum cold water pipe size to supply a hot water heater?

16- What preventive measure should be taken to protect metallic water pipes where hydrogen sulphide gas or other injurious elements exist?

17- Does corrosion affect only the *outside* of metal water pipe?

18- What causes galvanic corrosion?

19- What type of corrosion occurs as a result of weak electric currents?

20- Oxygen and iron in the pipe will cause what type of corrosion to occur inside a water line?

21- Name two other types of corrosion that are common within a water supply system.

22- What is the most common way of reducing the effects of corrosion in a plumbing system?

23- Why is it necessary to isolate ferrous water pipes from other metals within a building?

24- Where water velocity is high, what should be done to protect a water system from corrosion?

25- How does water containing large amounts of calcium and magnesium affect a water system?

26- Name two ways of reducing the formation of scale within pipes where water is unusually hard.

27- In climates where water service supply pipe is subject to freezing, how deep should the trench be?

28- What maintenance precautions should be taken where water pipes are not protected from the cold?

29- When water freezes within a pipe, what causes the pipe to rupture or burst?

30- When water freezes within a pipe, where is the pipe likely to rupture or burst?

31- Name two common insulation materials used to protect pipe against freezing.

13

Valves And Faucets

Valves

A valve is a device used to stop, start, or regulate the flow of a liquid or gas into, through, or from piping. Valves sized up to 2 inches in diameter are made of brass. The body of the valve has an opening through it which can be closed with a valve disc or washer. To shutoff faucet, this washer or disc is pressed tightly against a seating surface around or within the opening. There are many types of valves, but this chapter is limited to those commonly used in residential or simple jobs.

The *gate valve* (#1 in Figure 13-1) has a wedge-shaped or tapered metal disc which closely fits a similarly shaped, smoothly ground surface or *seat*. When the valve is open, the disc *gate* clears the opening formed by the seat and allows a straight line flow. When the flow must be stopped, the wheel is turned clockwise and the disc moves downward onto the seat Use gate valves on lines which are to remain either completely open or closed most of the time, as this limits wear on the metal gate and seat. The valve shown is a rising stem, single disc gate valve. Other common types are the double disc and the nonrising stem.

Gate valves are generally required in water service pipes (see Figure 12-1) and as the control valve for hot water heaters (see Figure 14-4). This is because an open gate valve obstructs the flow of water much less than an open globe or angle valve.

A *globe valve* (#2 in Figure 13-1) closely resembles a gate valve, but its interior is much different. The operating principle is similar to that of the compression faucet (Figure 13-2). A partition between the inlet and the outlet blocks the flow through the valve. The seat, which is at a right angle to the direction of flow, changes the direction of flow passing through it. The horizontal seat is a removable ground joint. A replaceable composition or fiber washer or metal disc at the end of the stem makes a tight fit against this joint. Use a globe valve where

you need either an adjustment of rate of flow or a tight shut-down. Globe valves are suitable for all kinds of service except where full flow is required. Install the globe valve in the line so that the inlet side will carry the pressure when the valve is closed. An arrow on the valve body usually shows the correct direction of flow.

The *stop and waste valve* (#3 in Figure 13-1) is similar to the globe valve except that it has a small drain opening on the outlet side of the valve body. When it is open, this outer valve can drain the building water system when the main valve is closed. This protects piping which may be exposed to freezing and expansion of the water within them. Most codes require that a stop and waste valve be installed near where the water service pipe enters the building. *Never install a stop and waste valve in a trench with water service pipe.*

An *angle valve* is a globe valve with the inlet and outlet at right angles to each other rather than in line. (See #4 in Figure 13-1). This valve controls the flow and changes the direction of a line.

A *check valve* checks or prevents reversal of flow in the line. (See #5 in Figure 13-1.) It permits a free flow of liquid through the valve in one direction but automatically closes if the liquid begins to flow the other way. The most common type in use is the horizontal swing check valve shown in Figure 13-1. Use this type on domestic or irrigation wells. Install the valve on the suction line between the well and the pump. Other types of check valves are the disc check, the ball check, the adjustable check, and the spring check.

Faucets

A *faucet* is a fitting which controls the flow of water at the end of a water supply line. The faucet most generally used for water outlets is the *compression type*. It can be made of chrome-plated or unplated brass, depending on its location in the piping system. The spout

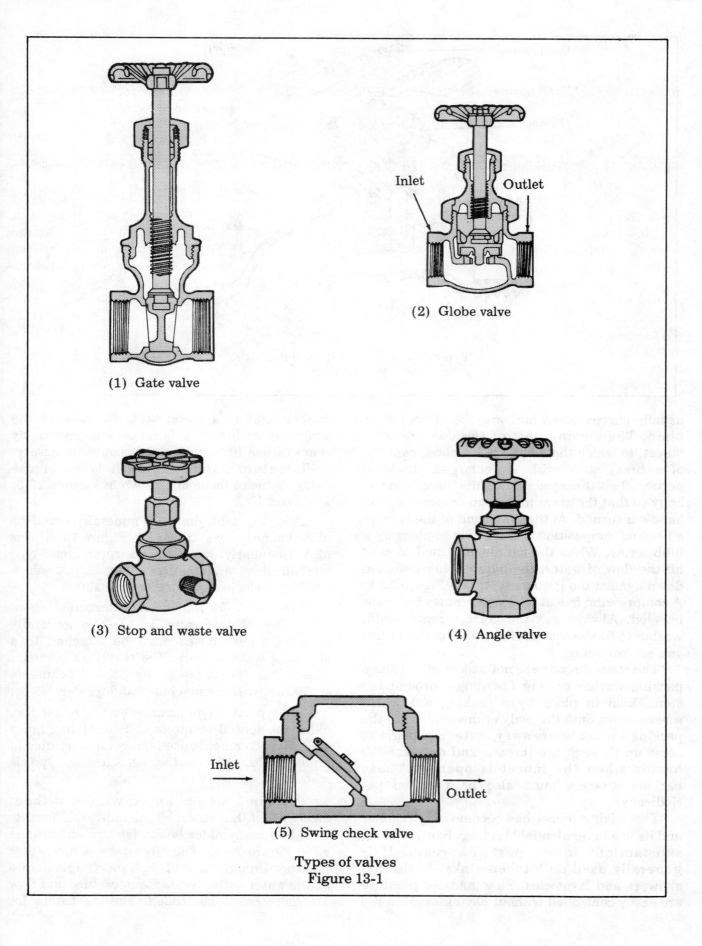

(1) Gate valve

(2) Globe valve

(3) Stop and waste valve

(4) Angle valve

Inlet →

Outlet →

(5) Swing check valve

Types of valves
Figure 13-1

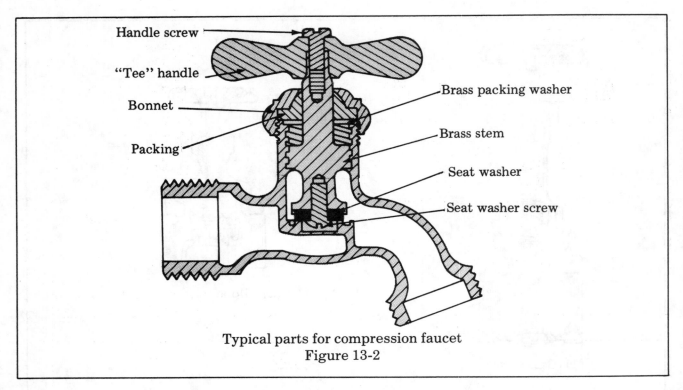

Typical parts for compression faucet
Figure 13-2

usually curves down and may be threaded or plain. The moving part of the compression faucet, to which the handle is attached, consists of a brass stem with an enlarged threaded portion. This threaded portion fits threads in the body so that the stem moves up or down as the handle is turned. At the lower end of the stem is a fibre or composition washer held in place by a bibb screw. When the handle is turned to shut off the flow of water, the bibb washer is forced down against the ground seat. See Figure 13-2. A compression faucet bibb washer may be flat or beveled. Always use the correct replacement washer to fit the seat configuration of the faucet you are servicing.

The stem threads are not watertight. Thus a packing washer or ring fits snugly around the stem, held in place by a packing nut which screws down onto the body of the faucet. As the packing washer wears away, water can begin to come up through the threads and out near the handle when the faucet is opened. These packing washers must also be replaced periodically.

The mixing faucet has become very popular and its use in residential buildings has increased substantially in the past few years. It is generally used on kitchen sinks, bathtubs, showers and lavatories. Flow and temperature are easily controlled without having to grind the

washer against a metal seat. Because of the simplicity of design, this faucet can give many years of use before repairs become necessary.

There are dozens of faucet types in use today. Some of these are shown in Figures 13-3, 13-4, and 13-5.

The *plain bibb faucet* is generally found on older kitchen sinks. (See #1 in Figure 13-3.) It is also frequently installed on commercial-type kitchen sinks and fixtures such as pot sinks, scullery sinks and service or slop sinks.

The *hose bibb faucet* has a threaded spout end. (See #2 in Figure 13-3.) It is generally installed where hoses must be attached to a source of water supply. The two most common uses of this faucet are as the hose attachments to clothes washing machines and garden hoses.

The typical *single lavatory faucet* is used on older residential lavatories. (See #4 in Figure 13-3.) It is commonly used today on lavatories in public toilets. Often the self-closing type is installed to conserve water.

The *combination faucet*, with or without swing spout, has replaced the individual faucets found on many older lavatories and sinks. (See #3 in Figure 13-3.) The advantage is that water passing through each of the faucets mixes in a single water outlet. Water temperature and flow are more easily controlled. The mechanics for

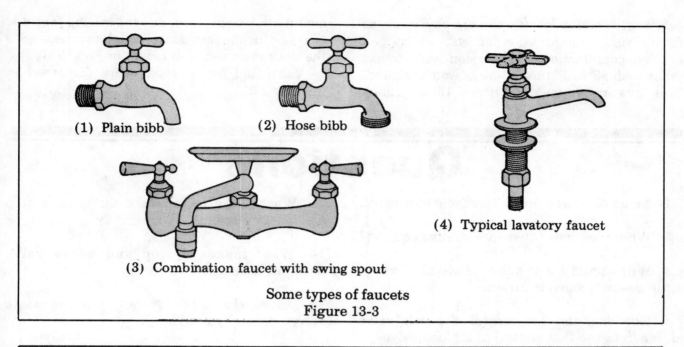

(1) Plain bibb   (2) Hose bibb

(4) Typical lavatory faucet

(3) Combination faucet with swing spout

Some types of faucets
Figure 13-3

Compression deck type lavatory faucet
Figure 13-4

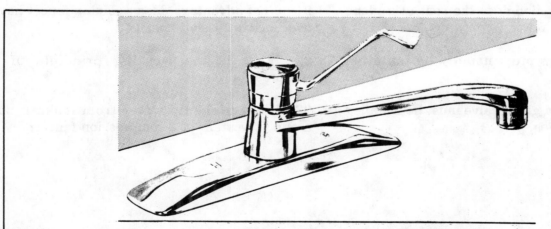

Mixing single lever deck type sink faucet
Figure 13-5

servicing combination faucets are the same as for individual compression faucets.

The combination compression sink faucet (shown as #3 in Figure 13-3) is wall mounted. The lavatory faucet in Figure 13-4 is deck mounted. A *single lever* or *knob mixing faucet* is common in modern sinks and lavatories. (See the deck mounted sink faucet in Figure 13-5).

Valve and faucet repairs are discussed in Chapter 21.

Questions

1- Name the three major functions of a valve.

2- Where are gate valves generally required?

3- Why should a gate valve be installed where little use of a valve is expected?

4- How does the flow control of a gate valve differ from the flow control of a globe valve?

5- Does a gate valve restrict the flow of liquids through it?

6- Which direction is the wheel turned to close any valve?

7- On the outside, a globe valve looks similar to what other valve?

8- Is the direction of the seat in a *gate* valve horizontal?

9- Is the direction of the seat in a *globe* valve horizontal?

10- In which direction does the seat of a globe valve change the flow?

11- Globe valves must usually be installed where?

12- Explain how a globe valve must be installed in the line it serves.

13- What is the main reason for using a stop and waste valve?

14- What makes a stop and waste valve different from other valves?

15- Under what conditions are stop and waste valves generally used?

16- Is it permissible to install a stop and waste valve in the same trench with the water service pipe?

17- Name the two main purposes an angle valve serves in a water line.

18- What is the working principle of a check valve?

19- Where is a check valve most generally used in residential construction?

20- What type of faucet is most commonly used for water outlets?

21- What advantage does a mixing faucet have over a compression faucet?

22- What is the working principle of a compression faucet.

23- What prevents water from leaking out around the stem of a compression faucet?

Hot Water Systems

The hot water system is the part of the plumbing installation that heats water and distributes it through pipes to fixtures dispensing hot water. The materials used in hot water systems are the same as or similar to those used in cold water supply systems. Galvanized steel pipe was once the most popular material, but in recent years copper hot water systems have become more common. Copper is especially good because of its ability to resist corrosion, which increases as the temperature of the water increases. Maintenance and repair of the hot water system is also similar to that of the cold water supply system.

The newer CPVC plastic and polybutylene (PB) plastic pipe are now accepted by most codes in both hot and cold water systems.

Some codes do not require a hot water system. The exception to this is any building designed for special occupancy. Even though some codes do not require a hot water system, most dwelling units have one. The code is quite specific in requiring proper installation of safety devices if there is a hot water system. Safety equipment is necessary to prevent damage to property and injury to persons using the plumbing fixtures.

The code does not specify design requirements for hot water distribution systems. It leaves the design of the system to the installer. The code simply states: "The sizing of the hot water distribution piping shall conform to good engineering practice." Since most plumbers are not engineers, this chapter is intended to help you design safe and functional hot water systems for residential and light commercial buildings.

Hot Water Distribution System

The hot water distribution system is made up of the pipe installations that carry the heated water from the storage unit to the plumbing fixtures. The installation begins with a water heating device and a main supply line connected to that device. Water for each fixture in the building is taken from the main hot water supply line as needed. Use Tables 12-3 and 12-4 to size the hot water main supply lines and the fixture supply openings.

Follow the same sizing procedure as for cold water systems, with one exception. The hot water main supply and branch lines may be one pipe size smaller, to a minimum of ½ inch. For example, the cold water main supply to a bathroom is ¾ inch; the hot water main supply to this same bathroom may be ½ inch. See Figure 12-2 for a typical sized hot and cold water distributing piping diagram.

Design Objectives

Two principal design objectives must be met in any good hot water system. First, the system must satisfy the hot water demand for a particular need. Second, safety features must be built into the system to prevent excessive pressure and temperature.

The design temperature for hot water in most plumbing fixtures is between 130 and 140 degrees F. A water heater thermostat setting of 150 degrees F is recommended where a dishwasher is used. This is hot enough for most purposes but cool enough to prevent scalding of the skin. The water heater thermostat is generally preset within the correct range at the factory and does not need additional adjustment. In all water heaters the energy transfer rate must be adequate for the quantity of hot water stored in the heater tank. All residential storage tanks must have insulation adequate to prevent excessive heat loss from the water stored.

There are three categories of hot water heaters in residential buildings.

- Electrical storage heaters

- Fuel-burning storage heaters

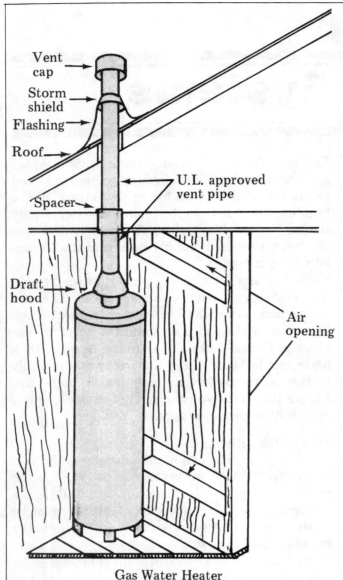

Gas Water Heater

The confined space must be provided with two permanent openings, one commencing within 12 inches of the top and one commencing 12 inches from the bottom. Adequate air for combustion, ventilation, and dilution of flue gases must be obtained.

Installation of fuel-burning water heaters
Figure 14-1

• Solar energy storage heaters

The most commonly used water heaters today are the electric and the fuel-burning storage tank types. Electric water heaters are clean and attractive and can be installed nearly anywhere within a building. But water heaters designed to burn gas or oil must be located in a well-ventilated area and must have flues to carry away combustion gases. See Figure 14-1. Tank locations for solar hot water heaters are also limited by the large size of the tank and by the type of system planned.

There is no need for constant circulation of hot water in a system designed for a one- or two-family dwelling. A small system has small pipes and short runs. The hot water in these pipes cools when water is not being used. But hot water is available again shortly after the faucet valve is opened.

Water expands when it is heated, but is otherwise relatively incompressible. Expansion should not be a problem in most residential and simple buildings with short piping runs.

Heater Capacities

Electric and fuel-burning units have small storage tanks because the heating capacity is usually adequate to maintain the water at 130 to 150 degrees Fahrenheit during the peak draw period. (The peak draw period is usually assumed to be one hour, although in most homes it is as little as 20 minutes.)

The amount of water used per person varies greatly with the habits of the users. The amount of water used daily is spread over a number of hours, and hot water is only a part of total water use.

When sizing hot water storage tanks, assume that only 75% of a tank's hot water supply is available during the peak draw period (one hour). Estimate the maximum hourly use and the number of hot water users. In general, 8 to 10 gallons is considered the maximum hourly use of hot water per person for dwelling units.

The *working load* of a water heater is the percentage of the *maximum load* that is expected to be used under normal conditions in any given hour. The working load factor is similar to the simultaneous use factor for cold water systems. The working load of single family residential type buildings is approximately 35%.

The maximum load (storage tank capacity) and the working load (percentage expected to be used) determine the amount of hot water needed per hour. The recovery rate per hour and the storage tank capacity determine the amount of hot water that can be drawn from the tank per hour during peak demand periods.

For a peak draw period of one hour, a

20-gallon water heater should provide 15 gallons (75% of the tank's capacity) plus the recovery rate of at least 50%. This should provide 22.5 gallons per hour.

Assume that for a one-bedroom dwelling a maximum of two persons each use 8 gallons of hot water in one hour. A 20-gallon water heater would be adequate. (This would even provide a backup of 6.5 extra gallons.) The number of bedrooms is the best indication of the quantity of water that will be used in the dwelling.

The following individual tank storage capacities are both economical and satisfactory for the average dwelling unit *when the peak draw period does not exceed one hour*. Special demand requirements which may last more than one hour might indicate that the next larger size should be used.

Recommended Tank Sizes

Number of bedrooms	Hot water storage tank capacity
1	20 gallons
2	30 gallons
3	42 gallons
4	52 gallons

For each additional bedroom above four, use the next larger size hot water heater. You may want to split the system and install two water heaters, especially if long pipe runs are required. (Central hot water systems for multi-family dwellings are explained and illustrated in *Plumber's Handbook* by this author.)

Water heaters located in closets or beneath kitchen counter tops should be installed so that the data plates giving the maximum working water pressure and other data is visible. The control and relief valves should be accessible. Locate the heater where it can be easily serviced or replaced without removing any permanent part of the building. See Figure 14-2.

Safety Devices

A combination pressure and temperature relief valve must be installed on all hot water heater storage tanks. The temperature relief valve must be the reseating type and must be rated by its B.T.U. capacity. *The B.T.U. rating of the temperature relief valve must be greater than the B.T.U. rating of the appliance it serves.*

In addition to the required temperature and pressure relief valves, an *energy shutoff device*

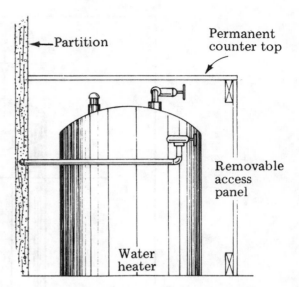

Installation of under counter hot water heater
Figure 14-2

is now required on all automatically controlled water heaters. This device will cut off the supply of heat energy to the water tank before the temperature of the water in the tank exceeds 210° F.

Domestic hot water heaters commonly used in residential buildings do not require separate pressure and temperature relief valves. A combination pressure and temperature relief valve is acceptable. *Fuse-type relief valves* are no longer acceptable on new construction and should be replaced on older hot water heaters with pressure and temperature relief valves of the reseating type. When the water overheats, the fuse melts and releases the excess pressure through the relief line. The water continues to run until the relief valve is removed and a new fuse reinstalled.

For relief valves to be most effective, they must be installed so that the temperature sensing element is immersed in the hottest water, which is in the top one-eighth of the tank. See Figure 14-3.

A check valve or shutoff valve should never be installed between the relief valve and the hot water heater storage tank. If the check valve fails to operate or if someone accidentally closes the shutoff valve, the relief valve would then be useless. The tank could then rupture or explode.

Relief valve drip pipes (popoff lines) should not be connected directly to any plumbing drainage or vent system. This could cause contamination of the potable water system. It could also conceal from view any continuous

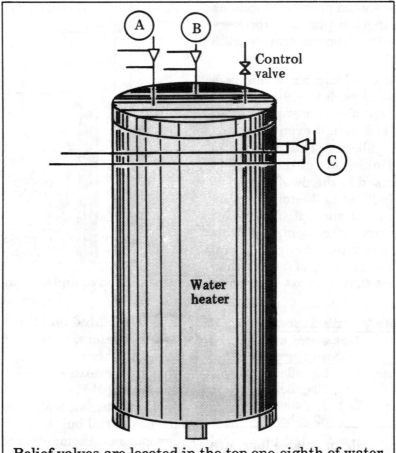

Relief valves are located in the top one-eighth of water heater. Positions illustrated at A, B, or C are acceptable

Figure 14-3

discharge. **Relief valve drip pipes should not terminate over any plumbing fixture or any other area where scalding might occur if the relief valve discharges.**

The terminus of drip pipes should extend to an observable point outside the building. It should be unthreaded to prevent connections and should reach to within six inches of the ground. It should never be trapped.

Relief lines are sized by the B.T.U. rating of the appliance. A 3/8 or 1/2 inch inside diameter pipe is adequate for sizing relief lines for single family residences. (All relief pipes are sized by their inside diameter.)

To install a relief line on a relief valve, first install a male flare or compression adapter (Figure 15-6) of the same size as the female threads in the relief valve. Spread pipe compound on the male threads and tighten the adapter into the female threads. Cut a length of

the soft copper tubing to extend to the outside of the building; allow enough tubing for the turn-down. Shape it by hand to fit the wall line and secure it to the flare or compression nut on the relief valve. Strap the tubing securely in place. See Figure 14-4, "Typical electric water heater piping detail," and Figure 14-5, "Typical combination temperature and pressure relief valve."

Hot water heaters must have the drain cock (valve) in an accessible location, both for flushing the tank of sediment and for emptying it for repairs or replacement.

The minimum size cold water supply pipe for a hot water heater, regardless of its capacity, is ¾-inch diameter. The cold water supply pipe must have a minimum ¾-inch shut-off valve in an accessible location. If a heater is located on a shelf in a utility room, for example, the shut-off valve should not be higher than six feet above the floor. The shut-off valve used must have a

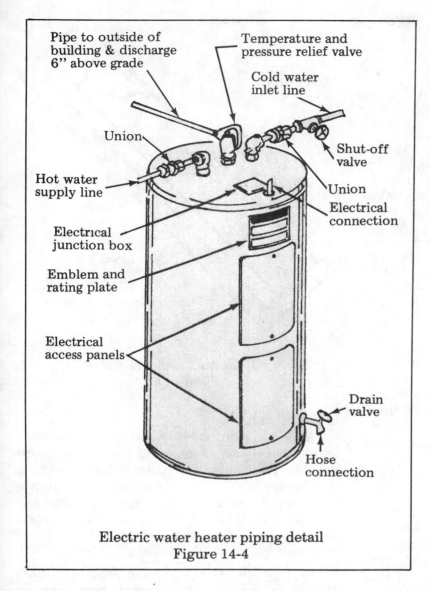

Pipe to outside of building & discharge 6" above grade

Temperature and pressure relief valve

Cold water inlet line

Union

Shut-off valve

Hot water supply line

Union

Electrical connection

Electrical junction box

Emblem and rating plate

Electrical access panels

Drain valve

Hose connection

Electric water heater piping detail
Figure 14-4

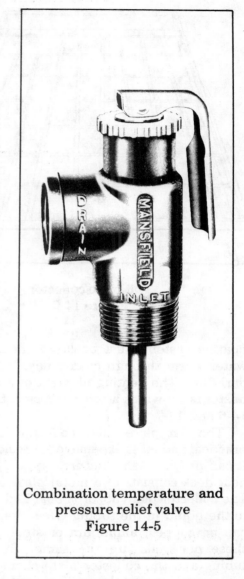

Combination temperature and pressure relief valve
Figure 14-5

cross-sectional area equal to 80% of the nominal size of the pipe in which it is installed.

Hot water piping materials and installation methods are the same as those for water distribution. See Chapters 15 and 16. The only difference in installation between the two is that hot water piping supports must permit the expected amount of expansion.

Solar Water Heaters

Solar energy is a practical means of heating domestic hot water. Solar energy hot water heating can be added to existing systems as well as to new construction at a relatively modest cost.

The use of solar energy for heating domestic water is not new. In many areas of the world solar heating has been used successfully since before 1900. Solar water heaters have been commonly used in Florida and California for

some 60 plus years. Solar water heating is technically feasible and within the cost of most any homeowner's budget. A family of four using an electric water heater should save as much as two hundred dollars a year by adding a solar water heater.

Solar Water Heater Components

There are at least three components in any solar heat collecting system: 1) the solar heat collector, 2) the circulation system, and 3) the solar storage tank. A fourth component used in many systems is a backup heat source to ensure that hot water is available even during periods of peak water use or low solar radiation.

In a conventional fossil-fuel water heater, the heating element can work continuously to produce hot water. The storage tank can be rather small, 30 or 42 gallons. The tank in a solar

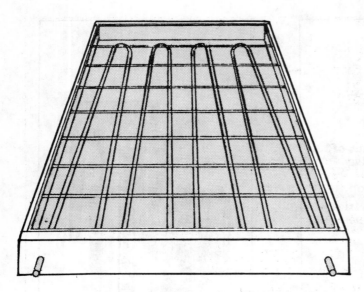

Flat plate collector
Figure 14-6

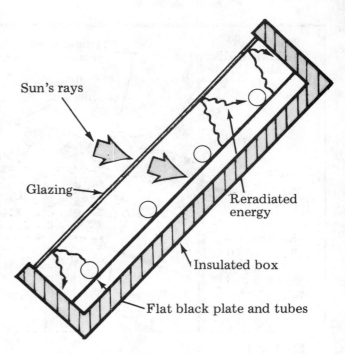

Collector cross section
Figure 14-7

heating system must be large enough to keep water warm through cloudy days and hours of darkness. The heating element of a solar water heater is known as a *solar collector*. See Figures 14-6 and 14-7.

The flat plate solar collector is the most practical and least expensive for residential use. It can produce temperatures up to 200° F. The heat deck consists of a metal plate and tubing. The collector plate absorbs heat and transfers it to the liquid in the tubing. Heat deck materials can be copper, aluminum or steel. Thermally, these materials are the same. But both the tubing and the collector plate should be of the same metal so that they will expand and contract at the same rates. Codes generally do not permit potable water to flow through aluminum tubing.

The heat collector box should be well insulated to shield the heat deck plate from the weather and to reduce heat loss. The heat deck plate and tubing are usually painted flat black to maximize absorption of the sun's energy. Light colors tend to reflect the sun's rays.

A transparent cover on the collector box permits the sun's rays to strike the metal collector plate and thus reduce the loss of reradiated heat back to the atmosphere. The glass used should have a low iron content to make it as transparent as possible to incoming rays. The glass admits solar radiation, but it is opaque to the long-wave radiation created when the sunlight hits a solid surface. This long-wave energy is trapped inside the box to heat the fluid in the tubing.

A clear plastic sheet is better than no cover at all, but glass is more efficient and more durable. Plastics tend to transmit both incoming and outgoing energy. And most available plastics deteriorate fairly rapidly when exposed to heat and moisture. If a more heat resistant plastic were available, it would be more desirable because glass breaks easily.

In cold climates a double layer of glass should be installed to prevent heat loss by convection when very cold air strikes the transparent surface.

A 4' x 12' solar heat collector heats approximately 80 gallons of water per day. This should be sufficient for the average family of four. Allow twelve square feet of heat collector surface per person to meet typical hot water needs of most families at any time of the year in most sun belt areas of the U.S. The minimum square footage of collector surface per person in northern states is of course higher.

The average family of four requires a storage capacity for solar heated water of at least 80 gallons. Larger tanks of 100 or 120 gallons do not cost much more than an 80 gallon tank and store extra hot water for use when demand is unusually high.

Collectors are manufactured by many res-

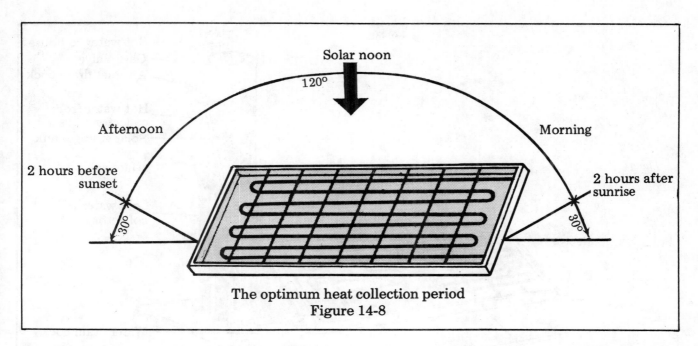

The optimum heat collection period
Figure 14-8

ponsible companies. It is your responsibility, however, to check for the approvals required by your local plumbing code before installing the unit.

Installing The Solar Heat Collector

Only about 30% to 65% of the solar energy which strikes the glass surface of a collector during the day actually heats water circulating through the tubing. The rest is lost back to the atmosphere through the glass plate.

The most efficient solar collector would be perpendicular to the sun's rays at all times during the day and at all seasons of the year. But this is possible only if the collector turns and tilts to follow the sun's path. Motorized collectors are far too expensive for installation in most homes. The best compromise is a flat plate collector tilted in the general direction of the sun's path across the sky at an angle equal to the latitude plus ten degrees. Fixed collectors should face *south*. However, collectors facing southeast or southwest work about 75% as well as those facing due south.

Locate the collector wherever it is most convenient and most attractive, as long as it is in full sun two hours after sunrise to two hours before sunset. Shading before or after these periods has little effect because at those hours most of the sun's energy is reflected back to the atmosphere. See Figure 14-8.

Mount collector as close as possible to the storage tank to reduce heat losses and friction in the pipes. Make sure that all piping is well insulated. In northern states, shield the transparent cover to protect it from winds that would otherwise cool the surface.

Collectors that are to be installed as awnings or as fixed overhangs on a residence should be approved by your local authority *before work commences*. This type of installation is usually not acceptable.

Collectors can be mounted on the ground almost anywhere, but they must be securely anchored as in Figure 14-9. Ground installations are subject to more repairs because the glass box cover is more exposed to accidental breakage and vandalism.

In climates subject to freezing temperatures, all tubes should be installed so they drain dry. Draining the system is the best and cheapest way to protect the water from freezing and bursting the pipes.

Mounting The Heat Collector

In many areas of the country, plumbers are responsible for mounting solar collectors. This is especially true in new construction where collector units are built into the roof as an integral part of the house. Built-in collectors look better because they can be designed to blend with the exterior and resist wind loads better. They can also be coordinated with roof surfaces to minimize leaks.

Roofing contractors make several recommendations for mounting collectors. They

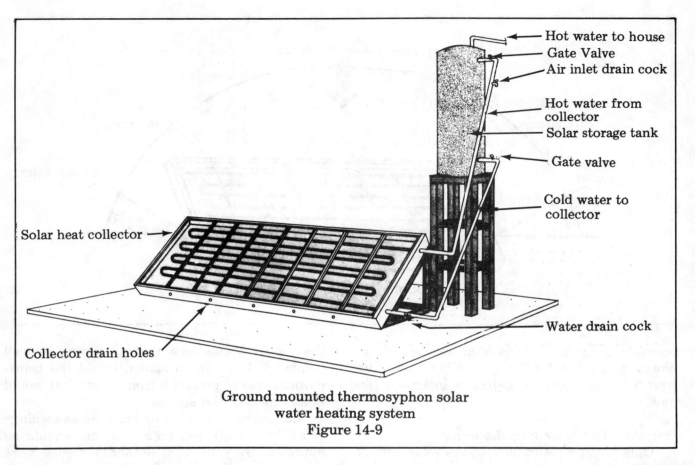

Solar heat collector

Hot water to house
Gate Valve
Air inlet drain cock

Hot water from
collector
Solar storage tank

Gate valve

Cold water to
collector

Water drain cock

Collector drain holes

Ground mounted thermosyphon solar
water heating system
Figure 14-9

suggest that the collector have a structural frame. The frame should be securely bolted to the main roof structure. Flashing and a rain collar should be put around the pipes and the collector. The collector is then fitted into the frame and anchored securely to it.

Roof leaks can be a serious problem with roof-mounted collectors. Seal every hole drilled through the roof membrane. If a hole is too large or not properly sealed, a leak will occur. To prevent leaks, bore a small hole and caulk around it carefully. A pitch pan is a practical solution on built-up roofs. If a pitch pan is used, seal it carefully before securing the pan and filling it with pitch. Otherwise, the pitch will heat up under summer sun and create a pocket. Water can then get into the pocket and remain there. This moisture will eventually rust the top of the securing bolt. Advise the owner to check his roof annually, as expansion, contraction, and wind can open fissures where the bolts and pipes penetrate the roof membrane.

A bearing plate should be installed under the cliff angle to prevent the collector from rocking. Screw the bolt all the way through the sheeting and then securely into the main roof structure.

Lag bolts fastened only to the sheeting will pull out under the load of the collector vibrating in the wind.

On roofs with asphalt shingles, apply a layer of plastic cement on top of the shingles and beneath the cliff angle. Apply another layer of plastic cement on top of the cliff angle. Then securely fasten the cliff angle to the main roof structure with lag bolts.

Water Circulation

Water circulates through the collector, moving hot water from where it is generated to where it is needed. A circulating pump can do this, but natural thermosyphon circulation moves the hot water through the system more simply. A thermosyphon requires no external energy source and no pumps, controls or other moving parts. The rate of movement of hot water from the collector to the storage tank, and cold water from the tank to the collector, is controlled by the intensity of the sunshine. A thermosyphon results when hot water, which is lighter, rises to the storage tank and replaces heavier cold water drawn into the collector. See Figure 14-10.

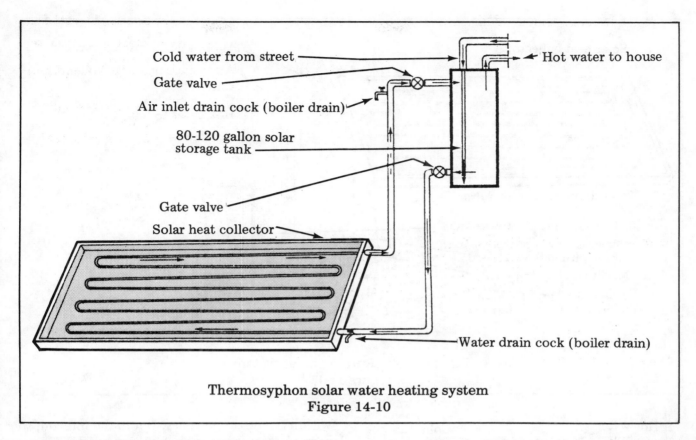

Cold water from street

Hot water to house

Gate valve

Air inlet drain cock (boiler drain)

80-120 gallon solar
storage tank

Gate valve

Solar heat collector

Water drain cock (boiler drain)

Thermosyphon solar water heating system
Figure 14-10

The thermosyphon will not work properly unless the storage tank bottom is at least two feet above the top of the solar collector. Locating a large, heavy storage tank higher than the solar collector may mean putting the tank on the roof or in the attic. This presents problems of weight, construction and appearance. A leak in an attic-mounted tank can cause considerable water damage inside the house. For this reason, solar energy codes usually require that an adequate-size drain pan with a safe waste pipe extended to the exterior be placed beneath all hot water storage tanks located above the ground floor.

In a thermosyphon system, at least ¾-inch inside diameter piping must be used both in the collector and in the circulation system. This reduces flow resistance. Be sure that the connecting pipe or tubing has a continuous fall with no sections that would permit the formation of an air pocket. An air pocket stops water circulation.

A pumped system uses the same basic piping as the thermosyphon system. But a pump forces hot water from the collector to the large storage tank, as in Figure 14-11. In a pumped system the heavy solar tank can be located in

any convenient place—the garage or utility room, for example. This avoids the design and potential leakage problems of a tank mounted on the roof or in the attic.

Half-inch copper tubing can be used in a pumped system. The pump controls must be set so that water circulates through the heat collector only when the water in the tank is cooler than the water leaving the collector. The pump and controls add to the total cost, but this may be offset by the lower cost of installing the heavy storage tank at ground level.

In a closed solar energy collection system a fluid such as antifreeze is heated in the collector. A closed system may have a built-in heat exchanger within the storage tank body or on the outside of storage tank, as illustrated in Figure 14-12. The fluid circulates through the solar collector and transfers its heat to water in the storage tank through a heat exchanger. This sytem, although more complicated and expensive, eliminates the possibility of freezing. A hard freeze can destroy a heat collector if it is not properly protected. If you expect freezing temperatures. install a closed system or make sure that the thermosyphon system can be drained dry in the winter.

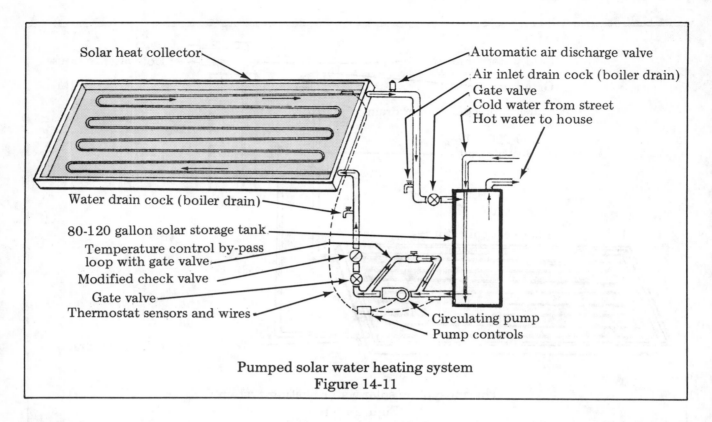

Pumped solar water heating system
Figure 14-11

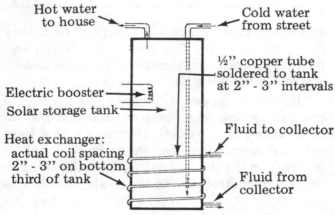

One type of heat exchanger
Figure 14-12

Materials and Installation

The rapid development of interest in solar energy has created a problem for many building code administrators. Building inspectors are faced with new installation problems, and current codes are not adequate to regulate these innovative installations. But until codes contain specific provisions for solar, the existing building and plumbing codes must be followed when you do solar work.

A standard solar energy code will be developed and adopted by most jurisdictions within the next few years. But in the interim, the rest of this chapter shows how existing codes and the *Uniform Solar Energy Code* apply to solar energy installations.

Pipe and fittings used for conveying fluids within a solar system must comply with the requirements for a potable water system. See Chapters 15 and 16.

Galvanized steel pipe and fittings and type K or L copper pipe and fittings are the most common materials used in a solar circulation system. *Plastic pipe must not be used for conveying heated fluids within a solar system*, since plumbing standards limit the use of plastic pipe where temperatures could exceed 180° F. *Aluminum tubing cannot be used in a potable water system*. Its use in solar heating systems is limited to the closed-type system with a heat exchanger, and only if it is first approved by the local authority.

Built-in heat exchangers within the body of the tank must not be of lighter weight pipe than type L copper tubing. The heat exchanger must have no seams, joints, fittings or valves. It must be constructed of double-wall material designed to prevent leaks that could result in a cross-connection with the potable water supply.

Fittings used in a solar piping system should be of the same material as the pipe. An exception is made for valves and similar devices. Where using dissimilar piping and fitting materials cannot be avoided, these materials should be electrically isolated using approved fittings which have been properly installed.

Valves installed in a solar piping system must be of brass or other approved materials. The fully opened valve must have 80% of the cross-sectional area of the nominal size of the pipe to which the valve is connected. Control valves must be installed so that they can isolate the solar system from the potable water supply. All control valves must be readily accessible.

Where excessive water pressure is possible, an approved type of pressure regulator must be installed. Since the solar energy system is an integral part of the building's water supply system, the pressure regulator can be the regulator installed in the water service pipe. In this case, a second regulator would not be necessary in the solar supply system.

Temperature relief valves must be installed on all equipment used for the heating or storage of domestic hot water. A combination pressure and temperature relief valve is acceptable under most codes. The temperature sensing element must be installed in the hottest water within the top one-eighth of the tank. See Figure 14-3. *Temperature relief valves must be the reseating type.* Fuse-type relief valves are not acceptable.

Relief valves located inside a building must have a full size discharge line. The line should discharge to the outside, turn down to within six inches of grade, and drain dry after use. Be sure that the line is securely strapped to the exterior of the building. The end of the discharge line must not be threaded. A discharge line can terminate at other locations if the installation is first approved by the controlling authority.

Some authorities may require a temperature relief valve located at the highest point of the solar piping system near the automatic air discharge valve. See Figure 14-11. Generally, codes do not require a separate discharge line from this relief valve to the ground but will accept the discharge of the solar system's relief valve onto the roof.

Automatic air discharge vents must be installed at all high points of a solar piping system. See Figure 14-11.

Storage Tanks

All storage tanks used for domestic hot water must meet the applicable A.S.M.E. requirements. Some codes require that any hot water storage tank located above the ground floor have an adequate-size pan and a minimum ¾-inch indirect (open) waste pipe which discharges to the outside of the building. This can eliminate interior building damage if the tank should leak.

All storage tanks must be equipped with an adequate and accessible drain cock. The tank and any devices attached to it must be accessible for repair or replacement.

Storage tanks must be permanently labeled with the maximum allowable working pressure and the hydrostatic test pressure which the tank is designed to withstand. These markings must be accessible to the inspector.

With special approval from the administrative authority, tanks designed and constructed to resist trench loads and corrosive soil effects can be buried underground. But no part of the tank can be covered or concealed prior to inspection and approval.

Installation Checklist for Solar Water Systems

- Only solar collectors approved by the local authority should be installed.

- Frames for securing solar collectors to the building must be of materials durable enough for outside use.

- Solar collectors must be anchored securely to withstand the expected dead, live and wind loads.

- Joints around pipes, ducts, bolts, or anything which penetrates the roof, must be made watertight.

- Collector panels that are not an integral part of the roof must be mounted a minimum of 3 inches above the roof surface.

- Collector panels and related piping must be installed so that they drain dry.

- Collector boxes should have drainage holes at the low point for draining rain or other liquids that might collect in the box.

- Glass used in the collectors should have a low iron content. Only tempered glass should be used.

- All piping in a solar system must be installed to allow for expansion, contraction, and normal pipe movement.

- All piping which carries heated water, fluids or gases from a collector or heat exchanger to a storage tank must be insulated to minimize heat loss. The insulation must be thick enough to limit maximum heat loss to 50 B.T.U.'s per hour per linear foot of pipe.

- Storage tanks must be insulated to limit heat loss to no more than 2% of the stored energy in a 12-hour period.

- Threaded, soldered and flare joints must be made as though for water piping. (See Chapter 15.)

- Piping in the solar heating system must be isolated so that gases, fluids or other substances can not enter any portion of the potable water system.

- The entire solar heating system should use only approved plumbing items and UL approved electrical components.

- A permit is usually required to install, repair or alter any solar energy system.

- After installation, the solar system must be tested and proved tight under a water, fluid or air pressure test. The system must be able to withstand 125 psi without leaking for 15 minutes.

Questions

1- Define a hot water system.

2- Why is copper pipe and tubing the most popular material for hot water systems?

3- What type of plastic pipe may be used for hot water piping, under limited circumstances?

4- Why is safety equipment necessary for a hot water system?

5- What governs the sizing of hot water distribution piping?

6- With what does a hot water installation begin?

7- What are two principal objectives in designing a good hot water system?

8- What is the recommended temperature setting for the hot water heater thermostat when a dishwasher is used in a residence.

9- What are the two most common sources of energy for residential heating units?

10- Name two other types of energy source for residential heating units.

11- Why are electric water heaters the most popular of all hot water heating units?

12- What provision must be made for gas- or oil-fired water heaters that is not necessary when installing other heating units?

13- Why are the possible locations of solar hot water storage tanks limited?

14- Why is it not necessary to insulate hot water lines for small jobs?

15- In designing a hot water piping system, what special consideration should be made for securing pipes?

16- Residential hot water heaters may be grouped into what three categories?

17- What are the two most common types of water heaters?

18- In most dwelling units, the peak draw period is usually assumed to be how long?

19- What percentage of a hot water storage tank capacity is used in sizing the tank for the peak draw period?

20- What is considered the maximum hourly use of hot water per person?

21- What is meant by the "maximum load" for a hot water heater?

22- Describe what is meant by the "working load" of a hot water heater.

23- For a peak draw period of one hour and a 50% recovery rate, how much hot water may a 30-gallon water heater be expected to supply?

24- What structural factor determines the quantity of hot water used in a residential unit?

25- When a special demand requirement is not considered, what is the recommended size of an individual storage tank capacity for a three-bedroom residence?

26- Water heaters must be installed where they are_____.

27- What determines the rating of a temperature relief valve?

28- What other special requirement must a combination pressure and temperature relief valve have?

29- Why is the older fuse-type relief valve not acceptable in new construction?

30- In what part of a water heater must a relief valve be installed to be the most effective?

31- Why is it against the code to install a check valve or shutoff valve between the relief valve and the hot water heater storage tank?

32- Why should relief valve drip pipes not be connected directly to any plumbing drainage system?

33- Where should a relief valve drip pipe terminate on a single family residence?

34- Why should relief valve drip pipes not be threaded?

35- What determines the size of relief lines?

36- What type of fitting should be used to secure a relief line to the female thread of a relief valve?

37- Where must the drain cock on a hot water heater be located?

38- What purposes does a hot water heater drain cock serve?

39- What is the acceptable minimum size cold water line to a hot water heater?

40- Why is solar energy for heating domestic water not considered new?

41- What are the three major components that make up a solar water heating system?

42- What are some of the obvious features that make the solar hot water system unique?

43- What type of solar collector is most practical for residential use?

44- What three materials may be used in building the heat deck for a solar collector?

45- Is there a thermal difference between heat deck materials?

46- Aluminum tubing may be used in a solar installation if it is limited to what type of system?

47- Why should the heat collector box be well insulated?

48- What is recommended for solar collector panels to prevent heat loss in cold climates?

49- How many gallons of water per day will a 4' x 12' solar heat collector provide?

50- What is the minimum size recommended for a solar storage tank for a family of four?

51- A flat collector should be tilted in what general direction to be most efficient?

52- Why is it important to install solar piping to drain dry?

53- What are some of the recommendations made by roofing contractors about mounting collectors?

54- Describe how a natural thermosyphon solar water heating system works.

55- What minimum size piping must be used in a thermosyphon circulation system?

56- Where may the hot water storage tank be located in a pumped solar water heating system?

57- What fluid is commonly used in a closed solar energy collection system?

58- Why is it not practical to use CPVC plastic pipe in a solar circulation system?

15

Water Pipe And Fittings

The plumbing code regulates the materials, sizing and installation methods for water piping. It requires a satisfactory supply of drinkable (potable) cold water to all plumbing fixtures so that they flush properly and remain clean and sanitary. The code establishes safeguards to avoid pollution of the water supply due to backflow or cross-connection. The many code requirements, limitations and restrictions applied to water piping will be discussed in this chapter.

Materials

Consider the water supply in your area before selecting the material and size for water supply pipes, tubing or fittings. The water in many communities corrodes or leaves deposits on the interior walls of some pipe. Some types of soil and fill can corrode the exterior of the pipe. (See ''Maintenance of Water Systems'' in Chapter 12.) Pipe that can leave toxic substances in the water supply must *not* be used for piping, tubing or fittings. And piping that has been used for anything other than a potable water supply system should *not* be reused in a potable water supply system. For example, pipe or fittings once used in a gas system should not be reused in a potable water supply system.

Many water piping materials meet code requirements. Some materials are acceptable for use in underground installation; others can be used above ground only. Some materials are acceptable both above and below ground. The materials considered in this chapter include only those commonly used in residential or simple installations. Materials and installation methods for larger buildings are found in *Plumber's Handbook* by this author.

Galvanized Steel Pipe

Galvanized steel pipe (coated with zinc inside and outside) is usually sold in 21-foot lengths threaded at each end. It comes in nominal inside diameter sizes of 1/8 inch and larger. In most sizes it is available in three weights or strengths: standard, extra strong and double extra strong. Standard weight pipe is adequate in most installations and is used in most plumbing jobs. Pipe of the heavier weights has the same outside diameter as the standard weight pipe but has thicker walls. Thus the inside diameter and the flow through the heavier pipe sizes are reduced.

Galvanized steel pipe usually has a pipe coupling screwed to one end of each length of pipe. This can be used to join two pieces of pipe unless a special fitting is required.

Of all the various kinds of pipe presently installed in water supply systems, galvanized steel pipe is by far the most common. Galvanized pipe has many advantages. It may be used to distribute both hot and cold water. It is considered superior as a building material because of its characteristic strength, durability, and resistance to trench loads. It is especially desirable for outside use in water service supply piping. Because of its resistance to trench loads, the depth of placement is not critical except as protection against freezing. Galvanized steel pipe may be expected to give many years of service. Buried underground it can last fifteen to thirty years. Within a building it can last fifty years or longer.

Galvanized steel pipe has several significant disadvantages. It is more subject to corrosion than copper or plastic pipe. Water with a high acid content will rust the inside of the pipe. Hard water, containing large amounts of calcium and magnesium compounds, forms scales or leaves deposits that harden in galvanized steel pipe more readily than in copper or plastic pipe. Where low water

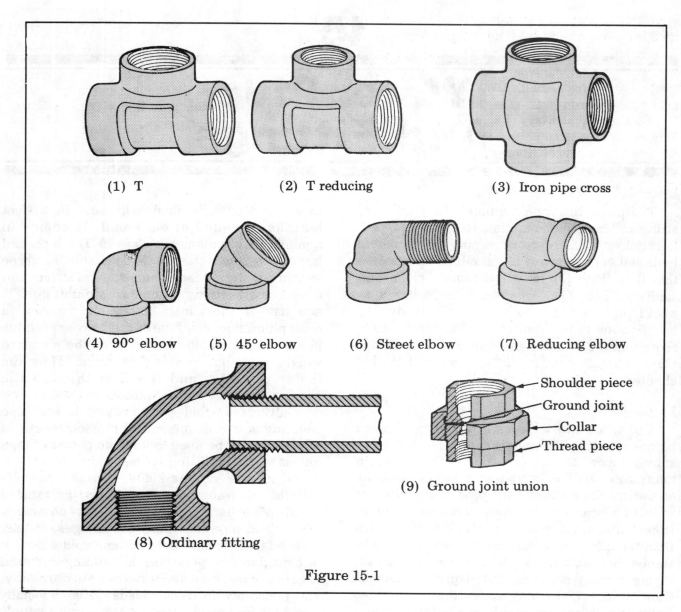

(1) T   (2) T reducing   (3) Iron pipe cross

(4) 90° elbow   (5) 45° elbow   (6) Street elbow   (7) Reducing elbow

Shoulder piece
Ground joint
Collar
Thread piece

(9) Ground joint union

(8) Ordinary fitting

Figure 15-1

pressure is a problem, the rougher inside surface of galvanized steel pipe creates more friction against moving liquid than the smoother interior of copper or plastic.

In installations *not* subject to these conditions, galvanized steel pipe and fittings may be used with confidence and long service may be expected.

Threaded Fittings

Galvanized malleable iron threaded fittings are usually found in water supply systems. They are pressure rated and are used to connect lengths of pipe, change the direction of flow, or reduce the pipe diameter. The correct fitting or combination of fittings is easy to work with. The name of the fitting usually describes its shape

and use. Some of the more commonly used galvanized fittings are illustrated in Figure 15-1.

Tees (t's) are used to provide an opening to connect a branch pipe at right angles to the through pipe or main pipe run. A *straight tee* (#1 in Figure 15-1) has the same size opening at all three connections. For example, if the main pipe run should be ¾ inch, and a ¾-inch side opening were necessary for a branch pipe to connect to the main pipe run, the tee would be called a ¾ inch tee.

When the branch opening is not the same size as the other two, the fitting is called a *reducing tee* (#2 in Figure 15-1). Specify a reducing tee by giving the straight-through (run) dimension followed by the side-opening (branch) dimension. Thus, a tee with run openings of ¾ inch and a branch opening of ½ inch is known as a ¾ by ½ inch tee. Assume you need a reducing tee with through (run) openings of 1 inch and ¾ inch and a branch opening of ½ inch. This tee would be called a 1 x ¾ x ½ inch reducing tee.

Tees may also have a run with male threads. These are known as *street* or *service tees*. A tee with two branch openings at right angles to each other is known as a *cross* (#3 in Figure 15-1). Crosses may also be straight or reducing.

Elbows (L's or ells) are used to change the direction of rigid pipe lines. Like tees, they are manufactured in many sizes and patterns. Specify elbows according to the angle in degrees between the two lengths of connected pipe. The more commonly used elbows have female threads for connecting to threaded pipe. They may be 90° or 45° (#4 and #5 in Figure 15-1).

For special situations street elbows are used (#6, Figure 15-1). These may be 90° (illustrated) or 45°. Reducing elbows are rarely used. They come in straight runs, or they may be of the reducing type (#7 in Figure 15-1). A cross-sectional view of a typical 90° elbow connected to a piece of screw pipe is shown in #8, Figure 15-1.

Unions are used where an additional branch line is installed in an existing screw pipe system or where certain types of appliances (a water heater, for example) must be freed for replacement. A union permits removing a section of pipe to install the necessary fitting for the branch opening without disturbing any of the existing pipe or fittings.

The *ground joint union* (#9 in Figure 15-1) has a shoulder piece with female threads, a thread piece with female and male threads, and a ring or collar with both an inside flange which matches the shoulder of the shoulder piece and a female thread which matches the male thread of the thread piece. The shoulder and thread pieces have a brass ground joint. The pipes to be connected are screwed to each end of the union pieces. Both pieces are then drawn together by the collar, making a watertight joint.

Couplings are used to join two lengths of the same pipe size when making a straight run. The coupling is a short fitting with female threads (#1, Figure 15-2). It is not used to join two lengths of pipe already installed and fixed in place; a union must be used in such cases. The reducing coupling (#2, Figure 15-2) is used to

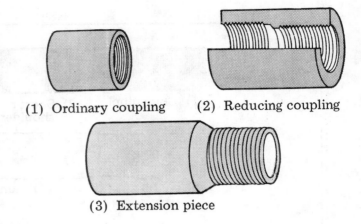

(1) Ordinary coupling   (2) Reducing coupling

(3) Extension piece

Iron pipe couplings
Figure 15-2

join two pipes of different sizes. Another type of coupling commonly used in water supply systems is the extension piece (#3, Figure 15-2), which has both male and female threads.

Among many other types of iron pipe fittings are the following:

Nipples are used to make an extension from a fitting or to join two fittings. A nipple is considered to be a piece of pipe 6 inches or less in length, threaded on both ends. A close nipple (#1, Figure 15-3), threaded its entire length, joins two fittings which must be very close to each other. A nipple threaded nearly its entire length with only a short unthreaded section in the center is called a short or shoulder nipple (#2, Figure 15-3). When the unthreaded portion is longer than this, the nipple is called a long nipple (#3 and #4, Figure 15-3). Long nipples are specified by lengths such as 3 inches, 3½ inches, 4 inches, up to 6 inches.

A *plug* is a short, generally solid length of galvanized metal, having a male thread and some means of being turned by a wrench. It is used to close openings in other fittings or to seal the end of a pipe. It has various types of heads. The square head plug (#1, Figure 15-4) is the most widely used; the slotted head plug (#2, Figure 15-4) is seldom used except in close spaces where a wrench can not be used.

A *cap* is a fitting with female threads. Its use is similar to a plug and is screwed to the threads of a piece of pipe or nipple. It may be used for plugging up water outlets when testing the system or to create an air chamber to eliminate water hammer. (#3, Figure 15-4).

A *bushing* has a male thread on the outside and a female thread on the inside. It is usually

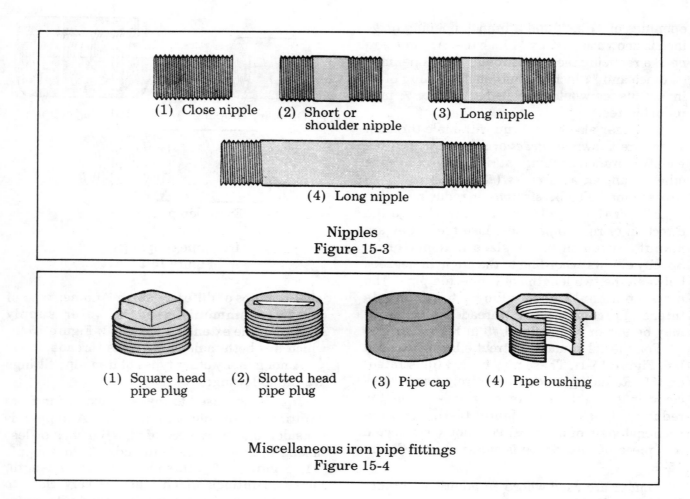

Nipples
Figure 15-3

(1) Square head   (2) Slotted head   (3) Pipe cap   (4) Pipe bushing
    pipe plug   pipe plug

Miscellaneous iron pipe fittings
Figure 15-4

used to reduce the opening of a fixed fitting to receive a smaller pipe. The ordinary bushing (#4, Figure 15-4) has a hexagon nut at the female end for screwing the bushing into the fitting. The faced bushing, without a hexagon nut, is prohibited by most codes.

Copper Pipe and Tubing

Copper pipe and tubing are used in both hot and cold water distribution systems. They can be used for both interior piping and underground installations including the water service line. Copper is light, easily installed, and resists corrosion. There are two types of copper water line: pipe and tubing.

Copper *pipe* is rigid, hard tempered, and comes in 20-foot lengths. It does not bend easily unless annealed or softened by heating. Changes in direction are usually made with fittings. Which are similar to those of many galvanized iron fittings, except that with copper the joints are soldered. The code requires lead free solder and flux in water distribution systems.

Copper *tubing* is flexible, soft tempered, and

comes in coils of 50 to 100 feet, depending on size. It is easily bent by hand while cold to form changes in direction. Since the soft tempered tubing may be easily bent, many fittings required for offsets with rigid pipe can be eliminated. Only 90° changes in direction, tees for branch line connections, and adapters must be used. Soldered fittings used for rigid copper pipe may also be used on soft copper tubing. In addition, flare fittings may be used with soft copper tubing if the need arises.

Copper pipe and tubing comes in three weights and wall thicknesses, as follows:

Type K has the heaviest or thickest wall and is generally used underground. It is available in hard and soft temper and in sizes useful for most piping installations.

Type L has a medium wall thickness and is most commonly used for water service and for general interior water piping. It is also available in hard and soft temper and in various sizes for most piping installations.

Type M has a thin wall, and many codes permit its use in general water piping installa-

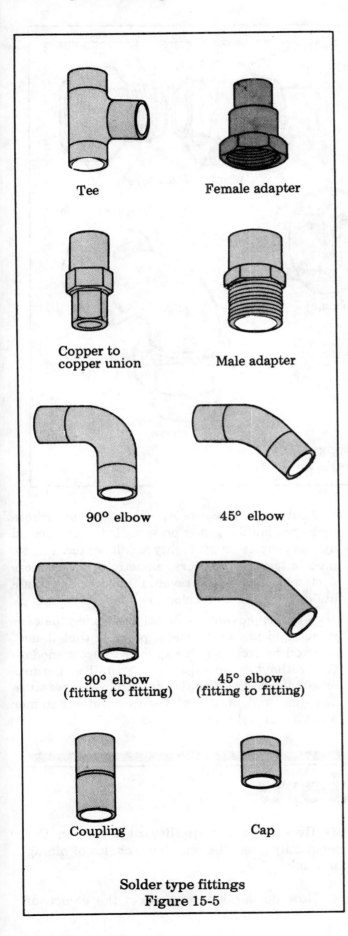

Tee

Female adapter

Copper to
copper union

Male adapter

90° elbow

45° elbow

90° elbow
(fitting to fitting)

45° elbow
(fitting to fitting)

Coupling

Cap

Solder type fittings
Figure 15-5

tions as with Types K or L. Check with your local authority for any restrictions placed on Type M before installation proceeds. It is available only in hard temper and comes in 20-foot lengths and in sizes for most piping installations.

Copper Fittings

There are two types of copper fittings: solder type and flare type.

Solder type copper fittings are more widely used than the flare. Solder fittings can be used equally well with either rigid or soft copper pipe or tubing. They are made in the usual iron pipe patterns such as tees, elbows and couplings. Figure 15-5 illustrates some of the many solder type fittings. In addition to those shown, solder type fittings include reducing tees, reducing ells, tees with male or female iron pipe adapter openings, iron pipe adapter ells, and others.

Flare type fittings (Figure 15-6) are used only with flexible soft copper tubing. Their principal use is limited to certain specialized installations because they resist vibration better than solder type fittings. They should not be used for general plumbing where the flare fitting would be sealed in the wall.

Plastic Pipe

PVC plastic pipe is rigid and comes in 20-foot lengths in sizes suitable for most piping installations. Where code permits its use, recommended weight is Schedule 40. PVC pipe and fittings can be used for water service supply piping, swimming pool piping and other applications located *outside* a building wall.

Joints and fittings are usually solvent welded with a liquid cement recommended by the pipe manufacturer. Fittings for plastic pipe are similar in size and shape to copper or iron fittings. Use only *pressure rated plastic fittings* in a water piping system. Installation of plastic pipe and fittings is described in Chapter 16.

Two plastics now accepted by most codes for water distribution above or underground are chlorinated polyvinyl chloride (CPVC) or polybutylene (PB) plastic pipe or tubing.

CPVC pipe comes in 20-foot lengths, is rigid, and suitable for both hot and cold water systems. The fittings are solvent welded to the pipe. CPVC comes in standard sizes for most hot and cold water piping installations.

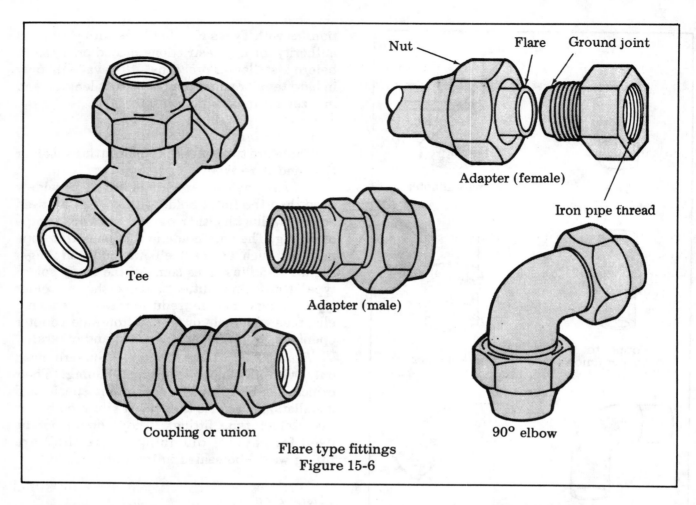

Flare type fittings
Figure 15-6

PB plastic comes in rigid 20-foot lengths or in flexible coils of varying lengths. Polybutylene tubing (coil) has one distinct advantage over rigid plastic. It bends easily, thus saving fittings. You can join polybutylene with compressions fittings, flare fittings, crimp type fittings, and by heat fusion.

All plastic pipe and fittings must carry ASTM identification numbers. All plastic pipe and fittings used in water distribution systems must have a minimum working pressure of 100 psi at 180° F for both hot and cold piping material.

Since plastic pipe is inert, it resists rust, rot, scale and corrosion. You can install it outside without fear of corrosion, even in soil where hydrogen sulfide gas is present. It is not affected by electrolysis or by highly acidic water. Plastic pipe interiors are very smooth. This reduces frictional loss and prevents hard water scale build-up which would clog the lines.

Another advantage of polybutylene pipe over some rigid plastic or metal pipes is that it isn't affected by freezing. It expands to accommodate ice without splitting. Its flexibility permits expansion and contraction caused by temperature changes. And, its "give" reduces water hammer and virtually eliminates noise.

Questions

1- What are three aspects of potable water piping that are regulated by the plumbing code?

2- What safeguards are required to avoid contamination of public water supply systems?

3- How does the quality of water in the community affect the plumber's choice of piping materials?

4- How do various soils affect the choice of

piping materials?

5- Why is it important not to reuse gas piping and fittings in a potable water supply system?

6- Steel pipe that is galvanized means that it has been coated inside and outside with what substance?

7- In what standard lengths is galvanized steel pipe manufactured?

8- What weight or strength of galvanized steel pipe is most often used in a plumbing water piping system?

9- Galvanized steel pipe usually has what type of fitting attached to one end?

10- Galvanized steel pipe is especially desirable for what particular use?

11- How does water with a high acid content affect galvanized pipe?

12- What substances does water contain that cause it to be considered ''hard''?

13- What three conditions limit the use of galvanized steel pipe?

14- Where are galvanized malleable iron threaded fittings usually used?

15- What does a tee provide in a water piping system?

16- A tee that has three openings of the same size is called a_____ tee.

17- A tee that has openings of different sizes is known as a_____tee.

18- Elbows are used to serve what purpose in a rigid pipe line?

19- What makes a street elbow different from a standard elbow?

20- What does a union accomplish when used in an existing line?

21- What is accomplished in a piping installa-tion by using a coupling?

22- What is the name of a coupling that has both male and female threads?

23- How are nipples used?

24- What is the name for a nipple that is threaded almost its entire length?

25- Name two common lengths for nipples used in a plumbing piping system.

26- Plugs are generally used for what purpose?

27- What plug is most widely used?

28- What is accomplished by the use of caps in a plumbing system?

29- What is accomplished when a bushing is used in a fitting?

30- Where is a faced bushing usually used?

31- Copper is known to resist what action?

32- Rigid copper pipe is usually manufactured in what lengths?

33- How do copper fitting joints differ from other fittings?

34- Flexible copper tubing is called by what name?

35- Name one main advantage in using flexible copper tubing.

36- Name the three weights or types of copper pipe or tubing used in piping systems.

37- What two types of copper fitting may be used in a water piping installation?

38- ABS and PVC plastic pipe may be used (where permitted) in what portion of a water supply system?

39- What minimum working pressure is re-quired for plastic pipe used in a water supply system?

16
Installation Of Water Systems

Most code sections which cover the installation of water supply systems make good sense if you take the time to understand them. Unfortunately, plumbers at the jobsite often overlook important points in their rush to finish the job on time. Think about each of the points in this chapter before you apply them on the job.

INSTALLATION METHODS

Water Service Supply Pipe

Water service supply pipe can be installed in the same trench as the building sewer. This procedure is common where the soil is hard or rocky or where there is inadequate space to separate the two lines. The following conditions must be met when a single trench is used for sewer and water service supply pipe:

• The water service supply pipe must be placed above the sanitary line on a solid shelf excavated at one side of the common trench. See Figure 16-1.

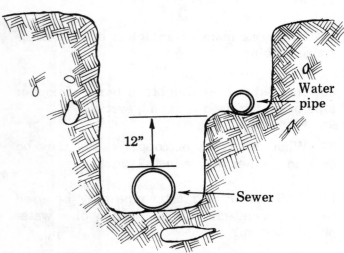

Placement of water service supply pipe
in sewer trench
Figure 16-1

• The bottom of the water service supply pipe must be at least 12 inches (some codes permit 10 inches) above the top of the sewer line. See Figure 16-1.

• The number of joints in the water service supply pipe must be kept to a minimum.

The water service supply pipe installed in a separate open trench usually must have a minimum separation of 5 feet from any sewer line. See Figure 16-2. Check with local authorities; they may require some other separation distance.

The water service supply pipe must have a minimum separation of 5 feet from any septic tank or drainfield.

Water service supply piping should be laid on a firm bed of earth for its entire length. It should be securely supported to prevent sagging, misalignment, and breaking. Installation in open trenches is recommended but not required. Tunneling with care under sidewalks or driveways is permitted. Young trees should not be planted near water service supply piping because the root system growth may cause the water line to buckle, bend, or break.

Since galvanized steel is characteristically strong, durable, and resistant to trench loads, it is considered superior as a building material. Galvanized steel is especially desirable for outside use in water service supply piping. Because of its resistance to trench loads, the depth of placement is not critical except as protection against freezing. Avoid backfilling the trench with large boulders, rocks, cinder fill, or other materials which might physically damage or encourage corrosion of the pipe.

The depth for copper pipe or tubing, type K, L, or M, is not established in most codes. Copper pipe and tubing is soft and easily damaged. Naturally, the pipe must be placed

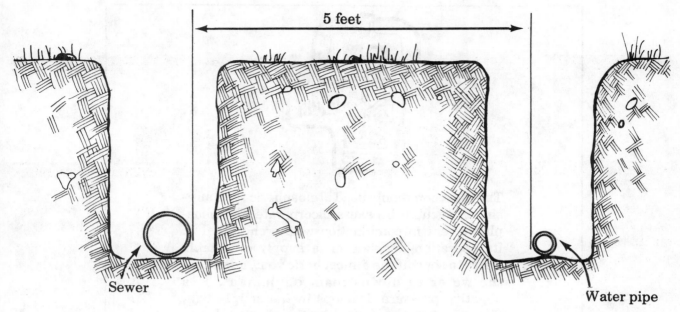

Placement of water service pipe
in separate trench
Figure 16-2

below the frost line in areas subject to freezing. In warmer climates, depth of placement is critical where there is likelihood of damage by edgers and other sharp tools. Select backfill material for the trench with care. Use fine rather than coarse materials. Avoid backfill materials such as cinder fill that promote corrosion of the pipe's exterior.

PVC plastic pipe is classed as fragile material and should be treated carefully during installation. The bottom quarter of the pipe should be supported continuously and uniformly on the trench bottom by 4 inches of fine, uniform material which should pass through a ¼-inch screen. The fine material should extend 4 inches on each side of the pipe and to a point 6 inches above the top of the pipe. The minimum allowable pipe depth below ground level is 12 inches. The installation method for plastic water service supply piping, when used in open trenches, is the same as for plastic pipe used in building sewers. See Figure 10-14.

Water service supply piping passing under the foundation of a building must have a clearance of 2 inches from the top of the pipe to the bottom of the foundation or footing. (Refer to Figure 10-27.) Water supply pipe passing through cast-in-place concrete such as a basement wall must be sleeved to provide ½-inch clearance around the entire circum-ference of the pipe. This avoids damage or breakage due to settling of the building or normal expansion and contraction of the pipe. This clearance also gives protection from the corrosive effects of concrete on metallic pipe.

In some geographic areas, such as near coastal regions, well water may be unsuitable for irrigation purposes. Buildings in these localities must depend on the public water supply to meet their irrigation needs.

When a lawn sprinkler system is connected to the potable water service supply pipe, certain installation methods must be used and preventive devices installed to eliminate the possibility of a cross-connection. The pipe to supply water to the irrigation system must extend at least six inches above the surrounding ground when it is connected to the water service pipe. There must be an approved backflow preventer located on the discharge side of the control valve. The backflow preventer (an atmospheric type vacuum breaker shown in Figure 16-3) must be located at least 6 inches above the highest sprinkler head and never less than 6 inches above the surrounding ground. See complete installation details in Figure 16-4.

Each building must have a separate water control valve installed in the water service supply pipe. This must be independent of the water meter valve or other interior control valves. The control valve must be accessibly

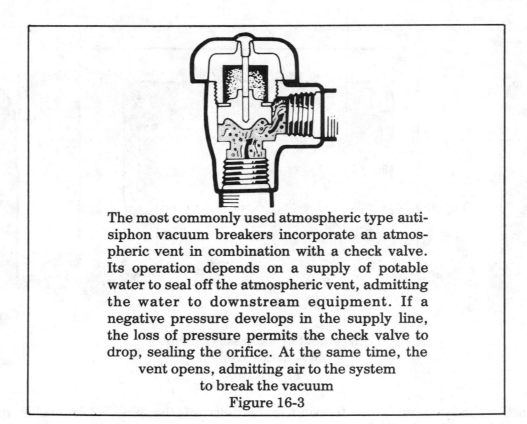

The most commonly used atmospheric type anti-siphon vacuum breakers incorporate an atmospheric vent in combination with a check valve. Its operation depends on a supply of potable water to seal off the atmospheric vent, admitting the water to downstream equipment. If a negative pressure develops in the supply line, the loss of pressure permits the check valve to drop, sealing the orifice. At the same time, the vent opens, admitting air to the system to break the vacuum

Figure 16-3

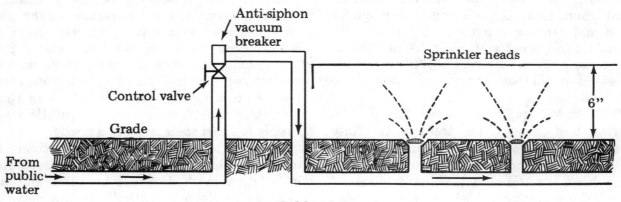

Table 16-4

located at or near the foundation line outside the building, either above ground or in a separate approved box with a cover. See Figure 12-1.

Some codes require an accessible control valve near the curb and a second control valve with a drip valve near where the water supply pipe enters the building. The drip valve is required in cold climates to drain off water during cold weather to prevent freezing and bursting the pipes.

Water Distributing Piping

Underground water distribution piping installed under a slab within a building must be firmly supported for its entire length on well-compacted fill. This prevents misalign-

ment, sagging, and the formation of traps or depressions in the pipe that could collect sediment or mineral deposits. Some mineral deposits solidify over a period of time, reducing the flow of water and causing premature failure of the system. Use fine rather than coarse backfill in the trench so that the pipe is not damaged when the soil is compacted.

Water distribution piping beneath a building constructed with wood joists should be strapped securely to the joists with pipe straps. Galvanized pipe straps should be used with galvanized pipe. Copper or copper plated straps should be used with copper pipe or tubing. Copper straps may also be used to support or

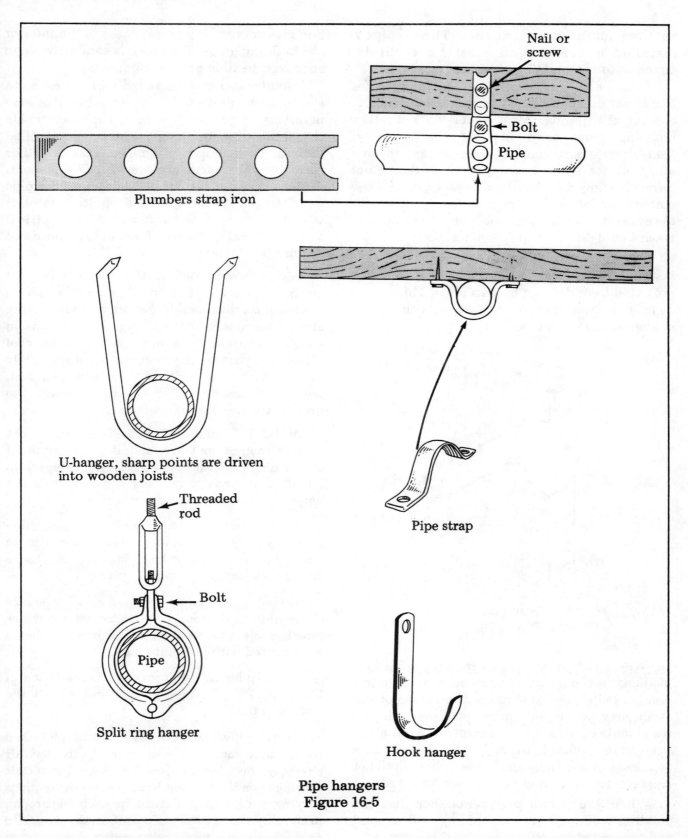

Pipe hangers
Figure 16-5

secure CPVC plastic pipe (where plastic pipe is permitted), as the outside diameter is approximately the same for the two materials. Other types of supports for small water distribution pipes are illustrated in Figure 16-5.

Water piping inside walls and partitions must be secured by pipe straps or some other approved method to prevent movement of pipes as they contract or expand. These pipes installed above ground must be securely attached to the building structure.

It will also be necessary in installing pipes inside wooden partitions to notch these studs to conceal the piping. Most codes do not allow notching more than 25% of the depth of *load-bearing* partitions. It is important to note also that *non-load-bearing* partitions should not be notched any deeper than necessary, so that the outside portion of the pipe may finish flush with the edge of the stud. Never under any circumstances notch a stud deeper than 40% of its depth.

Copper pipe, tubing, CPVC or PB plastic pipe installed in notched wooden partitions should be protected from damage by lath nails. This can be done by covering the notched opening with a metal stud guard as in Figure 16-6.

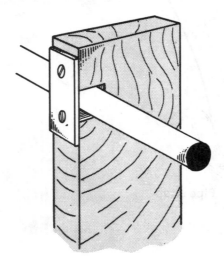

Stud guard
Figure 16-6

Where it is necessary to install piping in a load-bearing partition to meet code requirements, drilling holes through the center of the studs may be the only acceptable method. The disadvantage is that the pipe must be cut short enough to thread through the holes. This requires extra fittings and joints. A hole shall not be greater than 40% of the width of the stud.

When you install pipe remember that the building will settle and the piping will expand and contract with use. Make sure this movement will not damage the pipe. Water supply piping that passes through cast-in-place concrete floor slabs or bearing walls must be sleeved to provide ½-inch clearance around the entire circumference of the pipe. The sleeve installa-

tion gives crawling insects access to the interior of a building unless the space is caulked with an approved flexible sealing compound.

Metal water pipe installed under a concrete slab must be protected from corrosion. It is also necessary to protect pipe in any location inside the walls of a building constructed on filled ground if hydrogen sulphide gas or other corrosive substances are known to be present. Protection can be provided by painting the pipe and fittings with two coats of asphaltum paint. It is better still to install water pipe in the attic if you are working on filled ground. Some codes require this latter procedure.

Metal water distributing pipe must be electrically isolated from all ferrous pipes, electrical conduit, building steel, and reinforcing steel. This means that the pipe cannot come in physical contact with any good conductor of electricity. Where a physical contact is absolutely unavoidable, the pipe must be wrapped with heavy tar paper or other approved material to provide the necessary insulation.

Metal hot and cold water pipe must not contact each other when installed underground or within partitions. Contact or even close installation tends to transfer the heat from the hot water pipe to the cold water pipe.

Each plumbing fixture shall have accessible shutoff valves. Refrigerators with automatic ice makers must also be valved. Washing machines do not need separate shutoff valves.

Do not install water piping in crawl spaces or other unheated areas in climates where water pipe is subject to freezing, unless it is PB plastic or protected with adequate insulation.

Most codes now approve the installation of CPVC or PB plastics underground within the walls of a building.

Water pipe installations must be protected from water hammer with properly located air chambers or other approved devices. Install air chambers on the highest hot and cold water pipes serving a plumbing fixture in each bathroom, preferably the lavatory pipes. (See Figure 16-7.) Also install air chambers on water pipes serving individual fixtures such as kitchen sinks and clothes washing machines. Make up air chambers from a piece of capped pipe at least 12 inches long and one pipe size larger than the pipe it serves.

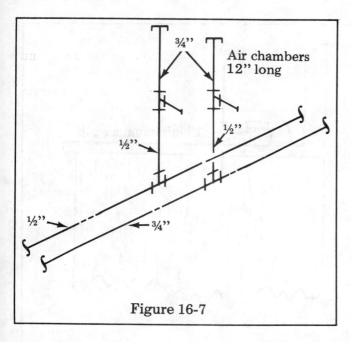

Air chambers 12" long

¾"

½"

½"

½"

¾"

Figure 16-7

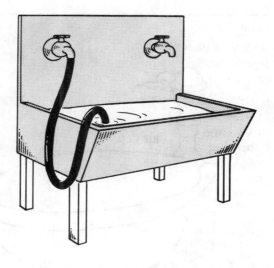

Cross-connection
Figure 16-8

What is potentially dangerous about an un-protected sill cock?

The purpose of a sill cock is to permit easy attachment of a hose for outside watering purposes. However, garden hoses can be extremely hazardous because they are left submerged in swimming pools, lie in elevated positions (above the sill cock) to water shrubs, are attached to chemical sprayers for weed-killing, etc., and are often left lying on the ground—which may be contaminated with fertilizer, cesspools, and garden chemicals.

Hose bibb vacuum breaker
for frost-proof hydrants

Hose bibb vacuum breaker

What protection is required for sill cocks?

A hose bibb vacuum breaker should be installed on every sill cock to isolate garden hose applications and thus protect the potable water supply from contamination.

Figure 16-9

Flush valve

Vacuum
breaker

6" minimum
air gap

1" minimum air gap

Air gap minimum 1"

**The three plumbing fixtures illustrated above
provide the minimum air gap permitted by code
Figure 16-10**

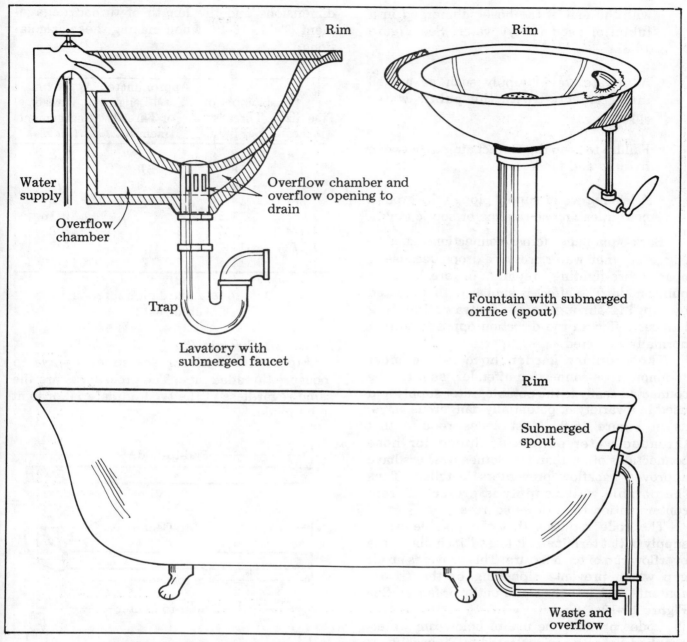

Fixtures with supply outlets below the overflow rim are prohibited. A stoppage in the waste line could cause a cross-connection. The liquid in the fixture bowl could rise to the flood-level rim and be siphoned into the water-supply system through the submerged water outlets

Figure 16-11

The entire water supply system should be installed so that it drains dry for repairs, for replacing lost air in air chambers, and to prevent freezing of the water in the pipes.

Cross-Connection

By definition, a cross-connection is a direct arrangement of a piping line that allows the potable (safe for drinking) water supply to be connected to a line which contains a contaminant. Such a condition is, of course, prohibited by code. There are many situations that might create cross-connections. Here are some of the most common:

- A garden hose attached to a hose bibb with the end of the hose lying in contaminated water.
- A garden hose attached to a laundry sink

with the end of the hose submerged in a tubful of used (dirty) water. See Figure 16-8.

- A private water supply such as a well interconnected with the public water supply.

- Failure to keep the correct air gap between fixtures and faucets.

- A leaking water supply pipe in the ground near a leaking sewer line or septic tank.

Back-siphonage (cross-connection) can be triggered when water pressure drops because of nearby fire-fighting, repairs, or breaks in a public main. The effect is similar to that created by sipping through a soda straw. Inhaling induces a flow in the direction opposite to that normally expected.

The ordinary garden hose is the most common cross-connection offender, as it can be connected easily to the potable water supply and used in a variety of potentially dangerous ways, as in Figure 16-9. Most codes require that threaded water outlets equipped for hose connections other than for clothes washers have approved backflow preventers installed. Thus the potable water supply is protected from contamination from these sources.

The code requires that the potable water supply outlet terminate at least 1 inch above the overflow rim of each fixture. This provides an air gap which prevents siphoning of the fixture contents back into the water outlet or faucet. See Figure 16-10.

Codes prohibit the use of below-rim water-supplied fixtures as illustrated in Figure 16-11. Back-siphonage of polluted water is possible in this type of fixture if the opening of the faucet becomes submerged in waste water and a negative pressure develops in the supply piping.

BASIC PLUMBING PROCEDURES

Measuring Pipe

Know the correct way to measure pipe before you attempt to cut and join it. There are two basic types of pipe measurement—the straight run and the offset—and several ways of measuring individual pieces. Be sure you know which measuring method is used in any given pipe measurement. You must know the fitting

dimensions and the length of thread engagement (Table 16-12) when making these calculations.

Pipe Size (inches)	Number of Threads per Inch	Approximate Total Length of Threads (inches)	Approximate Thread Engagement (inches)
1/4	18	5/8	3/8
3/8	18	5/8	3/8
1/2	14	13/16	1/2
3/4	14	13/16	1/2
1	11½	1	9/16
1-1/4	11½	1	5/8
1-1/2	11½	1	5/8
2	11½	1-1/16	11/16

Data for standard pipe threads
Table 16-12

Measuring Straight Pipe

Figure 16-13 illustrates several ways to measure threaded pipe. The procedures are the same regardless of the type of material or joint to be made.

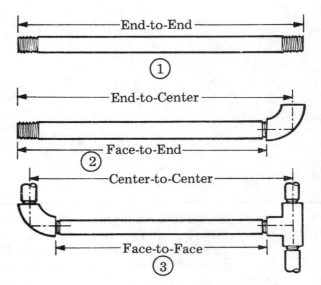

Threaded pipe measurements
Figure 16-13

End-to-End Measure (#1) Measure the full length of the pipe, including both threads.

End-to-Center Measure (#2) Used for a piece of pipe which has a fitting screwed on one end only. The pipe length is equal to the measurement, *minus* the end-to-center dimension of the fitting, *plus* the length of thread engagement.

Degree of offset	When C = 1, D =	When D = 1, C =	When C = 1, E =
60°	0.5773	1.732	1.1547
45°	1.000	1.000	1.4142
30°	1.732	0.5773	2.000
22-1/2°	2.414	0.4142	2.6131
11-1/4°	5.027	0.1989	5.1258
5-5/8°	10.168	0.0983	10.217

Finding the length of pipe to connect two
parallel runs of piping
Table 16-14

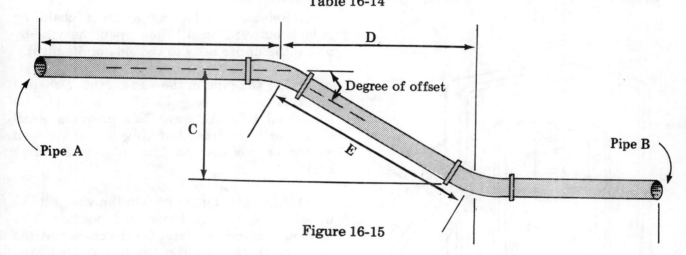

Figure 16-15

Face-to-End Measure (#2) Also used for a piece of pipe which has a fitting screwed on one end only. The pipe length is equal to the measurement *plus* the length of the thread engagement.

Center-to-Center Measure (#3) Used for a length of pipe which has fittings screwed on both ends. The pipe length is equal to the measurement, *minus* the sum of the end-to-center dimensions of the fittings, *plus* twice the length of the thread engagement.

Face-to-Face Measure (#3) Used in the same situation as center-to-center measurement. The pipe length is equal to the measurement *plus* twice the length of the thread engagement.

Measuring Offsets

In the layout and dimensioning of piping arrangements, the only place offsets are *not* a problem is where 90° fittings are used—that is, in *straight runs* as previously described. You can find the exact distance between centers of fittings of *offsets* by calculations using Table 16-14. For example, look at Figure 16-15, which shows a common pipe offset between parallel runs of pipe.

The center-to-center offset (C) is 10 inches. Assume that pipes A and B are to be connected with 45° elbows. Determine the length of pipe required for E using Table 16-14. The letters in the table refer to the dimensions in Figure 16-15. Find the number 1.4142 opposite the 45° offset. Multiply this number by distance C, in this case 10 inches. The length of pipe required would be 14.14 inches.

Use the same procedure to find the length of pipe required for other degrees of offset. For example, C is 10 inches. Assume the pipes A and B are to be connected with a 22½° offset. Find the number 2.6131 opposite the 22½° offset in Table 16-14. Multiply this number by distance C (in this case, 10 inches). The length of pipe required would be 26.13 inches.

The Pipe Vise

Vises are commonly used to hold pipe firmly during cutting, reaming and threading. They can also be used to tighten fittings to one or both ends of a threaded pipe. The two types of vises are described here.

The *hinged pipe vise*, known as a *yoke vise*, has a fixed toothed lower jaw and a movable toothed upper jaw. It opens on one side to receive the pipe. Turning the handle clockwise tightens the jaws on the pipe and holds it securely in place. See Figure 16-16.

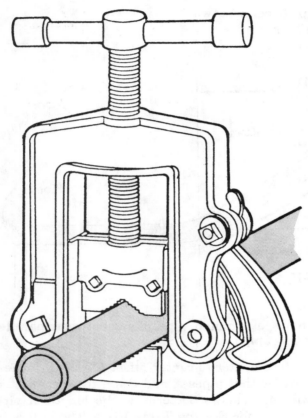

Pipe clamped in vise
Figure 16-16

The *chain vise* has a fixed toothed lower jaw and a chain that is looped over the pipe. The chain can then be hooked into slots on the opposite side of the vise body. Turning the handle clockwise causes the chain to tighten on the pipe and hold it securely in place. See Figures 16-17, 16-18, and 16-19.

Pipe vises are seldom used or needed to secure fragile or soft metal pipes. But if you must use one, protect the area of the pipe to be secured in the vise jaws. This prevents scratching or marring polished or chrome plated surfaces. Wrap the finished pipe with adhesive tape and tighten the vise with just enough pressure to hold but not damage the pipe.

Pipe vises may be mounted on a work bench, but they also come attached to a portable tristand. This three-legged stand can be folded for easy carrying or storing. When open for use, the tristand has a lower shelf for tools and pipe compound.

The procedure for using the pipe vise is the same for both the yoke and the chain vise:

1) Release the pressure on the jaws of the vise (Figure 16-16) by turning the handle counterclockwise.

2) Release the locking lever or chain by pulling outward on it. Then open the vise by tilting the upper body to the side on its pivot.

3) Lay the pipe in the vise on the V-shaped lower jaw. Long lengths of pipe should be supported in some way. This prevents small diameter pipe from bending under its own weight or the tristand from tilting or falling over.

4) After placing the pipe in the vise, tilt the upper body back to the closed position. The lever should lock in place. On the chain vise the upper body is looped over the pipe and is locked in the slots provided.

5) Turn the handle clockwise until the V-shaped upper jaw or chain presses firmly on the pipe. Tighten the handle only enough to keep it from turning as you thread the pipe.

CUTTING AND JOINING STEEL PIPE

Cutting and Reaming
Much pipe is wasted by careless cutting. Improper reaming causes a reduced flow of water through the pipes. Follow the procedures outlined below to avoid cutting waste and reduced water flow.

1) After measuring the pipe to determine the length needed, mark the spot with crayon or lead pencil.

2) Lock the pipe securely in the vise, as described above, with the cut mark up. Provide approximately 8 inches of clearance between the mark and the face of the vise. This distance should provide ample room for working the pipe

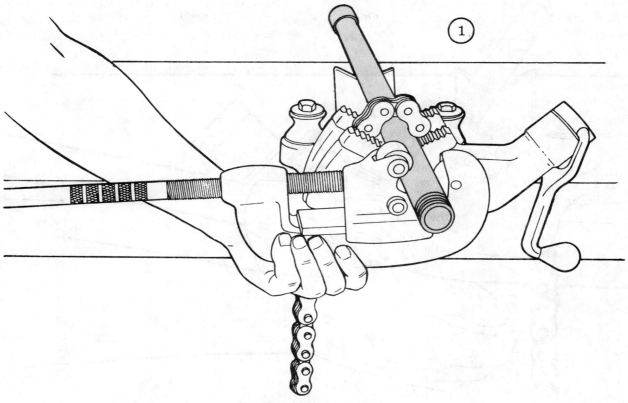

Cutting pipe
Figure 16-17

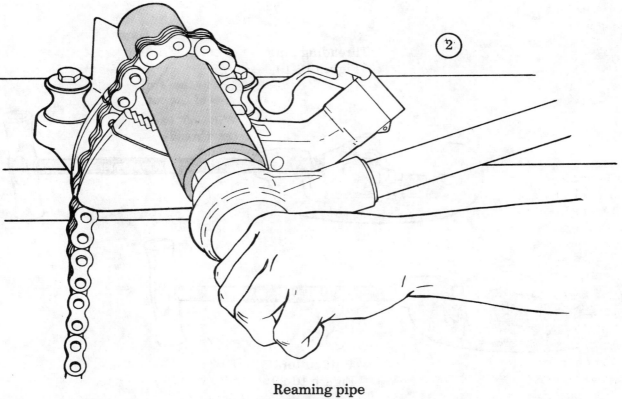

Reaming pipe
Figure 16-18

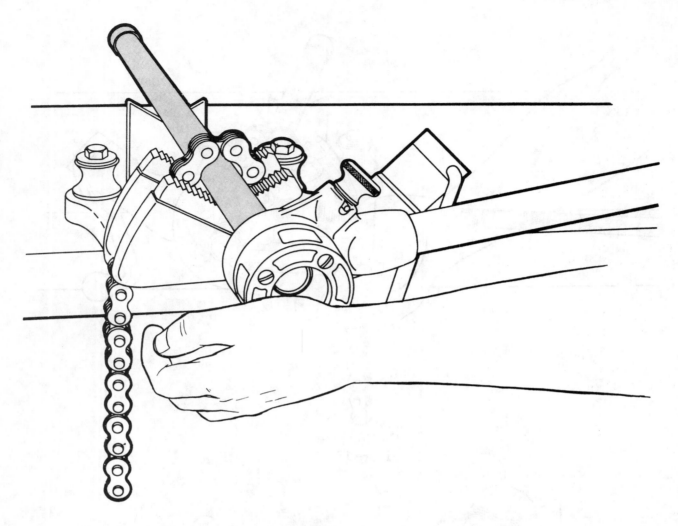

Threading pipe
Figure 16-19

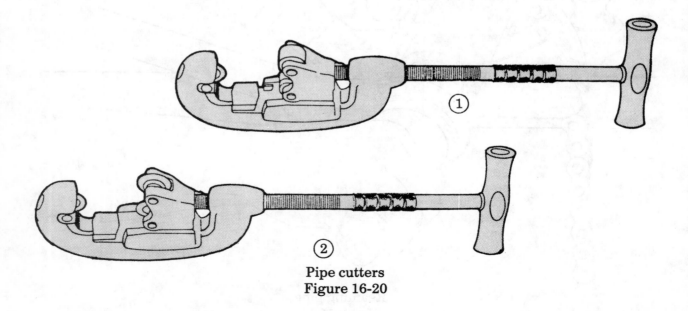

Pipe cutters
Figure 16-20

cutter and the space necessary for operating the die stock.

3) Open the pipe cutter by turning the handle counterclockwise until it clears the diameter of the pipe. Figure 16-20 shows the single-wheel cutter (#1) and the three-wheel cutter (#2). The single-wheel cutter is used more often and is the easier of the two models to operate. These are heavy duty pipe cutters and can be used to cut steel pipe.

4) Place the cutter around the pipe from underneath (Figure 16-17) with the cutting wheel (single-wheel cutter illustrated) exactly on the mark. The rollers of the single-wheel cutter ensure a straight cut. Begin each cut by lightly rotating the cutter completely around the pipe. This gives a "bite" or groove for the cutter wheel to follow.

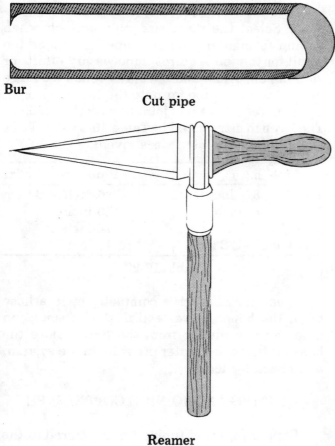

Bur

Cut pipe

Reamer

Figure 16-21

After each turn of the cutter wheel tighten the handle slightly. Tightening the handle too rapidly can break the cutter wheel or spring the

cutter frame, ruining the cutter. Be sure to use thread-cutting oil occasionally on the cutter wheel and rollers.

If a three-wheel cutter is used, place the two cutter wheels attached to the movable jaw on the mark and make sure that all three wheels lie in line at right angles to the center line of the pipe. Follow the procedure described above for a single-wheel cutter.

5) Rotate the cutter completely around the pipe, making a quarter turn on the handle for each complete revolution until the pipe is cut.

6) The pipe cutter wheel leaves a bur on the inside of the pipe. (See Figure 16-21.) Mineral deposits can collect at the bur and cause premature failure of the line. Use a pipe reamer as shown in Figure 16-18 to remove the bur before threading the pipe.

Threading Steel Pipe

The two most common hand operated pipe threading tools in use today are the *3-way threader* and the *ratchet threader*, shown in Figures 16-19 and 16-22.

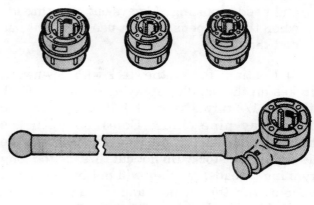

Ratchet stock and die set
Figure 16-22

The 3-way threader has three pipe-size dies as part of the stock, which has two handles for applying pressure as the pipe is threaded. This type can be used only where there is enough room to make a complete revolution. The stock may have die sizes of 3/8 inch, 1/2 inch, and 3/4 inch, or 1/2 inch, 3/4 inch, and 1 inch for threading these particular pipe sizes.

The ratchet threader is much easier to use and has dies that are interchangeable with a single stock. The dies can be 3/8 inch up to and including 1 inch. Ratchet threaders for larger pipe sizes are also available. The ratchet threader requires less pressure and can also be used to thread pipe in close places.

Pipe threaders cut standard pipe threads a taper of 3/4 inch per foot. This meets the standards adopted by the American Standards Association for pipe and fitting threads.

Procedure in Threading

1) Fasten the pipe securely in the vise.

2) Slide the stock over the end of the pipe with the guide on the inside (next to the vise). Push the die against the pipe with the heel of one hand. (See Figure 16-19.) Take three or four short, slow clockwise strokes, pressing the die firmly against the pipe. When enough thread has been cut so the die is firmly started on the pipe, apply plenty of thread-cutting oil on the threader dies. This prevents overheating of the dies. Occasionally check to be sure you are cutting clean threads.

Note: Chipped or torn threads indicate that the pipe at this location is too soft, too hard, has impurities in it, or that the dies are worn and need replacing. Always cut new threads when this happens. Do not use pipe that has bad threads.

3) Continue to turn the stock with downward strokes on the handle, applying oil often. Back off about 1/4 turn after each full turn forward to clear away pipe chips. Continue until the threaded pipe extends flush to 1/4 inch from the die end of the stock. Do not cut threads too long (running threads), as they will not have enough taper at the thread end and a bad joint will result. Place a shallow pan covered with a wire screen under the threader. The pan catches dripping oil for use again. The screen catches the pipe chips for safe disposal.

4) To remove the threader after threads are cut, turn the stock counterclockwise until the die is free of the cut threads.

5) Use a rag to wipe away the chips and excess oil that stick to the threads. The pipe is now ready to receive the fitting.

Joining Pipe and Fittings

Fittings are generally screwed to one end of

the pipe while it is still in the vise. The assembled pipe and fittings are then screwed into place in the installation. Follow these steps to make leakproof joints:

1) Inspect the female thread on the fitting and, if necessary, clean it with a wire brush. The thread should be free of rust or other foreign substances.

2) Repeat step 1 above for the male thread on the pipe if the thread is not new.

3) Apply pipe joint compound (known as *pipe dope*) evenly over the male threads of the pipe. Do not put pipe joint compound on the female threads of the fitting.

4) Hand screw the fitting on to the pipe. The fitting should turn easily for the length of about three threads.

5) Select the right size pipe wrench when joining fittings to threaded pipe. A wrench too small for the job requires unnecessary effort for your hands, arms, and back. A wrench too large forces the fitting too far onto the threaded pipe. This can result in a bad joint or a cracked fitting. The wrench sizes listed in Table 16-23 should be adequate for the pipe sizes given.

Pipe Sizes	Pipe Wrench Size
1/8, 1/4, 3/8 inch	6-inch or 8-inch
1/2 and 3/4 inch	10-inch
1 inch	14-inch
1-1/4 and 1-1/2 inch	18-inch

Table 16-23

When installation is complete, open a hose bibb, the house valve, and flush the sand and other foreign matter from the lines. Close the hose bibb, leaving water pressure in the system, and check for leaks.

CUTTING AND JOINING COPPER PIPE

Copper pipe and tubing are measured in the same way as threaded pipe (Figure 16-13) with one exception. Soft-tempered copper tubing is easily bent, and many fittings used for offsets with rigid pipe are not necessary. When using soft tubing, make allowance for fitting dimensions (solder or compression type) and for the distance the pipe or tubing is inserted into the fitting. Offsets for rigid copper pipe, on the other hand, are calculated in the same way as for threaded pipe. See Table 16-14 and Figure 16-15.

Cutting and Reaming

Both rigid and soft copper pipe and tubing may be cut with a tubing cutter (Figure 16-24) or

Tubing cutter
Figure 16-24

a hacksaw using a fine-tooth blade (24 teeth per inch). Follow the procedure for cutting iron pipe (Figure 16-17) when using the tubing cutter.

Usually no vise is required to cut copper tubing. Hold the tubing firmly with one hand and use the other hand to turn the cutter. A clean square cut is easier to make with a tubing cutter. Copper tubing cutters are similar in appearance to pipe cutters, though much smaller. Apply oil to the cutter wheel and rollers sparingly, as excessive oil must be cleaned off completely before soldering. Be careful to remove the bur the cutter wheel leaves inside the tubing (see Figure 16-18). Use a reamer attached to the tubing cutter frame for this purpose. When cutting tubing with a hacksaw, remove the rough edge with a round or half round file.

Soldered Joints

Soldering joins two metallic surfaces together with a fusible alloy (solder) which has a melting point lower than that of the pipe to be joined. The most common joint for copper tubing is the *sweat joint*. If this joint is properly made, it can last as long as the tubing.

The solder recommended for use in copper or brass potable water systems is about 95.5 percent tin, 4 percent copper and 0.5 percent silver. It can be used with any general purpose soldering flux on all materials except aluminum. It melts at 440 degrees, giving it excellent joint penetration and flow.

If the joint is properly cleaned and heated, surface tension spreads the solder to all parts of the joined surfaces. This results in a sound joint.

Soldering is an art in itself. It looks easy but requires care and precision. It takes much practice in soldering before you have the knowledge and skill necessary to solder copper water systems with confidence.

Remember to select a torch tip appropriate for the size of copper tubing you are joining. Too large a tip overheats and burns the tubing and fitting, preventing the solder from joining the two surfaces. Too small a tip heats the pipe and fitting unevenly and does not draw the solder into the joint. In either case, the result is a bad joint. Such a joint can leak when it is first tested or even months later after the piping has been enclosed by building walls.

Follow the procedure below in preparing and soldering copper pipe and fittings:

1) Inspect the end of the copper tubing to make sure it is round and free of burs. If the end of the tubing is out of round, cut it off.

2) Clean the inside of the copper fitting with a wire brush designed for this purpose or with emory cloth. The pipe end should be polished bright with emory cloth. All tarnish must be completely removed from the end of the tubing and the inside of the fitting so that the solder will tin all surfaces evenly. Use cleaned parts as soon as possible and do not handle them with dirty or sweaty fingers. The bright surface oxidizes very quickly and the cleaning process must then be repeated.

3) Apply a thin coat of non-corrosive soldering flux to the inside of the cleaned fitting and the outside of the cleaned pipe.

4) Push the tubing into the fitting and turn it a few times to spread the flux evenly. Fitting and tubing are now ready for soldering.

5) Heat the back side of the fitting evenly with a propane-type torch. (Professionals use a special soldering torch fired from a 20-pound tank of gas.) Do not apply the flame directly to the point where the pipe enters the fitting. See Figure 16-25.

6) When the flux begins to bubble, the

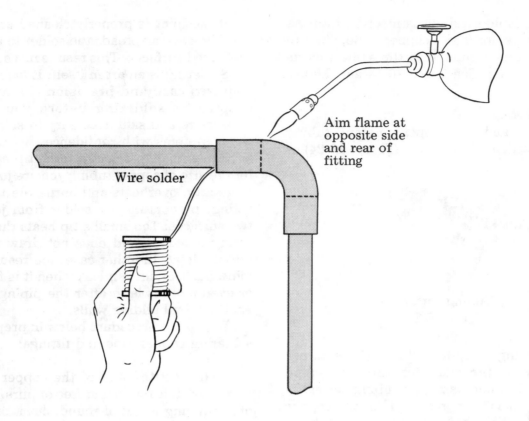

Soldering copper fitting
Figure 16-25

connection should be hot enough. Remove the torch and touch the wire solder to the point where the pipe enters the fitting. If the fitting and the tubing are hot enough, the solder will melt and be drawn into the joint. When a line of solder shows completely around the joint, the connection is filled with solder.

7) Do not move the tubing or fitting while the solder is cooling. Any movement may result in a faulty joint. Solder will harden in a minute or less if the joint has not been overheated. To speed the cooling process a wet rag may be placed over the fitting.

8) When soldering is to be done near wood or other combustible material, place an asbestos or metal sheet between the point to be soldered and the combustible material. Always keep a bucket of water available when installing copper in wood partitions.

If a leak develops after a system has been pressurized, it is much more difficult to re-solder because there is usually some water present in the pipe. The heat from the torch turns the moisture in the tubing to steam. This makes pinholes in the newly-applied solder before it can harden. Try applying heat to the tubing several inches to each side of the fitting to be soldered. This method usually dries the pipe long enough to make a good joint, but you must work quickly.

If you are unable to keep water away from the joint you are trying to solder, stuff plain white bread (remove the crust) as far into the pipe as possible in the direction from which the water is flowing. The bread absorbs the water, giving you time to make a good joint. When water pressure is returned to the system the bread flushes out easily, causing no trouble for the water system.

Use male-threaded or female-threaded brass adapters if you have to connect copper pipe or tubing to threaded pipe or fittings. Unions in a copper water supply systems should be brass. In galvanized water supply systems unions must have metal-to-metal joints with brass ground seats.

When installation is complete, open a hose

bibb, open the house valve, and flush sand and other foreign matter from the lines. After flushing, close the hose bibb and leave the water pressure on so the system can be checked for leaks.

Flare Joints

Occasionally a flare joint is required. This type of joint is used on soft flexible copper tubing when connecting certain types of fixtures and appliances. Hard (rigid) copper pipe can not be used to make flare joints as it is likely to split.

For example, a connector for a gas range or water heater might be joined with a flare fitting. The tubing is cut to the desired length. Then the flare nut that will make the connection is slipped on the cut tubing. A flaring tool (Figure 16-26) is

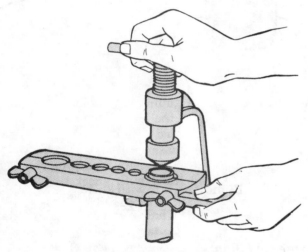

Using the flaring tool
Figure 16-26

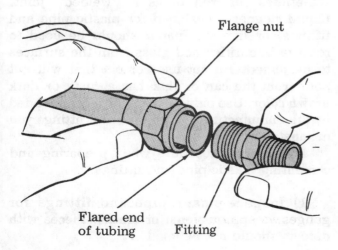

Flange nut

Flared end
of tubing   Fitting

Making up a flared fitting
Figure 16-27

used to flare the tubing ends to obtain a perfect fit. Slide the nuts to each end of the flare tubing and gently bend or shape the tubing by hand so that the ends fit together. The flare nuts must be screwed firmly by hand on the male thread of each fitting. (See Figure 16-27.) Tighten with a smooth jaw adjustable wrench of proper size until the fitting is snug. Then test the joint for leaks.

Flare fittings are not recommended for general copper installations and should be used only when necessary. They should never be concealed within walls or in any other inaccessible place. Flare fittings are made in ells, tees, adapters, unions, and other shapes. See Figure 15-6.

Compression Joints

Compression-type fittings are seldom used in a water supply system. They are most often used to connect fixture supply tubes to an ell or a cutoff valve located beneath the fixture. Although they are made in ells, tees, couplings, and other shapes, their use is limited. An example might be in connecting the water supply to an ice-maker unit in a refrigerator.

Compression type fittings are not flanged but have a threaded nut which forces a compression ring against a ground joint on the fitting. See Figure 16-28. The compression ring

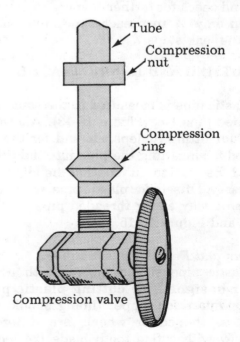

Tube

Compression
nut

Compression
ring

Compression valve

Figure 16-28

bites into the soft copper tubing, making a water tight connection.

Assembling a Compression Fitting

First, make any necessary offsets in the fixture supply line tube. Allow the tubing to fit fully into the fitting socket, marking the length with a pencil. Cut the supply line tube square using a small tubing cutter. A hacksaw will not do because it makes a rough cut and probably leaves the tubing end badly out of round and subject to leaking.

Remove the inside bur with a reamer attached to a tubing cutter frame. Slip first the nut, then the compression ring onto the tube. To help prevent a leak, apply a small amount of pipe compound over the compression ring.

Insert the tube end into the fitting. Slide the compression ring into the joint, making sure it is squarely aligned. Do not put any undue stress or strain on the tubing.

Slide the compression nut over the fitting thread and screw it down until it is hand tight. With a basin or other proper wrench, tighten the nut at the other end of the tube to the fixture faucet or other type of connection. Then tighten the compression nut firmly with a smooth-jawed adjustable end wrench. Over-tightening can cause a leak. Turn on the water and check for leaks. If there is any seeping, tighten ¼ turn *only* and check for further seeping, continuing to tighten only ¼ turn each time after each test until the leak stops.

CUTTING AND JOINING PLASTIC PIPE

Plastic pipe is measured in the same way as threaded pipe (see Figure 16-13). Allowance is made for fitting dimensions and for the length needed for inserting the pipe into the fitting. It should be cut to fit so that bending is not necessary. Offsets for plastic pipe are calculated the same way as for threaded pipe. See Table 16-14 and Figure 16-15.

Cutting and Reaming Plastic Pipe

Plastic pipe should be cut with a tubing cutter designed for cutting plastic pipe. A tubing cutter for copper tubing should not be used, as the cutter wheels are different. A hacksaw with a fine tooth blade (24 teeth per inch) may also be used. A square end cut is essential. A freehand cut with a hacksaw is an invitation to problems. Use a miter box when you cut plastic pipe with a hacksaw or handsaw. See Figure 10-17.

Be sure to remove the bur left by the cutter wheel inside the pipe. For this, use a reamer attached to the tubing cutter frame. A hacksaw does not leave a bur, but the inside and outside of the pipe are left with a rough ridge. Remove this roughness with a knife or small file. See Figure 16-29.

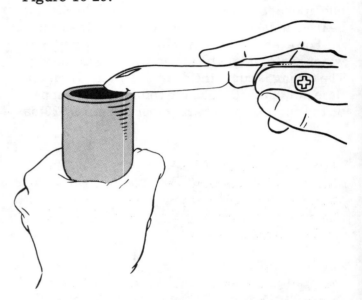

Removing bur with knife
Figure 16-29

Cemented Joints

Special plastic cement is always used to join plastic pipe and fittings. The joint created is sometimes referred to as a "welded" joint. Liquid cleaner (developed for plastic pipe and fittings) or fine sandpaper should be used to remove impurities and gloss from the surfaces to be joined. Do not use cement that will not pour from the can or that has a rusty or dark brown color. Use only the cement recommended by the manufacturer of the pipe and fittings you are installing.

Follow this procedure when preparing and cementing plastic pipe and fittings:

1) Inspect plastic pipe and fittings for gouges, deep abrasions, or cracks. Pieces with defects should not be used.

2) After the pipe is cut to the proper length and the burs are removed, check the dry fit

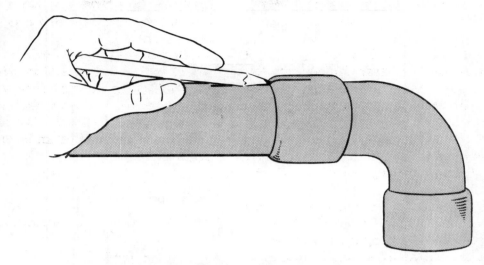

Mark both pieces with a pencil
Figure 16-30

before cementing. The pipe must enter the fitting socket smoothly but must not be so loose that the two surfaces do not make good contact. Test to see if the fitting will fall from the pipe. If it does, do not use it.

3) If the fit is correct, adjust the fitting to the position at which it will be cemented. Mark both pieces with a pencil (see Figure 16-30) so they can be quickly repositioned after cement has been applied. Cement sets very quickly once it is applied to plastic surfaces. For this reason, only one joint can be cemented at a time.

4) Clean the pipe end and the inside of the fitting with a clean rag to remove any grease or moisture.

5) Use liquid cleaner or fine sandpaper to remove surface gloss from the outside of the pipe end and the inside of the fitting socket.

6) Brush a light coat of cement on the outside of the pipe and the inside of the fitting socket, using the brush supplied with the can. Quickly brush a second coat of cement on the outside of the pipe.

7) Push the pipe fully into the fitting socket and give the pipe a quarter turn. Then set the pipe and fitting to the pencil marks and check the correctness of the position. Make any adjustments at once, as this is your last chance before the cement sets.

8) Wipe excess cement from the fitting with a rag and close the cement can immediately to prevent drying.

9) Hold the pipe and fitting together for approximately 15 seconds until the cement sets.

10) An even bead of cement should appear around the joint if it is properly made. If the bead is incomplete or does not show, not enough cement was used. Pull the pipe from the fitting immediately, apply more cement to the outside of the pipe only (not the fitting), and reinsert into the fitting.

Wait at least one hour after cementing the last joint before testing the system. If time is not critical, it is best to let the joints harden overnight.

If a leak occurs in a plastic system, cut out the bad joint and replace it with a new fitting. Remove enough pipe to allow room for two couplings and the fitting. Figure 16-31 shows how much pipe should be cut out if a joint at the elbow leaks. In this case 3 fittings, 2 short lengths of pipe, and 6 joints are required. Take your time preparing the pipe and fittings when you cement plastic piping. Plastic fittings cannot be reused if they don't work the first time.

When it is necessary to connect plastic pipe to iron pipe valves and fittings, use only approved male-threaded plastic adapters. Use only the thread compound recommended by the manufacturer for these threaded joints.

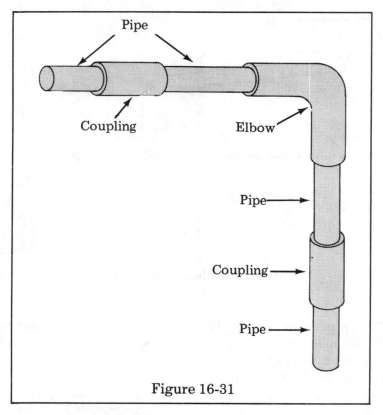

Figure 16-31

When the water system is complete, open a hose bibb to flush out sand, pieces of pipe shavings, and other impurities that may have collected inside the pipe during installation. This will save time in cleaning out faucets when fixtures are set and prevents the damage which sand and grit can do to washers.

Regardless of the type of material used, the entire water distribution system must be tested, inspected, and proved to be tight when completed. It must be tested at a water pressure not less than the maximum working pressure under which it is to be used.

Remember to install air chambers and secure all water supply pipes. A system with water shock or hammer will not pass the final inspection.

Questions

1- What two procedures are required when installing a water service pipe in the same trench with a building sewer?

2- What is the minimum separation distance between a water service pipe and a sewer line installed in separate trenches?

3- What is the minimum separation between a water service pipe and a septic tank or drainfield?

4- Why is it important to support water service supply piping for its entire length?

5- Why should young trees not be planted near a water service pipe?

6- What characteristics make galvanized steel pipe superior for use in outside trenches?

7- What should be avoided when backfilling a trench containing water service piping?

8- How deep should a water service piping trench be?

9- What is the code classification of ABS and PVC plastic pipe materials?

10- What precautions should be taken when backfilling over PVC plastic pipe in water service supply piping trenches?

11- What clearance is required between the top of a water service supply pipe and the bottom of a building footing?

12- What clearance is required around the circumference of a water supply pipe which passes through cast-in-place concrete?

13- When a lawn sprinkler system is connected to a potable water service supply pipe, what device is used to prevent a cross-connection?

14- This device must be located a minimum of how many inches above the highest sprinkler head?

15- To meet code requirements, what device must be provided on each building water service supply pipe?

16- Where must a water supply control valve be located?

17- Why is a drip valve required in cold climates?

18- What type of pipe straps should be used to secure copper water lines to wooden partitions?

19- When water line pipes contract and expand, what problem is likely to occur if they are not properly strapped?

20- What is the maximum depth a wooden partition may be notched to conceal water piping?

21- What is the standard acceptable method of providing openings to receive water piping in a load-bearing partition?

22- What should be provided to protect soft piping installed in notched wooden partitions?

23- What must be done to protect piping installed in filled ground where hydrogen sulphide gas is known to be present?

24- What must be done to protect water piping from other metals or ferrous pipes within a building?

25- What happens when hot and cold water piping within a building make contact with each other?

26- What must water closet supply pipes have that is not always required for other types of plumbing fixtures?

27- In what common areas of a building should water piping not be installed when subject to freezing?

28- Plastic water piping cannot be installed underground within the _____of a building.

29- What protects water pipe installations from water hammer?

30- Why must water supply systems be installed to drain dry?

31- What is a cross-connection?

32- Name two of the most common causes of cross-connection.

33- What triggers back-siphonage (cross-connection) in a water supply system?

34- Except for clothes washing machines, what must each threaded water outlet have to prevent cross-connection?

35- What is the minimum air gap required from water supply outlets to the overflow rim of each fixture?

36- Why are below-rim water supplied fixtures prohibited?

37- Why is it necessary to know the correct way to measure pipe before cutting it?

38- Name the two basic methods of measuring pipe for plumbing installation.

39- What is meant by end-to-center measure?

40- What mathematical figure must be known

to find the length of pipe between two 45° ells?

41- Name the two common types of pipe vises used to hold pipe.

42- In what direction do you turn a vise handle to tighten the jaws on the pipe?

43- When a vise is used to secure a chrome pipe, how should the finish be protected?

44- What is left inside a pipe when a pipe cutter is used?

45- To cut steel pipe square with a pipe cutter, how should you begin the operation?

46- What can happen when a pipe cutter handle is tightened too rapidly?

47- What adverse effect may occur in a water pipe if the bur is not removed?

48- What is the name of the tool used to remove a bur from the inside of a water pipe?

49- What are the two most common types of pipe threaders in use?

50- How do they differ?

51- Describe the type of pipe threader that can be used only where there is enough room to make a complete circumference.

52- What is the taper per foot of standard pipe threads?

53- What is the first thing you must do before threading a pipe?

54- Describe briefly the four steps in threading pipe.

55- How can used cutting oil be reclaimed?

56- Describe how joint pipe compound should be used when joining an elbow to a threaded pipe.

57- Why is it important to select the right size pipe wrench when joining fittings to threaded pipe?

58- What should be done after the installation of a piping system is complete?

59- What procedure is used in measuring copper pipe and tubing?

60- What makes soft-tempered copper tubing desirable for some uses?

61- What allowances must be considered when measuring pipe for cutting?

62- A fine-toothed saw blade is considered to have how many cutting teeth per inch?

63- What is the name of the most common joint used in a copper water system?

64- Name two other types of joints frequently used in a copper system.

65- Why is it important to solder copper pipe and fittings immediately after they are cleaned?

66- Why is it important to select the proper size tip when soldering copper pipe and fittings?

67- Describe the soldering process.

68- Solder generally used in a copper water pipe system is known as "50-50". What ingredients are found in this solder?

69- When the copper pipe and fittings are properly cleaned, what causes the solder to spread to all parts?

70- Describe briefly the seven steps necessary to prepare a copper joint properly for soldering.

71- What precaution should be taken when soldering is to be done near combustible material?

72- Why is it more difficult to resolder a joint that leaks than it was to solder it the first time?

73- When it is impossible to keep water away from a joint you are trying to solder, what "trick of the trade" should be tried?

74- What type of seat should a galvanized union

have in a water supply system?

75- What should you check for once a water system is complete?

76- With what type of copper should flare fittings be used?

77- What type of wrench should be used to tighten flare nuts?

78- Flare fittings are not recommended for general copper installations. When may flare fittings be used?

79- Flare fittings, when used, should never be _____.

80- Where are compression-type fittings most often used in a water supply system?

81- Describe how a compression-type fitting works.

82- Why should a tubing cutter be used to cut supply line tubes to fixtures?

83- Why should a freehand cut not be used when cutting plastic pipe?

84- What tools may be used to ream plastic pipe?

85- What piece of equipment should be used with a hacksaw to ensure a square cut for plastic pipe?

86- Cement used to create plastic pipe and fitting joints may be referred to as a_____ _____joint.

87- Liquid cleaner or sandpaper removes what from the surface of plastic pipe and fittings?

88- Describe briefly the ten steps necessary to create a good plastic joint.

89- After completion of a plastic system, what is the minimum waiting time recommended before water testing?

90- In a plastic system, what procedure must be followed in repairing a leaking joint?

91- What fitting must be used to connect plastic pipe to iron pipe valves and fittings?

92- What is the required test pressure for a newly completed water system?

93- What must *not* be present in a water system if it is to pass final inspection?

17
Private Water Wells And Sprinkler Systems

A recent survey by the Water Quality Association found that nearly one home in five is not connected to a public water supply. This means that about 47 million people depend on a private system for pure and wholesome water. Many suburban homes and nearly all rural homes use water from private wells. In some rural areas, lakes and streams make wells unnecessary.

The rapid spread of urban communities around large metropolitan areas has often outstripped the installation of public water distribution systems. And many smaller towns and cities have population densities low enough to make the cost of public water systems prohibitive.

Approval and inspection of potable water supply wells (not irrigation wells) is controlled by either the local health department or the Director of Environmental Resource Management (DERM). These authorities set guidelines on depth and separation distance of potable water supply wells (and in some cases irrigation wells) from sources of contamination. A permit is required for any drilled or driven well, regardless of whether the water is intended for domestic or irrigation purposes. Each well is inspected to ensure that the owner has complied with applicable regulations.

Well water is generally classified as hard water because of its high mineral content. Many people have to acquire a taste for this untreated water because it has a distinctly different taste than ''city water.'' It does not contain the chlorine or other chemicals to which city dwellers are accustomed. It also requires more soap to make a lather.

The minerals in untreated well water often stain the surfaces of plumbing fixtures, exterior building walls, and sidewalks to a dark reddish brown. These stains are virtually impossible to remove. Dissolved minerals can also cause a buildup of scale within water heater storage tanks and building water distribution pipes. The result of this buildup can be a premature failure of the system. In a domestic system which supplies plumbing fixtures, scale buildup and rust can be greatly reduced and the taste and smell of the water improved considerably with the installation of a water softener on the building water service line. See Figure 17-1. A well used for irrigation purposes and properly placed sprinkler heads can alleviate exterior rust stains.

The Source of Well Water

Well water is ground water that has filtered down through the soil to the water table level. It is known as *meteoric* water and makes up most of the estimated two million cubic miles of ground water in the upper crust of the earth.

Rain water soaks into the ground and moves slowly down to this underground water reservoir, which may be a few feet or hundreds of feet below the surface. Since it has been filtered through sand and rock, it is usually cool, low in harmful bacteria, and high in dissolved minerals.

When a well is driven or drilled, the bottom of the well casing must extend into the dry weather water table. Otherwise, during prolonged droughts the water table can fall below the level of the well. The result is a dry well. See Figure 17-2.

Most codes require that potable water supply wells, suction lines, pumps, and water pressure tanks be installed by certified and licensed professional well drillers or plumbing contractors. Irrigation wells may be installed by homeowners or other nonprofessionals.

The depth of a potable water supply well is determined by the local authority and by the

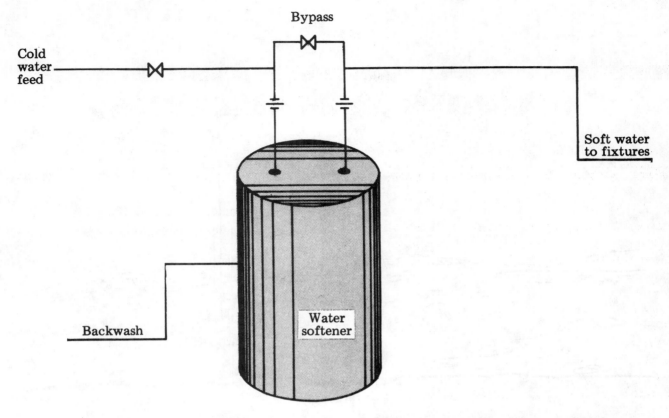

A full bypass should be installed on existing installations. The pressure drop through the softener will reduce the quantity of water available to operate all plumbing fixtures. A full bypass is not required on new installations. But the pressure drop through the softener must be considered when sizing the water pipe system. Backwash disposal methods should be approved by the local authority.

Water softener installed in building water service line
Figure 17-1

depth of the water table. Even though the underground water table may be within a few feet of the ground surface, the local authority can require a 30-foot minimum depth. The authority can also require a separation of the potable water supply well from a source of contamination: 50 feet from a septic tank, 100 feet from a drainfield. Irrigation wells do not have as strict depth and separation requirements as potable water supply wells. But it would be prudent to follow the stricter guidelines for irrigation wells too, if at all possible.

Wells which are dug or driven are classified as shallow wells. Dug or driven wells are used where the water table is within 22 feet of the ground surface. The well must penetrate deep enough below the existing water table to assure a dependable supply of water even in very dry seasons.

Wells which must be drilled are generally classified as deep wells. They may penetrate hundreds of feet into the earth. Water from deep wells is more desirable because there is less chance of contamination from a well 100 feet deep or more. The water level in these wells is little affected by seasonal rainfall or dry years.

There are two types of driven wells—those with an open end casing and those with a casing equipped with a well point.

An *open end casing* is commonly used for potable water supply as well as for irrigation purposes in areas where the water table is close to the ground surface and is in a good rock formation. In some areas rock or corrosion would clog the protective screening of a well point in a short time. When this happens, the well does not draw water and is useless. The open end well casing is preferable under these conditions.

If an open end casing is used, the pipe is

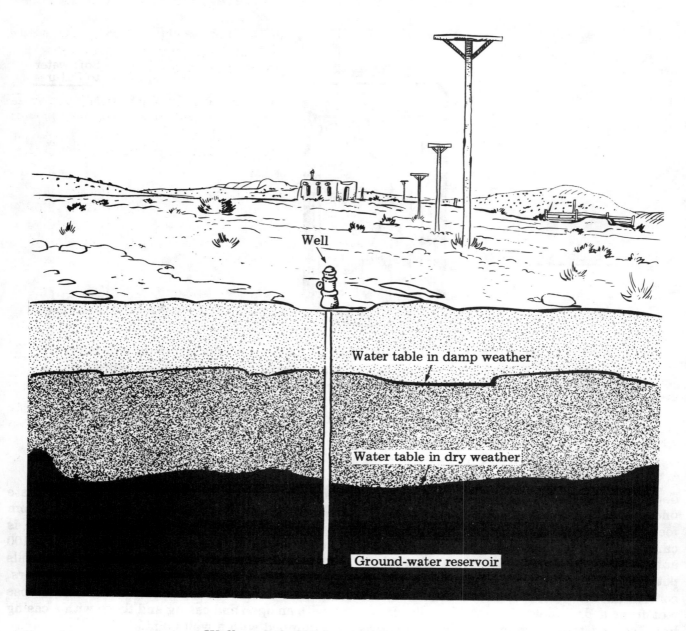

Well

Water table in damp weather

Water table in dry weather

Ground-water reservoir

Well extended to dry weather water table
Figure 17-2

driven to the desired depth and then loose soil and rock collection within the pipe are flushed from the driven section. Flushing for these shallow wells may be done by inserting a ½ or ¾ inch pipe into the larger pipe casing (usually 1½ or 2 inches in diameter). The smaller pipe has one end reamed and cut on an angle (see Figure 17-3) and is used to chip a pocket into the rock base. A garden hose is attached to the other end. The water is then turned on and the smaller pipe is worked up and down inside the larger casing. The water pressure flushes the loose debris out of the casing.

When most of the loose debris is flushed out and a good well has been installed, connect a 3 h.p. gasoline-driven centrifugal pump to the top of the well casing. Pump water out of the well until the water is free of rocks and sand. Check

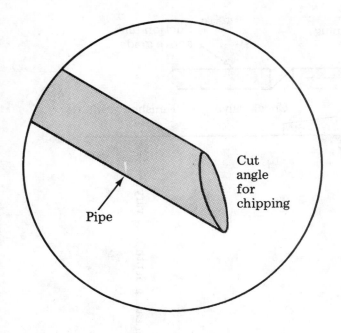

Cut angle for chipping

Pipe

Figure 17-3

the clarity of the water by catching samples in a glass jar. When no sand or rock fragments show up in the water samples, the well is considered good. The temporary pump may then be disconnected and preparations made to install the suction line to the permanent pump.

A *well point* is used in areas where the groundwater reservoir ends in loose shale or sand. The well point is screwed to the threaded well casing and is driven to the desired depth. Two types of well points are available in lengths that fit the use and water demand. One type has **fine perforations and a fine mesh screen and is** used in sand. The other type has larger perforations and a coarse mesh screen and is used in gravel or loose rock formations. No flushing is required if a well point is used. See Figure 17-4.

Potable water supply and irrigation wells must meet the following requirements to be acceptable by most codes.

- Unless specifically approved by the local authority, a well must not be located within any building or under the roof or projection of any building or structure.

- The well casing must be continuous, of new pipe, and must terminate in a suitable aquifer. Well casing pipe for a residential potable or irrigation water supply must be galvanized steel pipe.

- A concrete pad a minimum of 4 inches thick and 18 inches wide from the center of the well (or 36 inches across) must be poured around the top of the well casing. The concrete pad must be placed immediately below the tee if the suction line is above grade. This type of installation is usually used for irrigation water supply wells. See Figure 17-5. If the suction line is below grade, the concrete pad must be placed immediately above the tee and an extension piece with a plug must be extended to grade. This type of installation is usually used for potable water wells. See

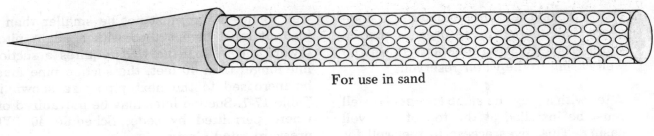

For use in sand

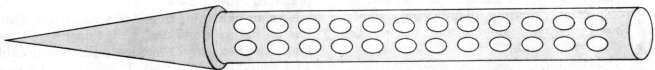

For use in gravel or loose rock formations

Well points
Figure 17-4

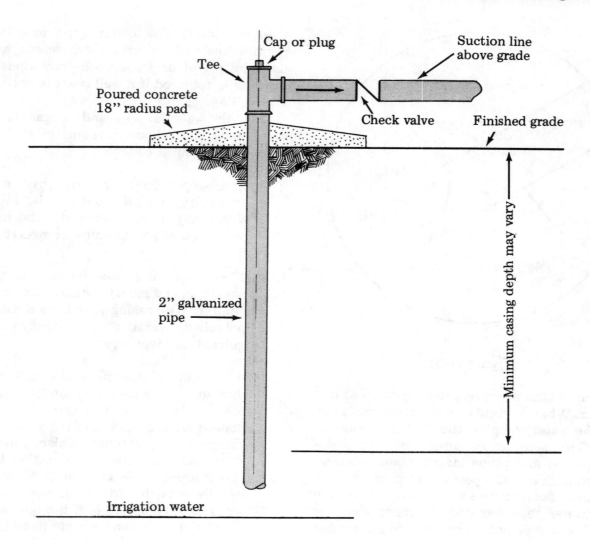

Supply well detail
Above grade suction line is generally used when
well location is very near pump
Figure 17-5

Figure 17-6. The concrete pad must slope away from the casing to prevent surface water from carrying pollutants down the well casing to the water reservoir.

• A tee with a plug the same size as the well must be installed at the top of the well casing. This gives access to the well for inspections, adding disinfecting agents, measuring well depth, and testing the static water level.

Suction Line for Potable Water Supply System

The suction line must be large enough to provide the water volume and pressure required to operate the plumbing fixtures in the building. The suction line or water service pipe from the well to the pump must not be smaller than 1 inch. It must be installed with a slight pitch toward the well. If the well requires a suction line longer than 40 feet, the suction pipe must be increased to the next pipe size shown in Table 17-7. Suction lines may be galvanized or, where permitted by code, Schedule 40 PVC pressure rated plastic.

As an example, consider a residence which has 30 fixture units and a suction line 65 feet long. The next larger suction pipe size in Table 17-7 would be 1½ inches. Therefore, a 1½-inch suction line pipe should be used.

Suction Line for Irrigation Supply Well

The suction line must be as large as the pump suction inlet, never smaller. For example,

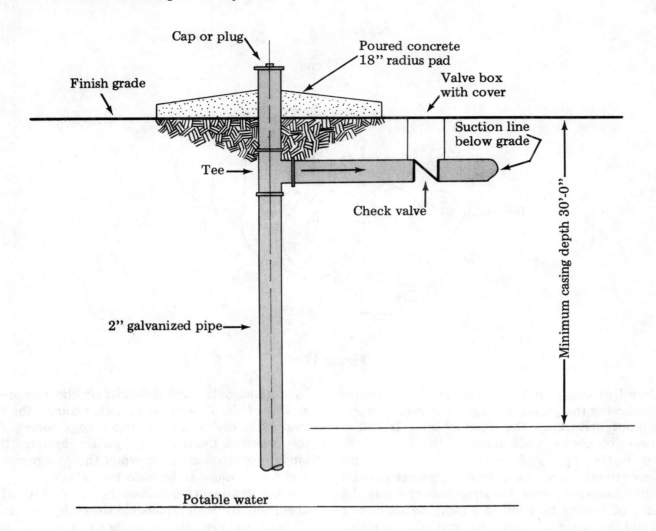

Supply well detail
Below grade suction line generally used when
well is located some distance from pump
Figure 17-6

Fixture units	Supply required G.P.H.	Diameter of suction pipe	Diameter pressure pipe	Diameter service pipe	Size of tank	HP	Well size
23	720	1	¾	¾	42	½	1½
30	900	1¼	1	1	82	¾	2
40	1200	1½	1	1	120	¾	2

(Predominantly for flush tanks)
Tank and pump size requirements
Table 17-7

a 2 horsepower pump would require a 2-inch suction line; a 1½ horsepower pump would require a 2-inch suction line. It must be installed with a slight pitch toward the well. Separation distance between pump and well is not a factor in sizing.

Suction lines of either type must have a soft seat check valve rated at 200 pounds which must be installed as close as is practical to the well. The check valve may be either the spring-loaded or flapper-type. The suction line must have a union or an approval slip coupling installed just

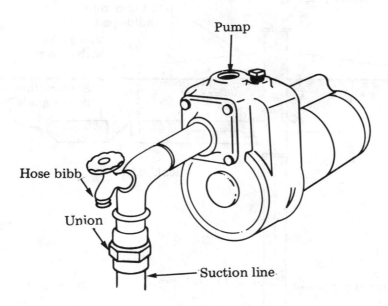

Figure 17-8

ahead of the pump. This provides for disconnecting the pump for repairs or replacement without disturbing the suction line. Install a hose bibb above grade ahead of the pump also. On both types of systems this can be conveniently used for priming purposes if the pump loses its prime. On irrigation systems the hose bibb may be used to supply pesticides or liquid fertilizer through the sprinkler system. See Figure 17-8.

Pressure Tank Water Systems

The water piping from the pump to the pressure tank should not be smaller than the discharge outlet of the pump. A gate valve with the handle removed should be installed in the piping between the pump and the tank if the tank has a capacity of more than 42 gallons. A minimum ¾-inch gate valve must be installed on the water piping on the discharge side of the tank. This serves as the house valve to control the water within the building.

There are generally two pressure tank systems in use today. The *hydropneumatic tank*, with its pressure switch, has been the dominant well water supply system for farm, residential, and other small systems since the 1920's. While the tank itself has given good service, unattended systems have given inconsistent service because of the failure of air chargers and air volume controls. The operation of a hydropneumatic tank depends on the compressibility of air. This compressibility causes the air trapped in the tank to act like a huge spring. As the pressure fluctuates within the system, the air is compressed again when the pump starts and forces more water into the tank.

Air in the pressurized tank is gradually absorbed by water passing through the tank. Unless the lost air is replaced, there is not enough in the tank to compress and expand. This causes excessive cycling of the pump in order to provide water for use.

Hydropneumatic tanks should be maintained with the necessary quantity of air by means of automatic devices, mainly air chargers and air volume controls. The tank can be manually drained of water, and this lets it fill with air at atmospheric pressure. As the pump fills the tank with water, it will again have a normal supply of trapped air to compress and expand for properly operating the water supply system. See Figure 17-9 for a typical pressure tank.

The hydropneumatic tank should be sized large enough to prevent excessive cycling of the pump. This depends on three things:

1) The draw-off capacity of the tank between the "cut-out" and "cut-in" limits. (The tank size should provide a draw-off of six gallons of water while maintaining an operating range of 20 to 40 psi water pressure.)

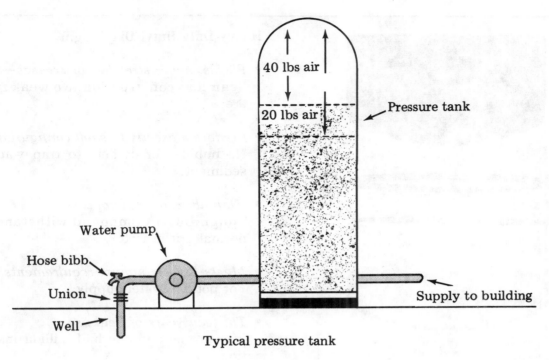

40 lbs air

20 lbs air

Pressure tank

Water pump

Hose bibb

Union

Well

Supply to building

Typical pressure tank

Figure 17-9

2) The pump capacity should be great enough to quickly refill the tank.

3) The minimum size hydropneumatic tank for a single family residence would be 42 gallons. Tanks are sized as to fixture units of a building. This is an indicator of the rate at which water will be used. For other sizes of tanks and specific requirements, see Table 17-7.

The second, the *breather-type pressure tank*, has been around for about thirty years. It has been approved for use by most codes. It has a built-in elastic diaphragm membrane which stores and releases water in response to pressure variations in the system. This pressure tank with its breather-type air-charging device is suitable for most small systems. It gives consistent service and requires no air to be injected into the tank. It gives effective system protection by keeping pump starts (excessive cycling) to a minimum. The tank volume for small systems may be as small as 14.0 gallons or as large as 36.0 gallons, depending on the water need. The sizing procedure covers updated water-use habits, increased off-peak demands, and the general increase in water usage that has occurred in the last twenty-five years. See Figure 17-10.

Install the pump and pressurizing system so

that equipment is reasonably accessible for repair or replacement.

Interior water piping, materials, and installation methods are the same as described in Chapters 15 and 16 for water distribution pipes.

INSTALLATION PROBLEMS

Inexperienced plumbers occasionally install pumps in ways which create problems that are not always obvious. In the majority of the cases where pumps were installed incorrectly, the instructions packaged with the new pump have not been followed. Any of the following circumstances could cause problems:

- connecting the suction line from the well to the discharge side of the pump.

- trying to use 115 volts on a pump wired for 230 volts. (Naturally the pump would not perform properly and if used too long will burn out.)

- trying to use a pump too small to supply a particular need.

- installing the well improperly.

At the end of this chapter is a trouble-shooting guide to well pump installation for the

Heavy-Duty Butyl Diaphragm

- *Flexes, never stretches or creases*—
  Seamless construction. No weak stress areas.

- *Conforms exactly to shell configuration*—
  No bubbles or corners to trap water or sediment.

- *High abrasion resistance*—
  Tough Butyl Compound withstands all normal grit in water.

- *Meets FDA minimum requirements*—
  For potable water supply.

- *Thirteen years of field use*—
  in over one and one half million installations.

Polypropylene Liner and Acceptance

- *Separate, rigid water reservoir*—
  100% non-corrosive. Not bonded to shell walls. Will never crack, pit, chip, craze or flake.

- *Mechanical "O" ring seals*—
  Exclusive mechanical clamping ring permanently bonds diaphragm and liner to shell groove.

  Swedged "O" ring seal on water side of acceptance fitting.

  Water never touches steel.

- *Copper-lined acceptance fitting*—
  Silver brazed for watertight seal.

- *Listed with NSF*—
  Polypropylene liner tested and accepted by the National Sanitation Foundation.

Figure 17-10

two most-used pumps for potable water supply wells, the *jet* and the *submersible*. It points out the mistakes commonly made by inexperienced installers, lists what to look for and the procedure to follow when checking out an installation, and explains how to make the necessary corrections once the problem is found. This guide can also be used to check similar symptoms for the centrifugal (irrigation) pumps.

LAWN SPRINKLER SYSTEMS

When planning and designing a lawn sprinkler system, consider first the number and type of sprinkler heads to be used. After selecting the heads, consider these factors:

- the amount of water required
- the horsepower of the pump (if any)
- the number of circuits necessary
- the pipe size

The two general types of sprinkler heads in use are the *flush spray head* and the *pulsating head*.

Flush Spray Heads

The flush spray head is generally the most popular for the average lawn sprinkler system. It is manufactured in many patterns, and this makes it easy to select heads for odd corners or hard to reach areas. This type of head provides a soaking spray. Therefore, the system does not have to operate for long periods of time. Some heads are provided with adjustment screws to control the distance of the spray (see Figure 17-11).

Adjustment screw

Adjustable spray head insert fitting
Figure 17-11

All spray heads provide about the same amount of water and coverage under the same installation conditions, regardless of type or of the material from which it is made. (See Table 17-12 for important data on flush spray heads.)

Type of Pattern	PSI	GPM	Diameter	Radius
Full head	20	3.4	24 feet	12 feet
3/4 head	20	2.2		12 feet
1/2 head	20	1.7		12 feet
1/4 head	20	.8		12 feet
End strip head	20	.6	3 x 12 ft. strip	
Full strip head	20	1.1	3 x 24 ft. strip	

Flush spray heads
Standard brass nozzle
Table 17-12

Pulsating Heads

The pulsating head is commonly referred to as a *rain bird* and is generally used to water large open areas. This type of head may pop up from the ground (as in play areas) or it may be secured to a piece of pipe. There are two kinds of rain birds commonly used—the full circle and the adjustable part circle.

These heads need more pressure to operate than spray heads, but each one covers about three times as much ground area as a spray head. Pulsating heads must be in operation for considerably longer periods of time to supply the same amount of water as the spray heads provide. (See Table 17-13 for important data on pulsating heads.)

Type of Pattern	PSI	GPM	Diameter	Radius
Full circle	25	2.9	44 ft.	22 ft.
Part circle	25	2.9	44 ft.	22 ft.

Pulsating heads
1/2'', 9/64'' nozzle
Table 17-13

Tables 17-12 and 17-13 show the approximate lawn area each type of head covers at 20-25 pounds per square inch (psi) and the number of gallons per minute (GPM) needed for their operation. Naturally, the greater the pressure the greater the coverage. More water per minute makes for faster soaking, thus less time is required for pump operation. Table 17-14 illustrates the various head patterns and degrees for flush spray heads.

Standard Patterns

	90°
	120°
	180°
	240°
	270°
	360°

Special Patterns

	End strip
	Ctr. strip

Flush spray head patterns
Table 17-14

Sample Plan for a Sprinkler System

Step 1 Assume you plan to use flush spray heads. First, determine the patterns required and the number and types of heads needed for complete coverage. Consult Table 17-12 for the coverage diameter of full heads, 24 feet. The coverage radius of all special pattern heads is 12 feet. Subtract 4 feet from the coverage of each type of head listed in Table 17-12. This allows for sufficient overlap of the spray and avoids dry spots. This means full heads can be placed 20 feet apart and special pattern heads (such as ¾, ½, ¼) 16 feet apart. The shape of the areas to be covered or the design of the head might require even closer spacing.

Step 2 Begin your layout of sprinkler heads from the wall line of the building or other permanent structure. Use small stakes driven into the ground to mark each head location.

Heads along a building wall would require half heads. Locate heads to spray out or away from the building wall to avoid rust stains. After laying out the special pattern heads around the building walls, driveways, walkways and the like, lay out the full heads.

The full heads should not be placed in a straight line with one-half heads or other full heads, but should be staggered off-center. This helps assure overlapping. See Figure 17-15.

You now have the number and types of heads required for the sprinkler system. Total

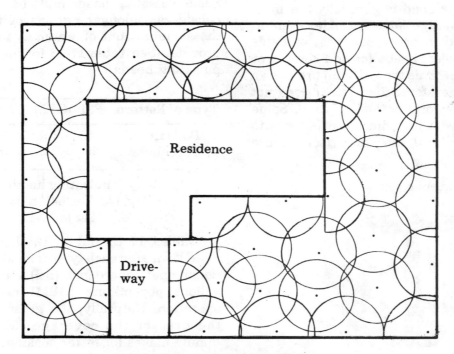

Flush spray head installation
Figure 17-15

the number of each type of head in Figure 17-15 to determine the amount of water needed for this landscape.

15 full heads	x 3.4 GPM =	51.0 gallons
2 ¾ heads	x 2.2 GPM =	4.4 gallons
27 ½ heads	x 1.7 GPM =	45.9 gallons
7 ¼ heads	x .8 GPM =	5.6 gallons
	Total gallons needed =	106.9 (or 107 gallons)

Step 3 If well water is to be used for irrigation, choose a pump to supply this volume of water. Consider the depth of the well when sizing the pump. The deeper the well, the larger the pump capacity must be to deliver the quantity of water needed.

The well in this example is 60 feet, an average depth for most sprinkler wells in most areas of the country.

Make a comparison of the three most widely used centrifugal pumps for irrigation purposes, using horsepower rating as a gauge.

Consider also the pressure (psi) required for the heads to function at their best. Flush spray heads need 20 psi, as in Table 17-12. This means that approximately 20% of the pump's capacity must remain uncommitted so that it can generate the necessary 20 psi.

Review each pump's capacity and select the appropriate one to supply the water required for the sprinkler layout in Figure 17-15.

Pump Size	Approximate Capacity 60' Well	Reserved For 20 PSI	Volume Available For Use
1 Horsepower	20 gallons	4 gallons	16 gallons
1½ Horsepower	60 gallons	12 gallons	48 gallons
2 Horsepower	94 gallons	19 gallons	75 gallons

Data on pumps
Table 17-16

Table 17-16 shows that an irrigation system requiring approximately 107 gallons of water per minute eliminates the 1 h.p. pump altogether. The 1½ h.p. pump could be used on three circuits by dividing the number of heads into three separate systems, each system controlled by a valve. With this pump, one third of the land area could be adequately watered at a time. Watering the landscape could thus develop into an all-day activity.

The ideal pump for this layout would be the 2 h.p. pump. The sprinkler system would operate efficiently on two circuits. The watering time would be cut by one third. The higher cost of the larger pump would be offset by a considerable long term savings from reduced pump wear and lower electricity use.

Materials and Pipe Sizes
Materials used in a potable water supply

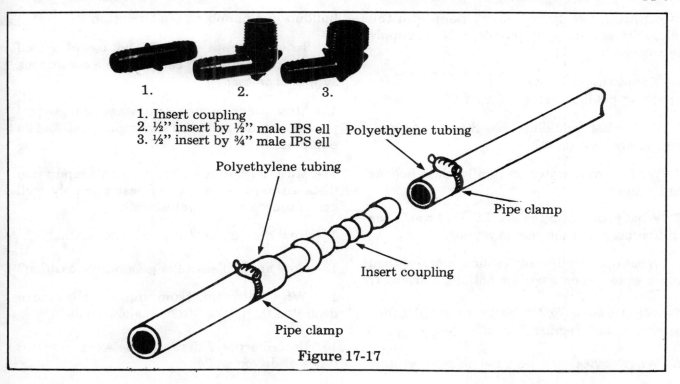

1. Insert coupling
2. ½" insert by ½" male IPS ell
3. ½" insert by ¾" male IPS ell

Polyethylene tubing

Polyethylene tubing

Pipe clamp

Insert coupling

Pipe clamp

Figure 17-17

system can also be used for piping and fittings in a sprinkler system. Note, however, that there are two types of materials used exclusively for sprinkler systems:

1) PVC Schedule 160 rigid pipe and fittings. This is a thinwall, lightweight plastic pipe which can be used only on open sprinkler systems. It is not designed to work under continual pressure. It is cut and assembled the same as PVC Schedule 40 pipe described earlier in this book.

2) Flexible polyethylene plastic tubing. This is obtainable in 100-foot coils and is cut to length with a hacksaw or knife. It is assembled with insert fittings held securely in place with special pipe clamps. See Figure 17-17.

0 - 5	GPM	½"
5 - 10	GPM	¾"
11 - 15	GPM	1"
16 - 25	GPM	1¼"
26 - 40	GPM	1½"
41 - 60	GPM	2"

Sizing guide for PVC and polyethylene pipe
Table 17-18

Rigid PVC plastic pipe should be sized from Table 17-18. Flexible polyethylene pipe should be sized one pipe size larger due to the restriction placed on the water flow by the insert fittings.

Place plastic sprinkler piping in the trench deep enough to protect it from sharp tools. At this depth a nipple 2 inches long screwed into the threaded fitting on the pipe should reach approximately one inch below the grass surface. On this nipple screw a flush head. This should leave the head flush with the grass surface. Backfill the trench with earth and replace the sod. Place a concrete collar around each sprinkler head for protection. Do not install the head at the end of each in-place run of pipe until the lines have been thoroughly flushed clean of sand and other foreign matter.

Circuits may be controlled by gate valves which have to be opened and closed manually as required. Special controls can be installed that change circuits automatically when the pump is shut off manually for a few seconds and then turned back on again. These special controls can also be equipped with electric timers so that the lawn may be watered automatically.

Questions

1- Approximately how many people in the United States depend on private water to supply their needs?

2- Name the two agencies responsible for approving potable water supply wells.

3- With what guidelines are these two agencies mainly concerned?

4- Why is well water generally classified as hard water?

5- What is the main reason that well water has a different taste than public water?

6- What does well water require more of than public water when used for bathing purposes?

7- What causes well water to stain other surfaces a dark reddish brown?

8- What device may be used to reduce scale

buildup in a plumbing piping system?

9- To what depth must the bottom of a well casing extend so that it can provide a continuous supply of water?

10- Most codes require that what professional person install a potable water supply well and its equipment?

11- Most authorities require what separation distance between a potable water supply well and a source of contamination?

12- How are dug or driven wells classified?

13- How are drilled wells generally classified?

14- Why is water from deep wells more desirable than water from shallow wells?

15- Describe the differences between the two types of driven wells.

16- Why is it recommended for some areas that wells with wellpoints *not* be installed?

17- Explain how to flush rock and sand from a shallow well.

18- How is a well checked to determine whether the water is free of sand?

19- What type of wellpoint is recommended when the water reservoir ends in sand?

20- What type of wellpoint is recommended when the water reservoir ends in gravel?

21- What material is required for well casing pipe used for residences?

22- What must be installed around the top of a well casing?

23- What purposes are served by having a tee screwed to the top of a well casing?

24- What is the minimum size suction line permitted for supplying water to plumbing fixtures?

25- In which direction must a suction line pitch?

26- For supplying water to a sprinkler system, what size must the suction line be?

27- What device must be installed on all suction lines as near the well as possible?

28- What device must be installed on all screw pipe suction lines as near the pump as possible?

29- What are some of the advantages of installing a hose bibb on the suction line above grade on an irrigation well pump?

30- What is the minimum size of the discharge pipe from the pump to the pressure tank?

31- What is the minimum size hydropneumatic tank permitted for a single family residence?

32- What controls the size of a hydropneumatic tank?

33- Name two advantages that the newer breather-type pressure tank has over the older hydropneumatic-type tank.

34- When initially installing the pump and equipment for a single family pressurized system, what important consideration must be made?

35- What may happen when a pump wired for 230 volts is connected to wires supplying 115 volts?

36- What are the two types of sprinkler heads that may be used for a sprinkler system?

37- What is the first factor to be considered in planning and designing a lawn sprinkler system?

38- What two factors determine the size pump to be used on an irrigation water well?

39- Where are pulsating-type sprinkler heads generally used?

40- What is the main disadvantage in using pulsating-type sprinkler heads in an average size residential yard?

41- What is accomplished when pressures exceed those listed in Tables 17-12 and 17-13?

42- What is the average radius coverage for a one-half flush spray head at 20 psi?

43- What spray pattern should flush heads have when installed along a building wall?

44- Why is it better not to install flush spray heads in a straight line with other heads?

45- Why is the depth of the well an important factor in sizing the well pump?

46- What type of well pump is usually installed for irrigation purposes?

47- What two types of materials are used exclusively for sprinkler systems?

48- What should be used to protect sprinkler heads?

49- What should be done to a newly installed sprinkler system before putting it in use?

50- What source other than a public water supply do many suburban and rural homes use for their water supply?

TROUBLE SHOOTING GUIDE — JET PUMPS

PUMP WON'T START OR RUN

CAUSE OF TROUBLE	HOW TO CHECK	HOW TO CORRECT
1. Blown fuse.	Check to see if fuse is OK.	If blown, replace with fuse of proper size.
2. Low line voltage.	Use voltmeter to check pressure switch or terminals nearest pump.	If voltage under recommended minimum, check size of wiring from main switch on property. If OK, contact power company.
3. Loose, broken, or incorrect wiring.	Check wiring circuit against diagram. See that all connections are tight and that no short circuits exist because of worn insulation, crossed wire, etc.	Rewire any incorrect circuits. Tighten connections, replace defective wires.
4. Defective motor.	Check to see that switch is closed.	Repair or take to motor service station.
5. Defective pressure switch.	Check switch setting. Examine switch contacts for dirt or excessive wear.	Adjust switch settings. Clean contacts with emory cloth if dirty.
6. Tubing to pressure switch plugged.	Remove tubing and blow through it.	Clean or replace if plugged.
7. Impeller or seal.	Turn off power, then use screwdriver to try to turn impeller or motor.	If impeller won't turn, remove housing and locate source of binding.
8. Defective start capacitor.	Use an ohmmeter to check resistance across capacitor. Needle should jump when contact is made. No movement means an open capacitor; no resistance means capacitor is shorted.	Replace capacitor or take motor to service station.
9. Motor shorted out.	If fuse blows when pump is started (and external wiring is OK) motor is shorted.	Replace motor.

MOTOR OVERHEATS AND OVERLOAD TRIPS OUT

CAUSE OF TROUBLE	HOW TO CHECK	HOW TO CORRECT
1. Incorrect line voltage.	Use voltmeter to check at pressure switch or terminals nearest pump.	If voltage under recommended minimum, check size of wiring from main switch on property. If OK, contact power company.
2. Motor wired incorrectly.	Check motor wiring diagram.	Reconnect for proper voltage as per wiring diagram.
3. Inadequate ventilation.	Check air temperature where pump is located. If over 100°F., overload may be tripping on external heat.	Provide adequate ventilation or move pump.
4. Prolonged low pressure delivery.	Continuous operation at very low pressure places heavy overload on pump. This can cause overload protection to trip.	Install globe valve on discharge line and throttle to increase pressure.

TROUBLE SHOOTING GUIDE — JET PUMPS

PUMP STARTS AND STOPS TOO OFTEN

CAUSE OF TROUBLE	HOW TO CHECK	HOW TO CORRECT
1. Leak in pressure tank.	Apply soapy water to entire surface above water line. If bubbles appear, air is leaking from tank.	Repair leaks or replace tank.
2. Defective air volume control.	This will lead to a waterlogged tank. Make sure control is operating properly. If not, remove and examine for plugging.	Clean or replace defective control.
3. Faulty pressure switch.	Check switch setting. Examine switch contacts for dirt or excessive wear.	Adjust switch settings. Clean contacts with emory cloth if dirty.
4. Leak on discharge side of system.	Make sure all fixtures in plumbing system are shut off. Then check all units (especially ballcocks) for leaks. Listen for noise of water running.	Repair leaks as necessary.
5. Leak on suction side of system.	On shallow well units, install pressure gauge on suction side. On deep well systems, attach a pressure gauge to the pump. Close the discharge line valve. Then, using a bicycle pump or air compressor, apply about 30 psi pressure to the system. If the system will not hold this pressure when the compressor is shut off, there is a leak on the suction side.	Make sure above ground connections are tight. Then repeat test. If necessary, pull piping and repair leak.
6. Leak in foot valve.	Pull piping and examine foot valve.	Repair or replace defective valve.

PUMP WON'T SHUT OFF

CAUSE OF TROUBLE	HOW TO CHECK	HOW TO CORRECT
1. Wrong pressure switch setting or setting "drift".	Lower switch setting. If pump shuts off, this was the trouble.	Adjust switch to proper setting.
2. Defective pressure switch.	Arcing may have caused switch contacts to "weld" together in closed position. Examine points and other parts of switch for defects.	Replace switch if defective.
3. Tubing to pressure switch plugged.	Remove tubing and blow through it.	Clean or replace if plugged.
4. Loss of prime.	When no water is delivered, check prime of pump and well piping.	Reprime if necessary.
5. Low well level.	Check well depth against pump performance table to make sure pump and ejector are properly sized.	If undersized, replace pump or ejector.
6. Plugged ejector.	Remove ejector and inspect.	Clean and reinstall if dirty.

TROUBLE SHOOTING GUIDE — JET PUMPS

PUMP OPERATES BUT DELIVERS LITTLE OR NO WATER		
CAUSE OF TROUBLE	**HOW TO CHECK**	**HOW TO CORRECT**
1. Low line voltage.	Use voltmeter to check at pressure switch or terminals nearest pump.	If voltage under recommended minimum, check size of wiring from main switch on property. If OK, contact power company.
2. System incompletely primed.	When no water is delivered, check prime of pump and well piping.	Reprime if necessary.
3. Air lock in suction line.	Check horizontal piping between well and pump. If it does not pitch upward from well to pump, an air lock may form.	Rearrange piping to eliminate air lock.
4. Undersized piping.	If system delivery is low, the discharge piping and/or plumbing lines may be undersized. Refigure friction loss.	Replace undersized piping or install pump with higher capacity.
5. Leak in air volume control or tubing.	Disconnect air volume control tubing at pump and plug hole. If capacity increases, a leak exists in the tubing of control.	Tighten all fittings and replace control if necessary.
6. Pressure regulating valve stuck or incorrectly set. (Deep well only.)	Check valve setting. Inspect valve for defects.	Reset, clean, or replace valve as needed.
7. Leak on suction side of system.	On shallow well units, install pressure gauge on suction side. On deep well systems, attach a pressure gauge to the pump. Close the discharge line valve. Then, using a bicycle pump or air compressor, apply about 30 psi pressure to the system. If the system will not hold this pressure when the compressor is shut off, there is a leak on the suction side.	Make sure above ground connections are tight. Then repeat test. If necessary, pull piping and repair leak.
8. Low well level.	Check well depth against pump performance table to make sure pump and ejector are properly sized.	If undersized, replace pump or ejector.
9. Wrong pump-ejector combination.	Check pump and ejector models against manufacturer's performance tables.	Replace ejector if wrong model is being used.
10. Low well capacity.	Shut off pump and allow well to recover. Restart pump and note whether delivery drops after continuous operation.	If well is "weak," lower ejector (deep well pumps), use a tail pipe (deep well pumps), or switch from shallow well to deep well equipment.
11. Plugged ejector.	Remove ejector and inspect.	Clean and reinstall if dirty.
12. Defective or plugged foot valve and/or strainer.	Pull foot valve and inspect. Partial clogging will reduce delivery. Complete clogging will result in no water flow. A defective foot valve may cause pump to lose prime, resulting in no delivery.	Clean, repair, or replace as needed.
13. Worn or defective pump parts or plugged impeller.	Low delivery may result from wear on impeller or other pump parts. Disassemble and inspect.	Replace worn parts or entire pump. Clean parts if required.

TROUBLE SHOOTING GUIDE — SUBMERSIBLE PUMPS

FUSES BLOW OR CIRCUIT BREAKER TRIPS WHEN MOTOR IS STARTED		
CAUSE OF TROUBLE	**HOW TO CHECK**	**HOW TO CORRECT**
1. Incorrect line voltage	Check the line voltage terminals in the control box (or connection box in the case of the 2-wire models) with a voltmeter. Make sure that the voltage is within the minimum-maximum range prescribed by the manufacturer.	If the voltage is incorrect, contact the power company to have it corrected.
2. Defective control box: a. Defective wiring.	Check out all motor and powerline wiring in the control box, following the wiring diagram found inside the box. See that all connections are tight and that no short circuits exist because of worn insulation, crossed wires, etc.	Rewire any incorrect circuits. Tighten loose connections. Replace worn wires.
b. Incorrect components	Check all control box components to see that they are the type and size specified for the pump in the manufacturers' literature. In previous service work, the wrong components may have been installed.	Replace any incorrect component with the size and type recommended by the manufacturer.
c. Defective starting capacitor (skip for 2-wire models).	Using an ohmmeter, determine the resistance across the starting capacitor. When contact is made, the ohmmeter needle should jump at once, then move up more slowly. No movement indicates an open capacitor (or defective relay points); no resistance means that the capacitor is shorted.	Replace defective starting capacitor.
d. Defective relay (skip for 2-wire models).	Using an ohmmeter, check the relay coil. Its resistance should be as shown in the manufacturer's literature. Recheck ohmmeter reading across starting capacitor. With a good capacitor, no movement of the needle indicates defective relay points.	If coil resistance is incorrect or points defective, replace relay.
3. Defective pressure switch.	Check the voltage across the pressure switch points. If less than the line voltage determined in "1" above, the switch points are causing low voltage by making imperfect contact.	Clean points with a mild abrasive cloth or replace pressure switch.
4. Pump in crooked well.	If wedged into a crooked well, the motor and pump may become misaligned, resulting in a locked rotor.	If the pump does not rotate freely, it must be pulled and the well straightened.
5. Defective motor winding or cable: a. Shorted or open motor winding.	Check the resistance of the motor winding by using an ohmmeter on the proper terminals in the control box (see manufacturer's wiring diagram). The resistance should match the ohms specified in the manufacturer's data sheet. If too low, the motor winding may be shorted; if the ohmmeter needle doesn't move, indicating high or infinite resistance, there is an open circuit in the motor winding.	If the motor winding is defective — shorted or open — the pump must be pulled and the motor repaired.
b. Grounded cable or winding.	Ground one lead of the ohmmeter onto the drop pipe or shell casing, then touch the other lead to each motor wire terminal. If the ohmmeter needle moves appreciably when this is done, there is a ground in either the cable or the motor winding.	Pull the pump and inspect the cable for damage. Replace damaged cable. If cable checks OK, the motor winding is grounded.
6. Pump sand locked.	Make pump run backwards by interchanging main and start winding (black and red) motor leads at control box.	Pull pump, disassemble and clean. Before replacing, make sure that sand has settled in well. If well is chronically sandy, a submersible should not be used.

TROUBLE SHOOTING GUIDE — SUBMERSIBLE PUMPS

PUMP OPERATES BUT DELIVERS LITTLE OR NO WATER		
CAUSE OF TROUBLE	**HOW TO CHECK**	**HOW TO CORRECT**
1. Pump may be air locked.	Stop and start pump several times, waiting about one minute between cycles. If pump then resumes normal delivery, air lock was the trouble.	If this test fails to correct the trouble, proceed as below.
2. Water level in well too low.	Well production may be too low for pump capacity. Restrict flow of pump output, wait for well to recover, and start pump.	If partial restriction corrects trouble, leave valve or cock at restricted setting. Otherwise, lower pump in well if depth is sufficient. Do not lower if sand clogging might occur.
3. Discharge line check valve installed backward.	Examine check valve on discharge line to make sure that arrow indicating direction of flow points in right direction.	Reverse valve is necessary.
4. Leak in drop pipe.	Raise pipe and examine for leaks.	Replace damaged section of drop pipe.
5. Pump check valve jammed by drop pipe.	When pump is pulled after completing "4" above, examine connection of drop pipe to pump outlet. If threaded section of drop pipe has been screwed in too far, it may be jamming the pump's check valve in the closed position.	Unscrew drop pipe and cut off portion of threads.
6. Pump intake screen blocked.	The intake screen on the pump may be blocked by sand or mud. Examine.	Clean screen, and when reinstalling pump, make sure that it is located several feet above the well bottom—preferably 10 feet or more.
7. Pump parts worn.	The presence of abrasives in the water may result in excessive wear on the impeller, casing, and other close-clearance parts. Before pulling pump, reduce setting on pressure switch to see if pump shuts off. If it does, worn parts are probably at fault.	Pull pump and replace worn components.
8. Motor shaft loose.	Coupling between motor and pump shaft may have worked loose. Inspect for this after pulling pump and looking for worn components, as in "7" above.	Tighten all connections, setscrews, etc.

TROUBLE SHOOTING GUIDE — SUBMERSIBLE PUMPS

PUMP STARTS TOO FREQUENTLY		
CAUSE OF TROUBLE	**HOW TO CHECK**	**HOW TO CORRECT**
1. Pressure switch defective or out of adjustment.	Check setting on pressure switch and examine for defects.	Reduce pressure setting or replace switch.
2. Leak in pressure tank above water level.	Apply soap solution to entire surface of tank and look for bubbles indicating air escaping.	Repair or replace tank.
3. Leak in plumbing system.	Examine service line to house and distribution branches for leaks.	Repair leaks.
4. Discharge line check valve leaking.	Remove and examine.	Replace if defective.
5. Air volume control plugged.	Remove and inspect air volume control.	Clean or replace.
6. Snifter valve plugged.	Remove and inspect snifter valve.	Clean or replace.

MOTOR DOES NOT START, BUT FUSES DON'T BLOW		
CAUSE OF TROUBLE	**HOW TO CHECK**	**HOW TO CORRECT**
1. Overload protection out.	Check fuses or circuit breaker to see that they are operable.	If fuses are blown, replace. If breaker is tripped, reset.
2. No power.	Check power supply to control box (or overload protection box) by placing a voltmeter across incoming power lines. Voltage should approximate nominal line voltage.	If no power is reaching box, contact power company for service.
3. Defective control box.	Examine wiring in control box to make sure all contacts are tight. With a voltmeter, check voltage at motor wire terminals. If no voltage is shown at terminals, wiring is defective in control box.	Correct faulty wiring or tighten loose contacts.
4. Defective pressure switch.	With a voltmeter, check voltage across pressure switch while the switch is closed. If the voltage drop is equal to the line voltage, the switch is not making contact.	Clean points or replace switch.

TROUBLE SHOOTING GUIDE — SUBMERSIBLE PUMPS

FUSES BLOW WHEN MOTOR IS RUNNING		
CAUSE OF TROUBLE	HOW TO CHECK	HOW TO CORRECT
1. Incorrect voltage.	Check line voltage terminals in the control box (or connection box in the case of 2-wire models) with a voltmeter. Make sure that the voltage is within the minimum-maximum range prescribed by the manufacturer.	If voltage is incorrect, contact power company for service.
2. Overheated overload protection box.	If sunlight or other source of heat has made box too hot, circuit breakers may trip or fuses blow. If box is hot to the touch, this may be the problem.	Ventilate or shade box, or remove from source of heat.
3. Defective control box components. (skip this for 2-wire models).	Using an ohmmeter, determine the resistance across the running capacitor. When contact is made, the ohmmeter needle should jump at once, then move up more slowly. No movement indicates an open capacitor (or defective relay points); no resistance means that the capacitor is shorted. Using an ohmmeter, check the relay coil. Its resistance should be as shown in the manufacturer's literature. Recheck ohmmeter reading across running capacitor. With a good capacitor, no movement of the needle indicates relay points.	Replace defective components.
4. Defective motor winding or cable.	Check the resistance of the motor winding by using an ohmmeter on the proper terminals in the control box (see manufacturer's wiring diagram). The resistance should match the ohms specified in the manufacturer's data sheet. If too low, the motor winding may be shorted; if the ohmmeter needle doesn't move, indicating high or infinite resistance, there is an open circuit in the motor winding. Ground one lead of the ohmmeter onto the drop pipe or shell casing, then touch the other lead to each motor wire terminal. If the ohmmeter needle moves appreciably when this is done, there is a ground in either the cable or the motor winding.	If neither cable or winding is defective—shorted, grounded, or open—pump must be pulled and serviced.
5. Pump becomes sand-locked.	If the fuses blow while the pump is operating, sand or grit may have become wedged in the impeller, causing the rotor to lock. To check this, pull the pump.	Pull pump, disassemble, and clean. Before replacing, make sure that sand has settled in well. If well is chronically sandy, a submersible should not be used.

TROUBLE SHOOTING GUIDE — SUBMERSIBLE PUMPS

PUMP WON'T SHUT OFF

CAUSE OF TROUBLE	HOW TO CHECK	HOW TO CORRECT
1. Defective pressure switch.	Arcing may have caused pressure switch points to "weld" in closed position. Examine points and other parts of switch for defects.	Clean points or replace switch.
2. Water level in well too low.	Well production may be too low for pump capacity. Restrict flow of pump output, wait for well to recover, and start pump.	If partial restriction corrects trouble, leave valve or cock at restricted setting. Otherwise, lower pump in well if depth is sufficient. Do not lower if sand clogging might occur.
3. Leak in drop line.	Raise pipe and examine for leaks.	Replace damaged section of drop pipe.
4. Pump parts worn.	The presence of abrasives in the water may result in excessive wear on the impeller, casing, and other close-clearance parts. Before pulling pump, reduce setting on pressure switch to see if pump shuts off. If it does, worn parts are probably at fault.	Pull pump and replace worn components.

Roughing-In

In a typical single-family residence there are approximately 300 feet of concealed piping beneath the floor and in the walls. This piping, discussed in previous chapters and known as the *rough plumbing*, includes the drainage and waste piping, the vents and the hot and cold water lines.

The *rough-in* is the portion of the rough plumbing system that brings the waste and the hot and cold water piping through to the wall and floor lines, where connections are made to the fixtures. Roughing-in the fixtures is possibly the most critical part of the installation. It is here that a plumber displays his skill or lack of skill. Plumbing knowledge and good workmanship are required to properly rough-in the waste and water outlets. The plumbing fixtures will be a vital part of the building all during the life of the building, so the plumber should do a sound professional job on this part of the installation.

Plumbing Fixture Clearance

One of the most critical parts of the roughing-in procedure is laying out and spacing the plumbing fixtures you intend to use. Every fixture must be spaced and installed so that it can be used for its intended purpose and is accessible for cleaning and repairs.

Observe the minimum clearances listed in this chapter so the plumbing systems you install will pass final inspection. These clearances are shown in Figure 18-1.

Water closets must be set a minimum of 15 inches from the center of the bowl to any finished wall or partition. Where a bidet is installed next to a water closet there must be a minimum spacing of 30 inches center to center. A water closet installed next to a bathtub must have a minimum of 12 inches from the center of the bowl to the outside edge of the tub apron. A water closet must also have a minimum clearance of 21 inches from the front of the bowl to any finished wall, door, or other plumbing fixture.

Bidets must have the same minimum spacing as water closets.

Lavatories are manufactured in various designs and widths, so center-to-center measurements do not apply. The minimum clearance is measured from the edge of the lavatory to the nearest obstruction. A lavatory must have a minimum clearance of 4 inches from its edge to any finished wall. It must have a minimum clearance of 2 inches from its edge to the edge of a tub. It must have a minimum clearance of 4 inches from its edge to the edge of a water closet tank and 21 inches from the front of the lavatory to any finished wall, door, or other plumbing fixture.

Shower openings must allow easy entrance and exit. The compartment or stall must have a minimum clearance of 24 inches from any finish wall, door, or other plumbing fixture. The minimum floor area for a shower stall is 1024 square inches.

Kitchen sinks and laundry trays are not governed by spacing requirements, although they should be spaced and installed to permit easy access for cleaning and repairs as well for the intended use.

Roughing-In Measurements

Plumbers must have a thorough knowledge of roughing-in measurements for various types of plumbing fixtures. Know the heights, distances, and locations of waste and water outlets for both wall-hung and floor mounted fixtures. Memorize these measurements for common fixtures or jot them down in a notebook which can be kept in your tool box. Complete roughing-in information is usually available from fixture manufacturers and distributors.

Roughing-in dimensions and installation information for the more commonly used fixtures are given at the end of this chapter. Included are measurements for bathtubs, water closets, lavatories, bidets, and kitchen sinks.

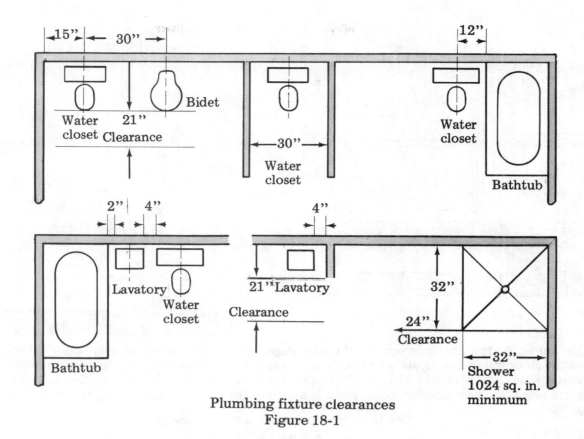

Plumbing fixture clearances
Figure 18-1

The roughing-in measurements are for American Standard fixtures but will be similar or identical to the measurements for like plumbing fixtures of other manufacturers. The author gratefully acknowledges the assistance of American Standard and Josam Manufacturing Company in supplying data and illustrations for this chapter.

Note that the standard roughing-in measurement for a floor mounted water closet is 12 inches from a finished wall. If you make an error and the water closet outlet is roughed in too close or too far from the finished wall, *the code does not permit the use of an offset closet flange to correct the mistake.* This would restrict the discharge flow from the water closet and could cause an overflow or a stoppage. Special water closets that can be set at 10 or 14 inches from the wall are available.

Plumbing Fixture Carriers

Bathrooms are the most susceptible of all rooms to unsanitary conditions. The bases of on-the-floor fixtures are natural areas for accumulated filth that is nearly impossible to remove. Bathroom floors made of wood tend to deteriorate next to and beneath toilet fixtures. Off-the-floor water closets can solve much of this problem and have been gaining popularity in residences.

Figure 18-2 shows two tank-type water closets, one for on-the-floor and the other for off-the-floor installation. Both of these are commonly used in residential buildings. The drawings clearly show the difference between the two installations.

Residential carriers have fewer parts and are therefore easier to assemble. They are designed to receive the waste from a single water closet or at most two water closets (when back-to-back installations are required, as shown in Figure 18-5). Of course, single family residences never have battery installations as do commercial buildings.

Residential carriers are designed to be compatible with newer piping materials. Figure 18-3 shows an off-the-floor residential water closet. Figure 18-4 shows the same closet carrier with the parts identified. Figure 18-5 shows units generally used in residential installations.

Some apprentices may never have an opportunity to work with off-the-floor plumbing fixtures because of the limited type of work their company performs. The following pages present some illustrations of the more commonly used residential carriers so that you may become familiar with their appearance and the methods of installation. (Carriers for commercial use are shown in the more advanced *Plumber's Handbook* by the author.)

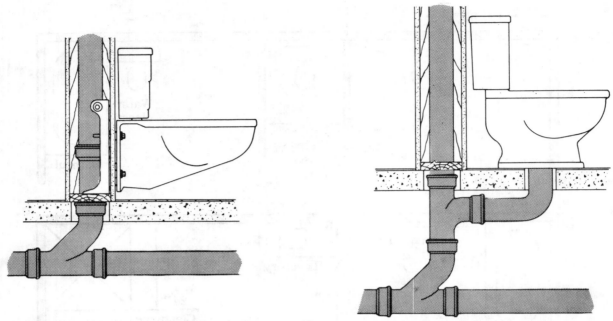

The above illustrations show some of the advantages of off-the-floor closet installation. With on-the-floor water closets, slabs or floor must be penetrated at each fixture to accommodate waste piping. Conversely, with off-the-floor fixtures, there is no slab penetration in toilet room areas. A clear, unobstructed floor is available for cleaning, and deterioration isn't a problem.

Two tank-type residential water closets
Figure 18-2

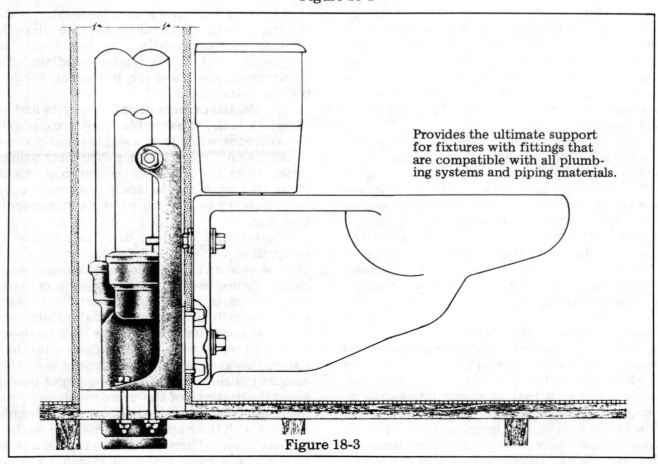

Provides the ultimate support
for fixtures with fittings that
are compatible with all plumbing systems and piping materials.

Figure 18-3

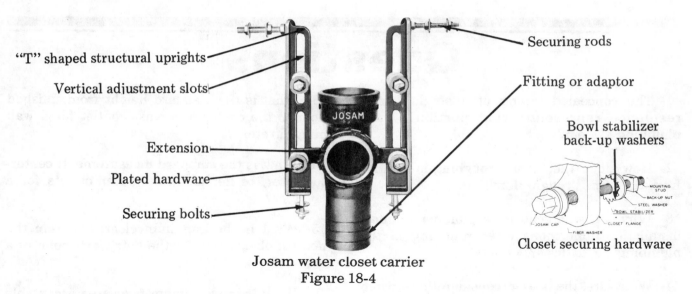

"T" shaped structural uprights

Vertical adjustment slots

Extension

Plated hardware

Securing bolts

Securing rods

Fitting or adaptor

Bowl stabilizer back-up washers

Closet securing hardware

Josam water closet carrier
Figure 18-4

3" Single, Copper, Adapter Type

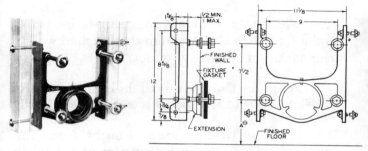

3" Single, P.V.C. Plastic, Adapter Type

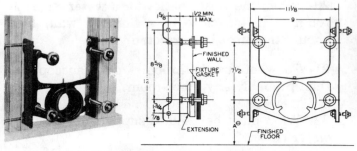

3" Back-to-Back, P.V.C. Plastic, Adapter Type

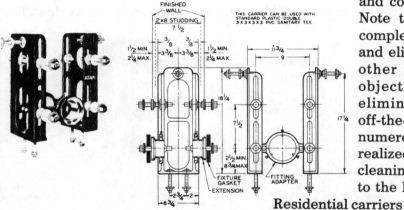

Residential closets This photo show the beauty and convenience of off-the-floor water closets. Note that the floor area under the closet is completely clear, which allows for easy cleaning and eliminates traditional maintenance. Tile or other floor surfaces are unobstructed, and objectionable crevices around the bowl are eliminated. The benefits derived from an off-the-floor residential closet installation are numerous. Easier and quicker installation is realized by the contractor and years of easy cleaning and minimum maintenance are offered to the home owner.

Residential carriers
Figure 18-5

Questions

1- The concealed piping of a single family residence represents what portion of the plumbing system?

2- How does the term "roughing-in" differ from the term "rough plumbing"?

3- Why is it so crucial for a plumber to know installation dimensions when roughing in the plumbing for fixtures?

4- What are the major considerations when installing plumbing fixtures?

5- How many inches should be allowed from the center of a water closet bowl to any finished wall?

6- What is the minimum center-to-center spacing for a bidet installed next to a water closet?

7- What is the minimum clearance between a water closet and the edge of a bathtub?

8- What is the minimum clearance from the front of most fixtures to any finished wall?

9- Why are minimum center-to-center spacing requirements not applicable to lavatories?

10- What is the standard height from finished floor to the overflow rim of a lavatory?

11- What is the distance from the center of most water closet bowls to the center of the water supply outlet?

12- What is the minimum distance from the edge of a lavatory to the nearest obstruction?

13- What is the standard height from finished floor to the center of a waste outlet for a wall hung lavatory?

14- What is the standard measurement, center-to-center, of hot and cold water outlets for a lavatory?

15- What is the minimum clearance from the nearest obstruction for the entry/exit point of a shower?

16- What is the minimum floor area acceptable for a shower?

17- What is the standard roughing-in measurement for a water closet from the finished wall to the center of the waste outlet?

18- What is the standard height from the finished floor to the top of a water closet bowl?

19- What is the standard height from the finished floor to the overflow rim of a kitchen sink?

20- When the waste outlet for a water closet is roughed in too close to a finished wall, how should the error be corrected?

21- To what are bathrooms more susceptible than probably any other room?

22- What must be used to support off-the-floor water closets?

23- Name two newer piping materials that are compatible with specially-designed residential carriers.

24- With what size drain is a bidet generally equipped?

Roughing-In Dimensions

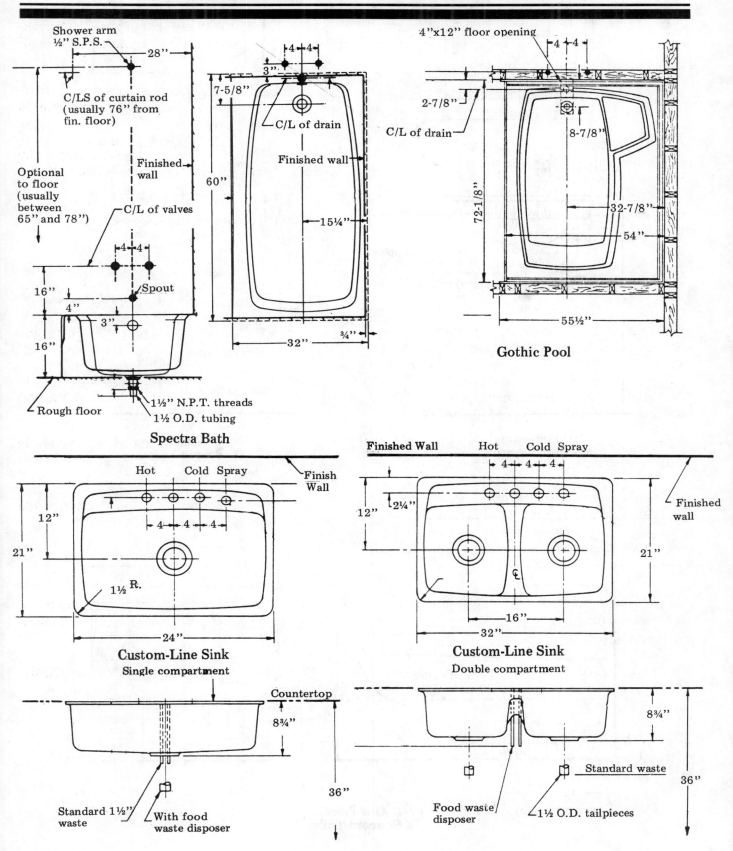

Shower arm
½" S.P.S.

28"

C/LS of curtain rod
(usually 76" from
fin. floor)

Optional
to floor
(usually
between
65" and 78")

Finished
wall

C/L of valves

16"

Spout

4"

3"

16"

Rough floor

1½" N.P.T. threads
1½ O.D. tubing

Spectra Bath

4 4

3"

7-5/8"

C/L of drain

Finished wall

60"

15¼"

32"   ¾"

4"x12" floor opening

4 4

2-7/8"

C/L of drain

8-7/8"

72-1/8"

32-7/8"

54"

55½"

Gothic Pool

Hot   Cold Spray

Finish
Wall

12"

21"

4 4 4

1½ R.

24"

Custom-Line Sink
Single compartment

Finished Wall   Hot   Cold Spray

4 4 4

12"

2¼"

16"

32"

21"

Finished
wall

Custom-Line Sink
Double compartment

Countertop

8¾"

36"

Standard 1½"
waste

With food
waste disposer

8¾"

Standard waste

36"

Food waste
disposer

1½ O.D. tailpieces

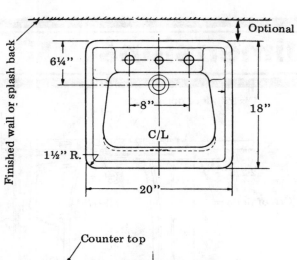

Finished wall or splash back

Optional

6¼"

8"

C/L

1½" R.

18"

20"

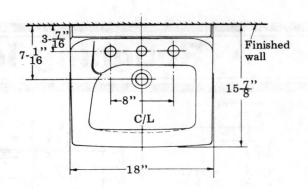

3-7/16"

7-1/16"

Finished wall

8"

C/L

15-7/8"

18"

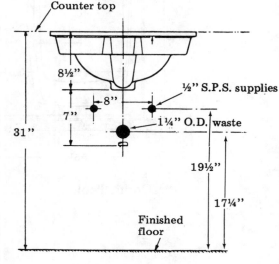

Counter top

8½"

½" S.P.S. supplies

8"

7"

1¼" O.D. waste

31"

19½"

17¼"

Finished floor

**Countertop
Merrilyn Lavatory**

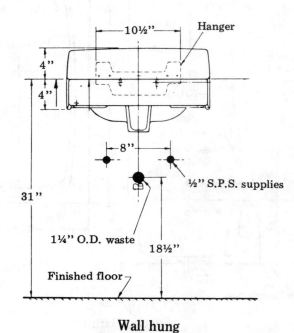

10½"

Hanger

4"

4"

8"

½" S.P.S. supplies

31"

1¼" O.D. waste

18½"

Finished floor

**Wall hung
Penyln Lavatory**

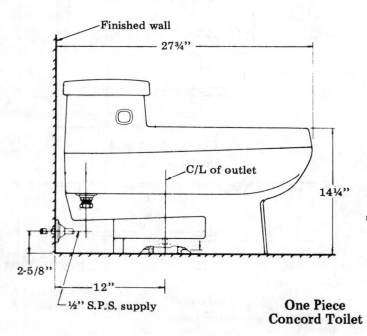

Finished wall

27¾"

C/L of outlet

14¼"

2-5/8"

12"

½" S.P.S. supply

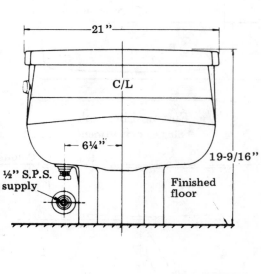

21"

C/L

6¼"

½" S.P.S.
supply

19-9/16"

Finished
floor

**One Piece
Concord Toilet**

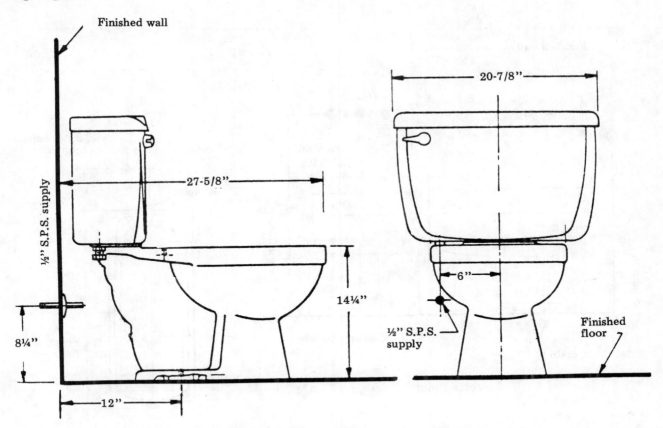

**Closed coupled combination round front
Plebe Toilet**

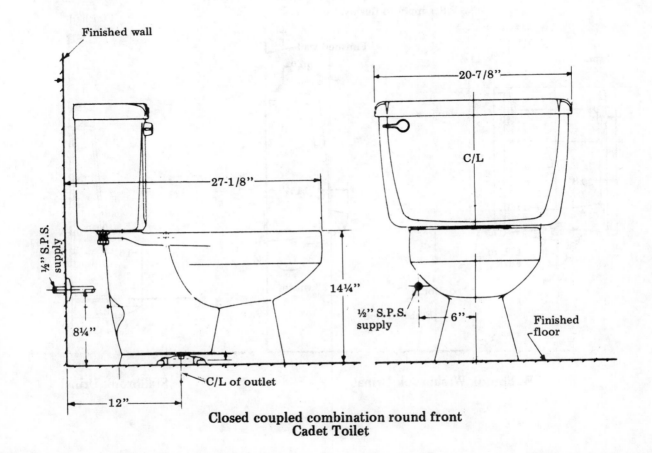

**Closed coupled combination round front
Cadet Toilet**

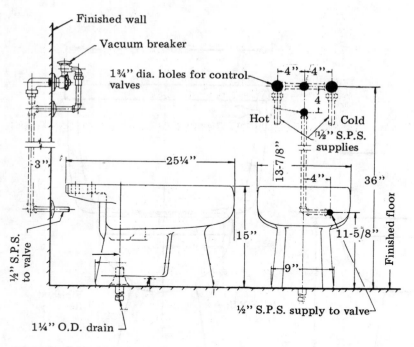

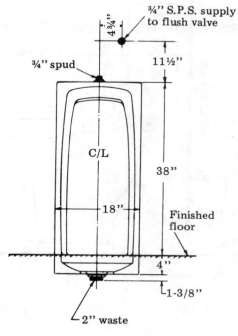

Luxette Bidet

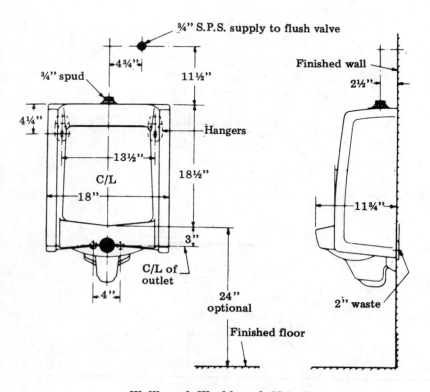

Wallbrook Washbrook Urinal

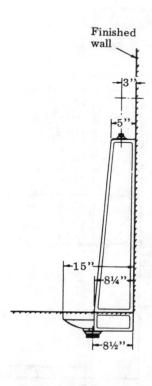

Stallbrook Urinal

Plumbing Fixtures

Important standards have been developed for plumbing fixtures over the past one hundred years. These code standards control the quality and design of all plumbing fixtures used today.

Plumbing fixtures are the end of the potable (drinkable) water supply system and the beginning of the sewage system. They are containers into which liquid waste flows before its release into the drainage system. The most commonly used residential fixtures are water closets, bidets, bathtubs, shower baths, kitchen sinks, and laundry trays.

By code, plumbing fixtures must be of high quality materials. They must be free of defects and concealed fouling surfaces. Fixtures must have surfaces that are smooth and non-absorbant and must be designed to allow all surfaces to be easily cleaned.

Fixtures designed to meet U.S. standards are commonly made of enameled cast-iron, enameled pressed steel, vitreous china, or stainless steel. Fixtures constructed of pervious materials (such as Roman baths or shower baths constructed of tile or marble) must not have waste outlets that can retain water.

Plumbing fixtures should be located in adequately lighted and ventilated rooms. If natural ventilation from a window is not available, a fan and duct are required. The lack of adequate lighting or ventilation promotes unsanitary conditions. Most codes prohibit locating fixtures in such locations.

Minimum Residential Fixture Requirements

The types and number of fixtures required by most codes for residences depends on the type of occupancy and the number of people expected to use the toilet facilities. Codes vary considerably (especially for commercial use) in the number of fixtures needed. For residential and light commercial buildings, the following fixture requirements should be adequate. Refer to your local code for exact requirements.

Single family residences - The minimum requirements are: 1 kitchen sink, 1 water closet, 1 lavatory, and 1 bathtub or shower unit. Provision must also be made for a clothes washing machine. Hot water is optional in some codes and required by others.

Duplex residential units - The minimum requirements are: 1 kitchen sink, 1 water closet, 1 lavatory, and 1 bathtub or shower unit for each dwelling unit. Provision must also be made for a clothes washing machine for each unit. One machine is adequate for both units if it is available to all residents. Hot water is optional in some codes and mandatory in others.

Fixtures for Places of Employment

The number of toilet facilities for light commercial buildings is based on the number of employees. The percentage ratio and the types of fixtures required for males and females may be changed by the plumbing plans examiner. The examiner in your area will consider altering the requirements in Table 19-1 if you can provide data showing that some other fixture ratio is more appropriate.

Consider an example: Assume that toilet facilities are needed for a storage warehouse employing 30 persons. Some building codes require a ratio of 50% male and 50% female facilities. Other codes use a percentage ratio of 75% male and 25% female for the same type of occupancy. Obviously, if these ratios were used rigidly an imbalance of toilet fixtures would result in many installations. In the example, the plumbing plans examiners could request a notarized letter from the owner giving the probable maximum number of male and female employees in the storage warehouse. If the letter stated that four females would be employed in the office and twenty-six males in other positions, then the correct number and type of plumbing fixtures could be determined from Table 19-1.

Males				Females		
Number of Males	Water Closets	Urinals	Lavatories	Number of Females	Water Closets	Lavatories
10-30	1	1	2	1-12	1	2
31-46	2	1	3	13-34	2	3
47-63	2	2	4	35-58	3	4

Places of employment
Table 19-1

More complete tables may be found in your local code or in the more advanced *Plumber's Handbook.*

Other essential requirements for light commercial buildings are noted below.

A drinking fountain must be provided for each 75 persons or portion thereof. The fountain must be accessibly located within 50 feet of all operational processes. Drinking fountains must not be located in any restroom or vestibule to a restroom.

Wash-up sinks may be substituted for lavatories where the type of employment warrants their use.

Manufacturing plants that may subject their employees to excessive heat, infection, or irritating materials must provide a shower for bathing for every 15 persons.

Small office buildings or similar establishments that have 10 or more offices or rooms and that employ 25 persons or more must provide a service sink on each floor.

Where more than one person can use the toilet facilities at a time, water closets must be separated from the rest of the room and from each other by stalls made of some impervious material.

Toilet rooms connected to public rooms or passageways must have a vestibule or must otherwise be screened or arranged to ensure decency and privacy. This vestibule must not be common to the toilet rooms of both sexes.

Toilet bowls must be of the elongated type having seats with open fronts.

Establishments employing 9 persons or fewer that do not cater to the public (such as storage warehouses and light manufacturing buildings) have less rigid requirements. Here, some codes require only one water closet and one lavatory for both sexes. But consider the following code conditions for such establishments:

- If the minority sex exceeds 3 persons, separate toilet facilities are required. For example, where 4 males and 5 females (or vice versa) are employed, separate toilet facilities must be provided.

- If the number of males employed exceeds 5, a urinal must be provided.

Retail Stores

Establishments frequented by the public must provide toilet facilities for the number of employees and the public reasonably anticipated. Retail stores of 1,000 square feet or less (allowing for displays and storage) are usually required to have one water closet and one lavatory to adequately serve both employees and the public.

Fast Food and Small Restaurants

The required toilet facilities for establishments such as fast food restaurants that have a countable seating capacity are determined by the maximum number of people that can be served at one time. The ratio observed by most codes is 50% male and 50% female.

Use Table 19-2 to determine the minimum toilet facilities for establishments where food and drink (but no alcoholic beverages) are served and consumed on the premises. This includes small town restaurants and restaurants serving hamburgers, fish and chips, barbecue, and the like.

More complete tables will be included in your local code and in *Plumber's Handbook.*

Here are other essential fixture requirements for fast food and small restaurants:

- Establishments serving drive-in custo-

Males				Females		
Number of Males	Water Closets	Urinals	Lavatories	Number of Females	Water Closets	Lavatories
1- 62	1	1	1	1- 30	1	1
63- 98	2	1	1	31- 62	2	1
99-138	2	2	2	63- 98	3	1
				99-138	4	2

Food and drink establishments
serving no alcohol
Table 19-2

mers must provide toilet facilities at the ratio of 1 person for each 100 square feet of parking area. (Use Table 19-2.) For example, a drive-in restaurant with 10,000 square feet of parking area would need to provide toilet facilities for 50 males and 50 females, plus employees.

• Public food service establishments that offer only take-out service need not provide guest toilet facilities. Toilet facilities are required here only for employees.

• The floors and walls of *public* toilet rooms in fast food and small restaurants must be covered with tile or other impervious materials to a height of 5 feet. These toilet rooms must have easy and convenient access for both patrons and employees. The restrooms must be located within a 50-foot line of travel from the nearest exit to the dining room or food service area. Toilet rooms must be located on the same floor as the area they serve.

• A dishwashing machine or suitable 3-compartment sink must be installed in food or drink establishments where dishes, glasses, or cutlery are to be reused.

In establishments where food or drink are prepared or served, a hand sink must be installed for employees' use. Lavatories in adjoining toilet rooms may *not* serve this purpose.

• Water closets must be separated from the rest of the room and from each other by stalls made of an impervious material. A privacy lock is not permitted on the entrance door to a public toilet room.

• Toilet rooms connected to public rooms or passageways must have a vestibule or must otherwise be screened or arranged to ensure decency and privacy. This vestibule must not be common to the toilet rooms of both sexes.

• Toilet bowls must be of the elongated type and must have seats with open fronts.

• Note that all public toilet rooms must have adequate built-in provisions for the handicapped.

PLUMBING FIXTURES

Fixture Overflows

Bathtubs and lavatories are two of the most common fixtures provided with overflows. The code does not require that certain plumbing fixtures have overflows, however. Thus the current trend in fixture design is to omit overflows on lavatories. Integral overflow passageways do provide secondary protection against self-siphonage of the fixture trap seal. They also let excess water escape below the flood-level rim of the fixture.

When installing fixtures which have overflows, remember that the waste pipe must be designed to prevent water from rising into the overflow when the stopper is closed. The waste pipe must also prevent water from remaining in the overflow when the drain is open for emptying.

The overflow pipe or passageway from a fixture must be connected on the *inlet* side of the fixture trap. This prevents sewer gases and

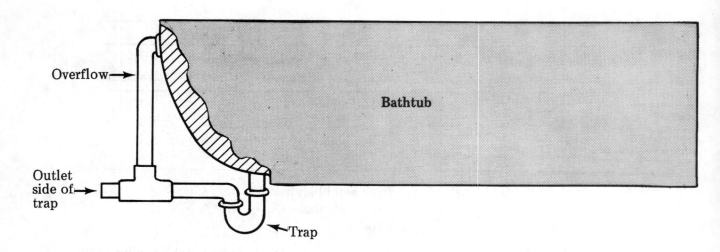

Overflow →

Outlet
side of →
trap

← Trap

Prohibited overflow connection
Figure 19-3

odors from entering the room through the overflow. In fact, the code prohibits connecting a fixture overflow to any other part of the drainage system. See Figure 19-3.

Fixtures must have durable strainers or stoppers. (An exception is made for fixtures with integral traps.) The strainer or stopper must not prevent rapid drainage of the fixture. The strainer should not be smaller than the fixture waste outlet it serves and (except for fixed strainers) should be easy to remove for cleaning.

Bathtubs

Bathtubs come in many styles, designs, and colors. The enamel coating is generally acid-resistant and has an easy-to-clean glasslike surface. Bathtubs are manufactured in enameled pressed steel, enameled cast-iron, or in gel-coated fiberglass.

Recessed tubs are built into the floor and walls at two ends and the back. Bathtubs that recess into tile or other finished wall materials must have waterproof joints. Such tubs are either right-hand or left-hand. Both are shown in Figures 19-4 and 19-5.

Corner tubs are built into the floor and walls at one end and at the back, as shown in Figure 19-6.

Modern bathtubs have slip-resistant surfaces for safe tub bathing and a secure standing area for showering. See Figure 19-7.

The minimum size waste and overflow for

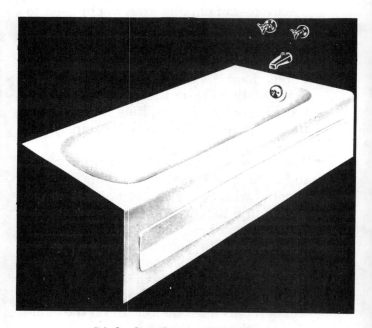

Right-hand recessed bathtub
Figure 19-4

bathtubs is 1½ inches. There are several types of approved tub wastes and overflows. Figure 19-8 shows the three most common types in use today.

Bathtub waste and overflow (A in Figure 19-8). This is the most trouble-free of the three wastes shown. In this type there are no internal moving parts to wear or break, and collected hair or other debris can not cause a stoppage.

Drain plugs for this waste are manufactured in two types. One is the *tip-toe* drain plug. If you wish to close the drain to retain the water, press

Left-hand recessed bathtub
Figure 19-5

the stopper down with your toe. To release the water, press the stopper down a second time and the drain opens.

The second type of stopper (drain plug) operates like the basket strainer in a kitchen sink. You lift with your fingers and turn to open; turn and let it drop to close. If these parts ever need replacing, unscrew the stopper with your fingers (the only tool necessary) and replace it with a new stopper of the same type.

Bathtub waste and overflow (B in Figure 19-8). This type is known as a *trip waste*. It has several disadvantages, and some codes will no longer permit its use. The water for the tub is controlled by a heavy brass cylinder attached to a small adjustable rod that connects to a lever on the overflow plate. The overflow plate is the large chrome disk near the top of the tub. The waste opening at the bottom of the tub is covered by a perforated strainer. When the lever is pressed to the "down" position, the rod lifts the cylinder and the waste water drains away. To close the waste opening and retain the water, pull the lever up, lowering the cylinder to a ground seat.

One disadvantage of the trip waste is that hair, lint, and other debris adhere to the cylinder surfaces. This prevents the cylinder from retaining water in the bathtub and keeps water from draining away freely. To clean the cylinder, remove the two screws in the overflow plate and pull it up and out. Clean the cylinder of foreign matter, cover it with petroleum jelly and replace as before.

Another disadvantage of the trip waste is that the rod must have threads and a small lock nut for adjustment. If the cylinder is adjusted downward too much, the waste water will not

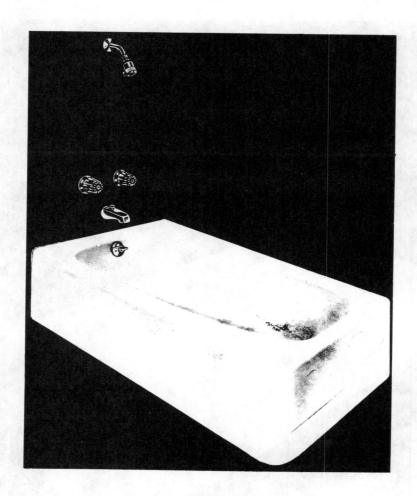

Left-hand corner bathtub
Figure 19-6

drain away freely. If the cylinder is adjusted upward too much, the tub will not retain water long enough for a leisurely bath. Very accurate adjustments are required for it to function properly.

Because the rod is approximately 1/8 inch in diameter and tends to deteriorate over the years, the cylinder may break loose from the rod. It could then drop to the bottom and partially block the opening. Unless there is an access panel, the only way to retrieve the cylinder short of breaking into the bathtub wall is to remove the overflow plate. Then fashion a hook at one end of a length of strong wire such as a clothes hanger and ''fish'' for the cylinder. With patience the cylinder can usually be removed.

Bathtub waste and overflow (C in Figure 19-8). This type is known as a *pop-up*. It is the most popular waste and overflow now in use.

The pop-up stopper and the trip waste operate similarly. Each has a lever attached to an adjustable rod, which is in turn attached to the overflow plate. Turn the lever to the left to close and to the right to open the drain. A coiled spring is attached to the rod instead of a cylinder. Pressing this spring down raises the stopper and opens the drain. Raising the spring closes the drain opening. An advantage of the *pop-up* is that the adjustment on the rod does not have to be as accurate to give maximum performance. A disadvantage is its tendency to collect hair, lint, and other foreign matter. When this occurs, clean the pop-up by raising the tub stopper to an open position and pulling it out with your fingers. Foreign substances can then be removed and the stopper can be worked back to its original position. Then remove the two screws located on the overflow plate, pull the mechanism up and out, and clean the spring. Replace as before.

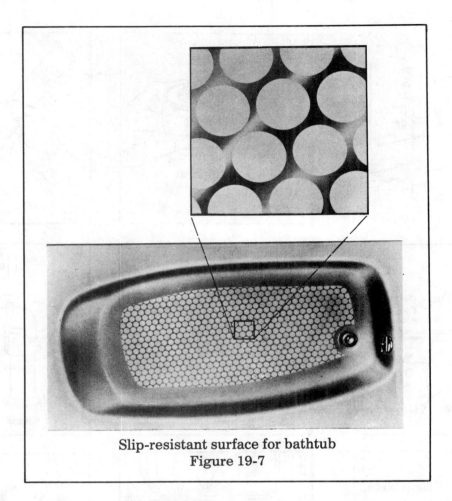

Slip-resistant surface for bathtub
Figure 19-7

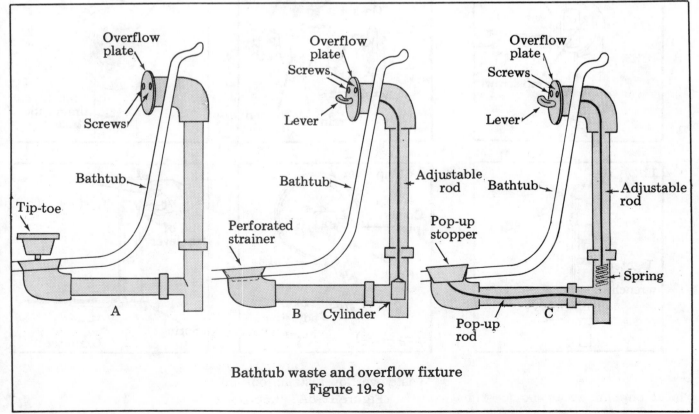

Bathtub waste and overflow fixture
Figure 19-8

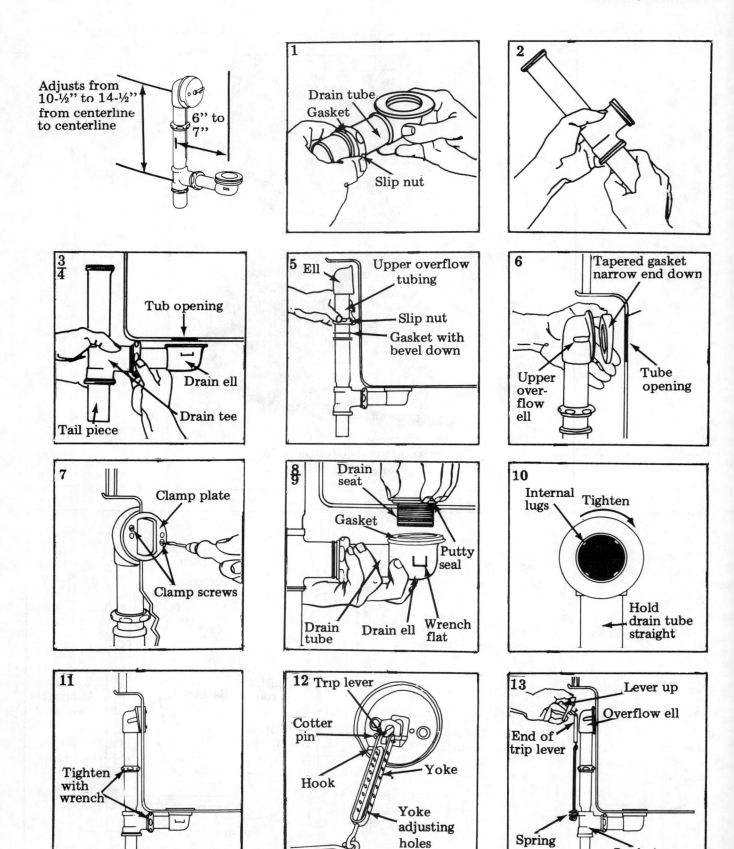

Waste overflow installation
Figure 19-8A

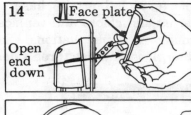

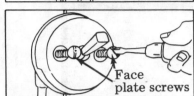

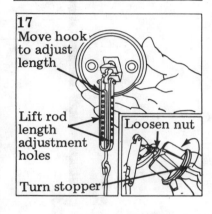

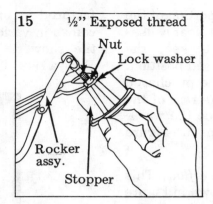

1. Place slip nut and gasket on drain tube.

2. Screw the tailpiece up into the bottom of the drain tee. Use pipe dope on threads. Make tight.

3. Place drain tube into the drain tee. Hold the drain against the bottom of the tub at the opening and tighten slip nut hand tight.

4. This positions the waste assembly so that the connection to the waste pipe opening can be determined, whether by slip nut, direct connection to a 1½'' waste pipe thread, or any other method of your own choosing.

5. Place the slip nut and gasket on the overflow tube. The bevel on the gasket should' face down. Slip the upper tube into the lower tube.

6. Center the upper overflow ell with the tub opening. Place the tapered gasket between the flange on the ell and the tub opening, narrow side down.

7. Center the clamp plate over the tub opening and thread the brass clamp screws through it into the overflow ell. Use the 4 o'clock and 10 o'clock positions. Tighten screws making sure all parts are aligned.

8. Next, install the waste seat. Thoroughly clean the tub around the drain opening. Apply a liberal amount of plumber's putty around the bottom of the flange on the drain seat.

9. Position the drain ell and tube underneath the tub, with the gasket on the top of the flange. Use the thinner gasket if the tub is cast iron. Use the thicker gasket if the tub is steel.

10. Thread the drain seat into the drain ell. Hold the ell on the wrench flats. Tighten the seat using a block of wood or a spud against the internal lugs provided for this purpose. Make sure the drain tube is straight.

11. Now tighten both slip nuts, the one on the overflow tubes and the one on the drain tube. Test all joints with water.

12. Remove the cotter pin in the top of the yoke in the lift assembly. Assemble the lift lever to the trip lever using the cotter pin.

13. Position the trip lever UP, the lift assembly DOWN, and hold the lift assembly against the overflow tubes. Hold the end of the trip lever in the center of the overflow ell. The end of the spring should be in the center of the drain tube and tee. If it is not, adjust the lift length by moving the hook in the yoke adjusting holes.

14. With the open side of the face plate DOWN, push the lift assembly down into the overflow tube. Thread the chrome plated face plate screws into the overflow ell, 9 o'clock and 3 o'clock.

15. Remove the brass nut and star lockwasher from the end of the rocker assembly. Replace, putting on nut first and lockwasher last. Leave about ½'' of thread exposed on the end. Thread on stopper and tighten the nut against it.

16. Thread rocker assembly into the drain tube. It will be necessary to work it in to clear the tubing joint.

17. Adjust lift rod length until the lever opens and closes freely. Adjust stopper screw. If stopper raises too far and won't drop down, loosen the nut and screw the stopper down a couple of turns at a time until it drops freely.

18. Fill the tub. Drain and check all joints.

Waste overflow installation (Continued)
Figure 19-8A

The spring and the raised back portion of the overflow plate wears with use. This wear causes a loss of the tension necessary to keep the stopper or cylinder in an open or closed position. Replaceable overflow plates may be purchased and installed without disturbing the rest of the waste and overflow mechanism. See the waste and overflow installation directions in Figure 19-8A.

The *old-fashioned waste and overflow.* This type of waste and overflow (not shown in Figure 19-8) is the old chain and stopper type. The stopper must be manually placed in the waste opening to retain water and manually removed to release water. This type is used today only as a replacement in the older leg tubs still found in some homes.

Shower Baths

Shower baths installed in homes usually are located against a wall at one end of the bathtub. See Figures 19-5 and 19-6. This type usually has a diverter valve or diverter spout that supplies water to the tub and can transfer the flow to the shower head. This type of shower may use two separate valves and a diverter spout as shown in Figure 19-6, or a mixing valve and a diverter spout as shown in Figure 19-5. The user adjusts the temperature of the water from the spout and then pulls up the knob located on the tub spout to divert water to the shower head.

Shower Stalls

A shower stall installation has many advantages. It takes up little space and uses less water than a bathtub. Stall showers are generally manufactured from one of two materials—prefabricated porcelain steel or fiberglass.

The porcelain steel shower stall is a prefabricated unit with three sides and a base. The sides and base are assembled at the job site. These are usually very small units just meeting minimum space requirements of the code. The sides are made of thin sheets of steel grooved so that they fit together and make a watertight joint. The base may be of a heavier porcelain steel or of precast concrete or other approved material. The metal sides tend to rust and it is difficult to keep the joints clean. Usually a shower curtain closes off the entrance.

The newer fiberglass shower stall is manu-

factured in one piece, including the base. It comes in various sizes and colors and replaces tile showers in many new installations. See Figure 19-9. The fiberglass stall has three sides.

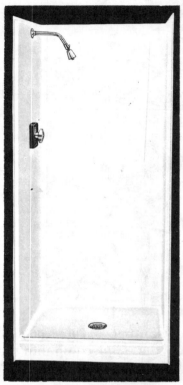

Fiberglass shower stall
Figure 19-9

The entrance may have a glass or plastic door or shower curtain. It has a smooth watertight finish which is easily cleaned.

Stall shower bases of both types usually come with the shower drain attached to an opening in the base. Accuracy is critical when roughing in stall showers. They are usually installed after the floor is poured and the rough partitions are in place.

Tiled Showers

Tiled showers have tile or marble walls on three sides and a floor of similar material. These showers can be designed and sized to the owner's taste. Tiled showers can be constructed level with the existing bathroom floor or recessed to the most usable depths. Recessed tiled showers are called *sunken* showers. Tile showers may have shower curtains or, more likely, they can be equipped with doors of plastic or safety glass. The waste opening may be installed in any location within the shower as

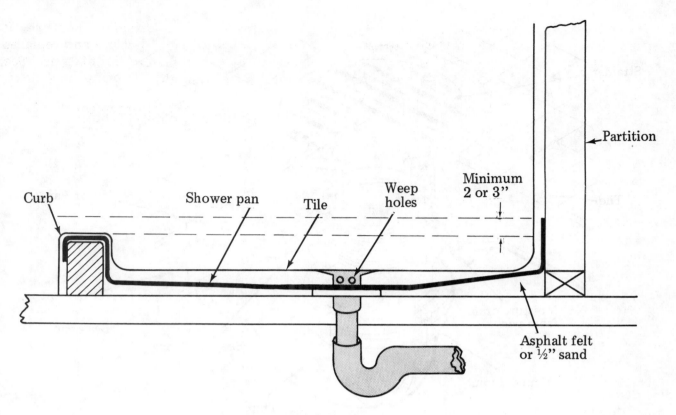

Curb   Shower pan   Tile   Weep holes   Minimum 2 or 3"   Partition

Asphalt felt or ½" sand

Figure 19-10

long as it is low enough so that the shower floor slopes toward the drain from all angles. Roughing in the waste opening for a tiled shower is not as critical as for a stall shower.

Completely waterproofing the shower is most important. The walls of a tile shower must be waterproof, smooth, non-corrosive, and non-absorbent to six feet above the floor. There must be an impervious waterproof base under the tile on the shower floor, as water standing on the tile surface can slowly seep through and leak out into adjoining rooms.

Other Shower Requirements

Waste outlets for shower compartments must be a minimum of 2 inches in diameter. After a shower, water should drain from the shower floor without puddling. The free area of the shower strainer must be a minimum of $3\frac{1}{2}$ square inches. The strainer must be removable to allow easy cleaning of the shower trap. Shower traps must not be smaller than the waste outlet pipe used in the shower compartment.

Shower compartments need a minimum floor area of 1,024 square inches. This requires a minimum 32-inch span between walls, a space considered adequate for adult use. Floors of

shower compartments must be smooth and sound.

Shower pans of lead, copper, or other approved materials should be used when such pans are required. Lead pans should not weigh less than 4 pounds per square foot. Copper pans must not weigh less than 12 ounces per square foot. Shower pans of lead or copper must be painted with asphaltum paint inside and outside to protect the pan from corrosion where it joins concrete or mortar.

The carpenter must have already installed the rough partitions and the curb before the shower pan is installed. If the building has wood floors, the carpenter must provide a solid base of subflooring or plywood on which the shower pan is to rest. In buildings with concrete floors, a layer of 30-pound saturated asphalt felt or a ½-inch layer of sand should be placed under the pan. This is absolutely necessary to protect the pan against rough surfaces that could cause accidental puncturing of the pan before it is given the protection of the finished floor material.

Shower pan material is soft and flexible and must be supported by an adequate backing secured to the partition studs. This should keep

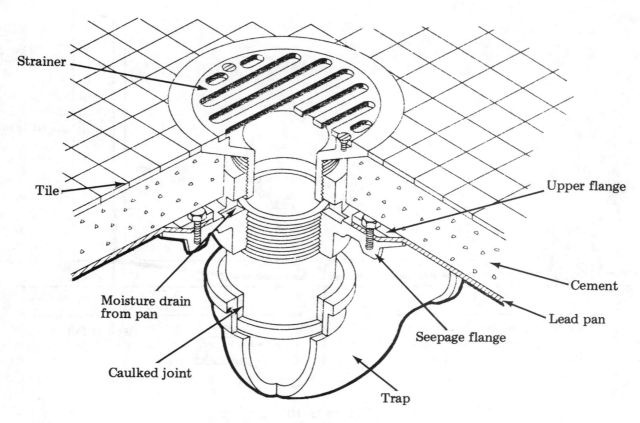

Strainer

Tile

Moisture drain
from pan

Caulked joint

Upper flange

Cement

Lead pan

Seepage flange

Trap

Shower pan installation
Figure 19-11

the pan sides from sagging until the interior of the shower compartment is in place to hold the pan rigid. If the shower pan material must be punctured to secure it in place, the penetration can not be lower than 1 inch from the top of the pan's turn-up. Figure 19-10 shows an installed pan.

Cut the shower pan large enough so that when properly folded to fit the shower compartment it has an upturned edge on all sides at least 2 inches above the finished curb, or 3½ inches above the rough curb. Cut a hole where the drain is located and then put the shower pan in place. Securely fasten the shower pan to the shower strainer base at the invert of the weep holes.

Paint the top of the shower base and the bottom side of the clamping ring with a pipe joint compound. Place the clamping ring on top of the shower base and screw them together. The pan material is clamped between the two flanges, thus forming a watertight joint between the shower waste outlet stub and the pan. Screw the strainer portion of the shower drain down into the threaded flange body to the desired height of tiling (generally, 1 inch above bottom

of pan at this point). See the shower pan installation detail in Figure 19-11.

The pipe between the trap and shower drain can be threaded as illustrated or it can be made of other code accepted drainage materials.

Each shower pan must be tested for inspection. Remove the strainer plate and plug the waste outlet. Fill the pan with water. The pan must be full and ready for inspection during the tub and water pipe inspection. Otherwise, the contractor you work for may have to pay for a reinspection. While shower pans are not required for prefabricated shower stalls, each stall requires approval by the plumbing inspector for watertightness.

Shower pans can be omitted in shower compartments built on a concrete slab on the ground floor, provided the bottom, sides, and curbs of the shower compartment are poured at the same time the floor slab is poured. A curb one inch higher than the existing slab must be poured around three sides of the shower compartment. This usually keeps the water level below the height of any surrounding wood plates or studs and helps keep the compartment watertight. See Figure 19-12.

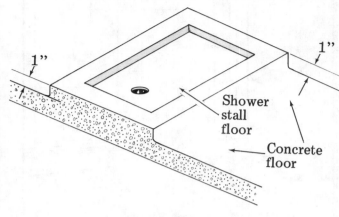

Figure 19-12

Shower Rods

Shower rods are generally installed 6'-6'' above the finished floor. The rod should be centered over the bathtub rim or shower entrance curb so that the shower curtain drains splashed water back into the fixture. Cut the shower rod with a hacksaw to fit the opening. The end flanges which support the rod and shower curtain must be held securely in place by the screws supplied with the flanges. The screws must pass through the finished wall and penetrate the backing material installed during the rough partition work.

Shower Enclosures

In many parts of the country shower enclosures are now installed by plumbers. They are not difficult to install and the tools needed are a hacksaw, a screwdriver, a level, and a masonry hand drill or a ¼ H.P. electric drill with a masonry bit.

The enclosure comes in a kit which generally includes a can or tube of sealing compound with instructions on how to use it. The kit should contain screws and shields for fastening the uprights. Enclosures are manufactured with either plastic or safety glass doors.

Enclosures are designed for installation either on the rim of a bathtub or on the curb of a shower stall. Since most bathtubs are 5 feet long, no measurement of dimensions between walls is necessary. For shower enclosures, measure from finished wall to finished wall along the top of the curb.

Manufacturers are usually generous with the length of the bottom and top track or rail they supply. This rail must be cut with a hacksaw at least ¼ inch shorter than the overall length

between the end walls. This is necessary so that the uprights (which are a standard height) fit snugly between the bottom track and the vertical tile wall.

Install the bottom track first with the drain holes facing into the shower compartment. Place a generous amount of sealing compound on the center line of the bathtub rim or shower stall curb the full length of the bottom track. Then press the bottom track firmly down and against the rim or curb. (See A, Figure 19-13.) Place the uprights against the end walls and plumb them straight with the level as shown in B, Figure 19-13. Use a pencil to mark the tile through the preformed screw holes in the uprights. Remove the uprights so the holes can be drilled in the tile wall. The masonry bit should be the same size as the enclosed shields. Before drilling, punch a small hole in the glazed tile finish with a nail punch or large nail. This gives the bit a bite and keeps it from slipping.

Drill the holes—there are usually three in each upright—and insert the shields. (See C, Figure 19-13.) Spread sealing compound on the wall side of the uprights and set the uprights in place. Secure them to the wall with screws. Drop the top rail into place as shown in D, Figure 19-13.

When attaching the door rollers to the top rail, install the *smooth* side of the glass or plastic doors toward the *inside* of the bathtub or shower stall to permit easy cleaning.

Lavatories

Lavatories are commonly made of enameled cast iron, enameled pressed steel, vitreous china, stainless steel, or the newer acrylics. Lavatories are manufactured in various designs, widths, and sizes. The bowl may be round, square, or oval and may hold from one to two gallons of water. Most lavatories are supplied with fittings for both hot and cold water. Wall-hung, pedestal type, and countertop lavatories are available. See Figure 19-14.

The wall-hung lavatory hangs from a steel or cast-iron wall hanger secured to the wall with screws. It is used more often in older homes and in commercial buildings.

The pedestal lavatory is supported by both a wall bracket and a pedestal base. Once commonly used in residences, this type has been almost entirely replaced by the countertop lavatory.

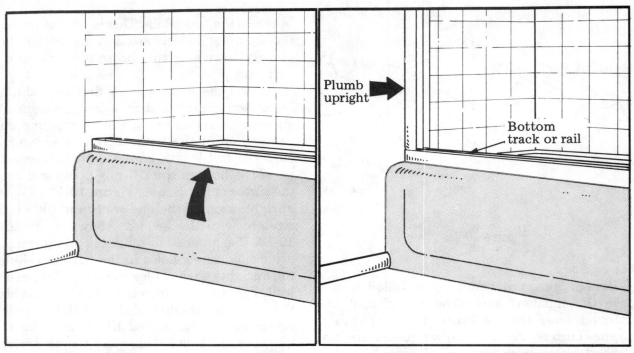

Attach bottom track to tub with sealing
compound.

A

Plumb uprights with level and mark
holes for drilling.

B

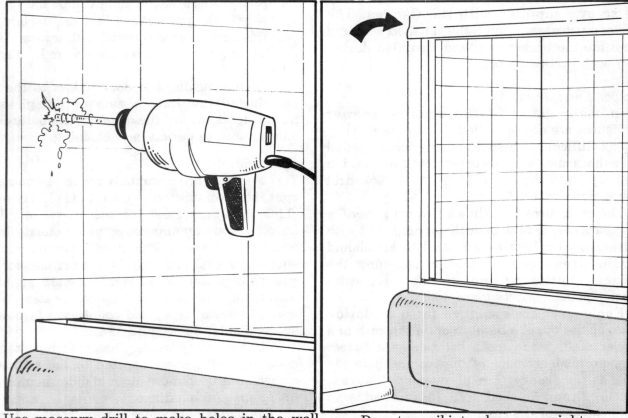

Use masonry drill to make holes in the wall.
Insert shields in holes and attach uprights to
wall with screws. Seal uprights to tile

C

Drop top rail into place on uprights
and hang doors.

D

Figure 19-13

Self-rimming lavatory

Stainless rim lavatory

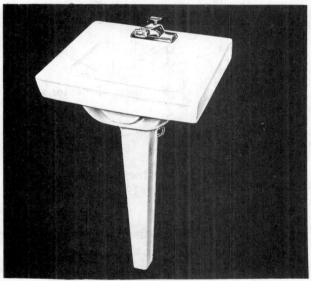

Pedestal lavatory

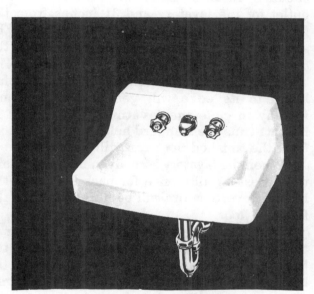

Wall hung lavatory

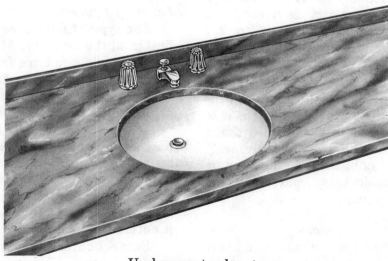

Undercounter lavatory

Figure 19-14

The countertop lavatory is the most popular of the three for new construction. It is designed for both above-counter and under-counter installations. The new acrylic lavatory is available as either a basin alone or as a one-piece unit with counter.

Installing The Wall-Hung Lavatory

Wall-hung lavatories for residential and commercial use are generally supported by a hanger screwed securely to wooden backing material fastened to the bathroom partition studs. The mounting board, usually a 1 x 6, must be installed flush with the face of the studs during the rough carpentry work. The board should be installed above the center of the waste outlet for the lavatory and 31 inches from the finished floor to the center of the board. The standard height from the finished floor to the overflow rim of the lavatory is 31 inches. The extra width from the center of the mounting board to its edge will generally allow some leeway in case the lavatory back varies in height. Thus the standard height of the lavatory from the finished floor can still be maintained.

When the lavatory is ready to be hung, make the necessary allowance for back height so the lavatory overflow rim will be 31 inches from the finished floor.

Place the hanger directly over the center of the waste pipe and use a pocket level to level the hanger. With a pencil mark the holes on the backing through those provided in the hanger. (Use two at each end and one in the center.) If the wall is of tile or other hard surface, use an electric drill with a masonry bit to drill through the finished wall to the mounting board. Use wood screws long enough (usually 2 to 2½ inches) to penetrate mounting board at least one inch. See Figure 19-15.

Wall-hung lavatories must be provided with enough wall support so that no strain is transmitted to the fixture pipe connection or to the finished wall. There is little possibility that a lavatory so secured can pull away from the wall. The weight of the countertop lavatory is transferred to the cabinet top and places no strain on the fixture piping.

Installing The Countertop Lavatory

Many types of clamp assemblies are available to secure lavatories to countertops. The installation depends entirely on the

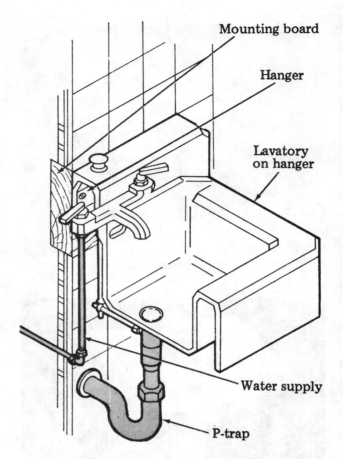

Wall-hung lavatory installation
Figure 19-15

manufacturer of the fixture, the type of fixture, and the construction of the cabinet. Shown in Figure 19-16 are two types of clamp assemblies now in use. Follow closely the instructions accompanying the clamp assembly you choose.

Lavatory Faucets

The type of faucet installed depends on the design of the lavatory. Many wall-hung lavatories in residences have factory installed faucets as shown in Figure 19-14. Wall-hung lavatories for commercial use are usually designed for separate hot and cold water faucets.

Many commercially used lavatories have only a single cold water faucet with the opening for the hot water faucet concealed by a faucet hole cover. A lavatory which has two single faucets for hot and cold water is less desirable as this makes it impossible to mix the water before it enters the lavatory.

Some commercial lavatories have self-closing faucets. The faucet has a spring which snaps the handle closed when released, shutting

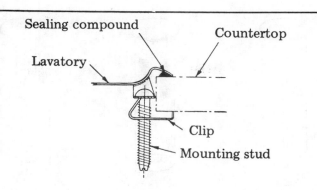

Apply sealing compound around edge of hole. Drop lavatory into hole. Install clips as shown and tighten firmly. Remove excess sealing compound.

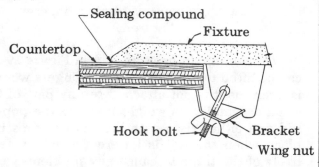

Apply sealing compound to underside of lavatory rim. Place lavatory over cut-out. Make final fitting connections and align lavatory. Install clamp assemblies (see illustration). Tighten wing nuts. Add or remove excess sealing compound around base of lavatory rim. (In 20'' countertop, also add sealant on top of back edge of lavatory). Form fillet with wet finger or sponge. Carefully wipe excess compound from countertop and lavatory rim with damp cloth or sponge.

Two types of clamp assemblies
Figure 19-16

off the supply of water. A disadvantage to this type of faucet is that one hand must remain on the handle to hold it open as long as flow is desired. And the strength with which the faucet snaps closed when released can set up a water hammer in the pipes.

The combination faucet with either a mixing valve or separate hot and cold water valves is most frequently used in residences. The combination lavatory faucet is usually made to allow either a 4-inch or an 8-inch distance between the centers of the handles. This has partially standardized the holes in lavatories. The combination faucet has a chamber in which

hot and cold water are mixed before coming out a single central spout. The advantage of this is that the user can adjust the temperature of the water leaving the spout.

Lavatory Drain Assemblies

There are two types of drain assemblies for lavatories.

1) The chain and rubber stopper pull-out (P.O.) plug type is generally used in commercial buildings. It consists of an open drain with a flange on the upper end and a threaded tube projecting from the lower end. The rubber stopper is inserted in the drain opening to plug it so that water can be contained in the lavatory. The stopper is removed to release the lavatory contents to the drainage system. It is inexpensive, usually trouble free and can be used on lavatories that have single hot and cold water faucets. See Figure 19-17.

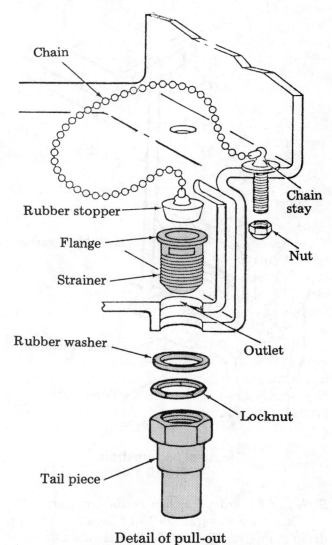

Detail of pull-out
Figure 19-17

2) The pop-up drain stopper has a combination faucet and a lift rod connected to a pop-up rod in the waste. When the lift rod is pressed down, the pop-up stopper is pushed up, releasing the lavatory's contents to drain away. When the lift rod is pulled up, the pop-up stopper closes to seal the opening in the drain. See Figure 19-18. The pop-up drain assembly is more sanitary than the pull-out plug drain with chain and rubber stopper.

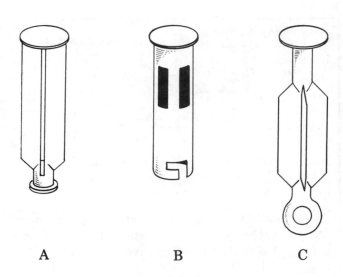

Lavatory pop-ups (drain stoppers)
Figure 19-19

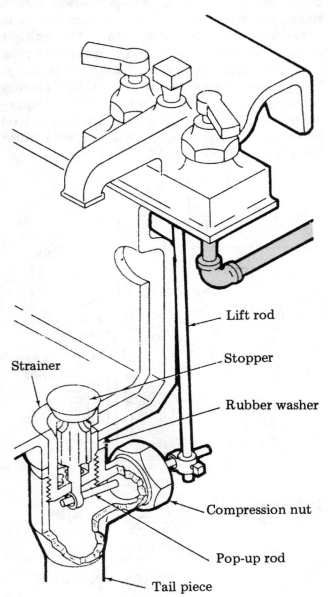

Detail of pop-up drain
Figure 19-18

Styles of Lavatory Pop-Up Drain Stoppers
There are several styles of lavatory pop-up drain stoppers. Figure 19-19 illustrates three of the more common types.

The pop-up drain stopper, A in Figure 19-19, can be lifted up and out with the fingers when it is open, as it is not attached to any part of the drain assembly. Its weight rests on the pop-up rod and either rises with the rod or drops into place when the rod is lowered. It is the least likely of the three to cause any problems, as it will not collect drainage debris. The rod that extends into the drain assembly on which the stopper rests may catch long hairs or strings. These can easily be removed with tweezers.

The pop-up drain stopper, B in Figure 19-19, is more difficult to remove as it has a slot in the side that locks directly to the pop-up rod. To remove it, push the lift rod down all the way. This in turn will raise the stopper. Twist to unlock it and lift it up and out. This style of pop-up stopper is a round, hollow tube and hair tends to wrap around the outside of the tube. This makes stopper operation difficult and slows the flow of waste water. When this happens, remove all hair and other foreign matter from the stopper and rod. Then replace as before.

The pop-up drain stopper, C in Figure 19-19 is the most difficult of the three to remove for replacing or cleaning. This style of pop-up stopper is designed with a hole on the opposite end through which the pop-up rod extends. See Figure 19-18. Loosen the compression nut with pliers so the rod can be pulled toward the rear of the lavatory bowl, freeing it from the stopper. The stopper can now be removed for cleaning. Reassemble as before, tighten the compression nut, and run water through the lavatory waste. Check the compression nut for leaks.

Installing Lavatory Trim

Faucets and waste assemblies should be fitted to the lavatory prior to installing it on the wall, as this job is more easily done when the lavatory is on the floor. Protect the lavatory from scratches by placing it on rags or cardboard. Place a ring of putty on the undersides of single or combination faucets before securing them in place. This prevents water from seeping under the faucet body through the preformed holes in the lavatory and leaking on to the floor or in the cabinet. Also put a ring of putty under the drain assembly flange before it is secured in place. A mack washer seals the underside of the drain assembly to the bottom of the lavatory. The mack washer is held in place by a metal washer and lock nut. The rest of the drain assembly can now be screwed to the threaded tube projecting below the lavatory body. The rubber stopper and chain may be secured to a chain stay fastened to the lavatory. See Figure 19-17. The lift rod may be secured to the pop-up rod and adjusted to properly operate the pop-up stopper. See Figure 19-18.

The wall-hung lavatory can now be placed on the hanger and pushed down into place. Counter-top lavatories are securely fastened to the counter top by special rim clips and are made watertight with a caulking compound or other recommended adhesive. Wall-hung lavatories must be sealed with white cement or other suitable material at their point of contact with finished wall surfaces.

A lavatory supply tube may be used with a compression ell or stop to connect the water supply rough-in to the faucet. Lavatory supply tube is soft and flexible and can be bent to any offset by hand and then cut to the desired length. The top of the supply tube is secured to the faucet by a coupling nut supplied with the faucet. The bottom of the tube is usually secured to the ell or stop by a compression nut and ring. See Figure 16-28. The coupling nut is tightened by a basin wrench, which is used to reach less accessible places. See Figure 19-20. The compression nut may be tightened into place with a six or eight inch adjustable smooth-jawed wrench.

Waste outlets for lavatories must be a minimum of 1¼ inches O.D. A 1¼ inch P-trap installed in the waste line extending through the finished wall completes the installations. The P-trap and the waste line must be of a similar material.

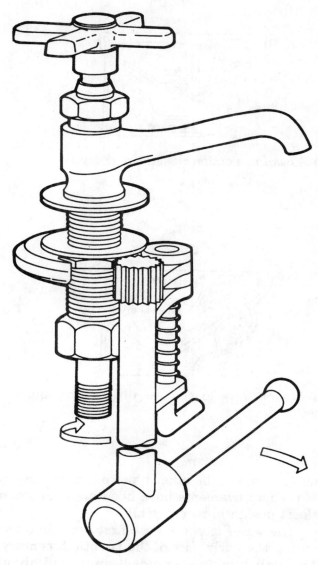

Basin wrench
Figure 19-20

Water Closets

Credit for inventing the first water closet goes to an Englishman, Joseph Bramah. He developed the valve closet in 1788. It worked with water pressure supplied by a pump. About 1830 the pan closet was introduced, followed by the long hopper water closet, around 1850. The plunger type water closet came into use about 1870 and was more popular.

The English pioneered the design of plumbing fixtures until the early 1880's. By 1890 the washdown water closet, which originated in the United States, was introduced. Its design included the siphonic action, greater water coverage of interior bowl surfaces, and complete scouring of all interior bowl surfaces with each flushing. The washdown water closet

Standard

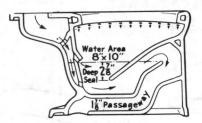

(1) *Washdown* action closets are noisy.

Quiet

(2) *Reverse Trap* closets are quieter than the washdown models.

Quieter

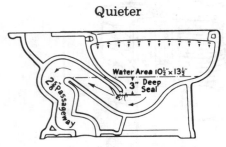

(3) *Siphon Jet* closets are the next in quietness.

Quietest

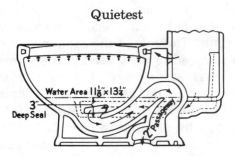

(4) *Siphon Action* closets are the quietest toilets.

Figure 19-21

made obsolete the other types of water closets. Many improvements have been made in water closet design in recent years.

The water closet we use first came into use during the early part of the twentieth century. The wall-hung water closet appeared about 1905. Around 1915 the water tank came down from the wall with the introduction of the low-down flush tank combination water closet. By the early 1920's the reverse-trap and siphon jet water closets were introduced. In these the flushing action was quieter. In the 1930's the one-piece water closet was introduced. Improvements continue to be made in water closet design to make them quieter, more sanitary, and easier to repair.

Types Of Water Closets

The water closet is a fixture used to carry body waste solids to the drainage system. It is constructed of vitreous china and is available in various designs and colors. The different types of water closets include the common washdown bowl, the reverse trap bowl, the siphon jet bowl, and the siphon action bowl. Figure 19-21

illustrates each of these. A separate trap is not needed, since all of them have a built-in trap to provide a seal. Although all of these water closets are installed in the same way, they differ in flushing action.

1) **Washdown bowl** - This is the least expensive and simplest type of water closet. The trap is at the front of the bowl and the bowl is flushed by small streams of water running down from the rim. Some codes now prohibit use of the washdown bowl water closet.

2) **Reverse trap bowl** - This type is similar to the washdown bowl except that the trap is at the rear of the bowl, making the bowl longer. This bowl holds more water than the washdown bowl and operates more quietly.

3) **Siphon jet bowl** - This looks like the reverse trap bowl but flushes differently. The unit has a small hole in the bottom that delivers a direct jet of water into the trap. This, combined with more water from the flushing rim, starts a siphoning action with each flush.

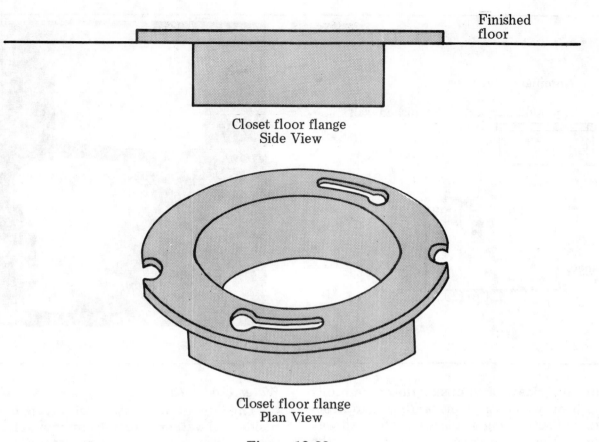

Closet floor flange
Side View

Closet floor flange
Plan View

Figure 19-22

4) Siphon action closet - This is the most efficient, the quietest, and the most expensive of the water closets. It looks like the siphon jet bowl but holds almost a full bowl of water. Almost all the inside surface of the bowl is covered when the water is at rest in the bowl.

Water closet bowls 1, 2, and 3 in Figure 19-21 are two-piece. That is, the tank is separate from the bowl. They are usually referred to as "closet combinations." Water closet 4 in Figure 19-21 is of one-piece construction.

Bowl Installation

The type of piping material used in a drainage system determines the type of closet floor flange used to hook up the water closet. Where lead stubs are used to secure the fixture to the drainage system, a brass or hard lead flange must be soldered securely to the stub. In a copper drainage system where copper stubs are used to secure the floor flange, a brass flange should be soldered securely to the copper stub. In a cast iron drainage system a cast iron flange with a lead and oakum joint secures the floor flange to the cast iron stub. In a plastic drainage system a plastic flange with a cement welded connection is used to secure the floor flange to the plastic stub.

Slip the water closet flange (Figure 19-22) over the stub and slide it down until the flange portion rests level with the finished floor.

Make up the joint as with any other joint of like material. Cast-iron stubs that extend above the top of the closet flange must be broken off with a hammer and chisel. Copper, lead, and plastic stubs that extend above the top of the closet flange may be cut flush with the top of the flange with a hacksaw.

Put two brass closet bolts in the closet flange slots provided, threaded ends up. If the water closet bowl needs four bolts as some do, place it properly on the flange and mark the spots on the floor with a pencil for the two additional (front) bolts. For wood floors, screw the closet screw into the floor, leaving the machine threads to secure the front of the bowl to the floor with closet nuts and washers. For tile or concrete floors, drill holes into the floor where marked and insert the proper size and length lead shield into the hole. Screw the closet screw into the lead shield as described above for wood floors.

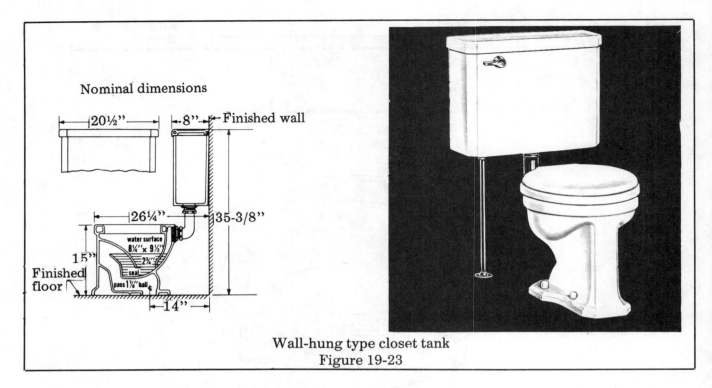

Wall-hung type closet tank
Figure 19-23

Turn the closet bowl upside down, carefully setting it on wood strips or cardboard to avoid breaking or scratching it. Place a preformed wax setting seal over the discharge opening of the horn.

Turn the water closet bowl right side up and set it on the flange with the horn projecting down into the closet stub. Guide the two closet bolts up through the bolt holes on either side of the base of the water closet. Place one hand on each side of the closet rim, rock the bowl gently, and press firmly down at the same time.

Place the closet washers and nuts on the closet bolts and tighten them. Use a two-foot level alternately across the width and then the length of the bowl. The bowl must be level in each direction. Alternate this tightening procedure from one closet nut to the other until the closet bowl is firmly set on the water closet flange. The wax seals the installation, making it gasproof and waterproof against leakage. Do not overtighten the bolts, as this could crack the base of the closet bowl.

When you are satisfied that the bowl is secured in place, grout the bowl with white cement or another suitable material to provide a watertight seal at the joint where it contacts the floor. This prevents the accumulation of odor-causing materials, avoids other unsanitary conditions, and keeps roaches and other insects away from these ideal breeding areas.

Water Closet Tanks

The closet tank is most often used on residential water closets. A minimum amount of water flushes the water closet quietly and effectively with this type of tank. Because of the longer time needed to refill the tank after each flushing, closet tanks can not be used in certain commercial installations where many flushing actions are performed in a short time. Closet tanks and bowls are manufactured of vitreous china. Specify the closet tank you need by the height at which it is to be installed above the water closet bowl.

There are two types of low closet tanks; the *wall-hung type* (Figure 19-23) and the *close coupled type* (Figure 19-24). The wall-hung type is usually found in installations made more than 25 years ago. They still exist in many older buildings. The close coupled type has been used almost exclusively for new construction in recent years. It can be installed easily and is not subject to shifting or sagging on the wall. The tank is supported entirely by the water closet bowl, to which it is attached by two tank bolts.

Water closets with tanks designed to use ballcocks should refill after each flushing and then close tight when the tank is full. The ballcock must have a refill tube reaching and turning down into the overflow tube. Water from this tube automatically restores the closet bowl water seal. An anti-siphon valve must be

Close coupled type tank closet
Figure 19-24

must be automatic after manual activation. Each tank must have an overflow tube adequate to prevent tank overflow and remove excess water at the rate it enters the tank. Consider what would happen if the flush ball is securely in place on the flush valve seat and the ballcock should become locked in an open position. The flush valve seat must be a minimum of one inch above the rim of the bowl.

The flushing device and the connection between the tank and the bowl should have enough flow capacity to allow the water to flush all surfaces of the bowl.

Flushometer Valves

The diaphragm type flushing valve (Figure 19-25), also called a flushometer, is a compact and efficient device for delivering water under pressure directly into the water closet bowl. The flushing action is quick and automatic and the amount of water delivered can be adjusted. This valve has an automatic device which shuts off the water after approximately 10 seconds. It is generally used in toilet rooms serving large numbers of people.

built into the unit to prevent contamination of the water supply system. The flush valve is operated manually but the flushing operation

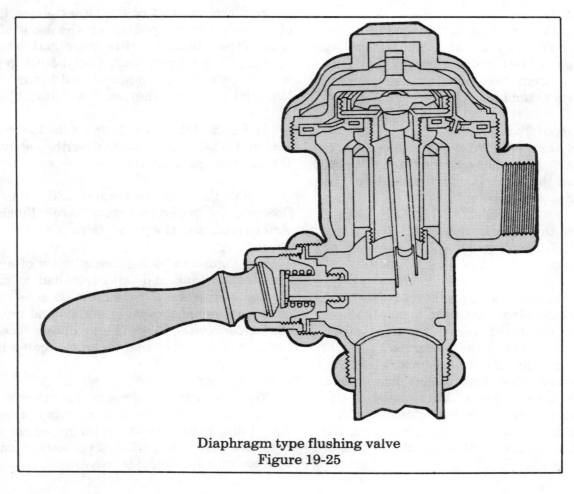

Diaphragm type flushing valve
Figure 19-25

The flushometer valve must have a vacuum breaker a minimum of 6 inches above the rim of the bowl. See Figure 16-10. This prevents back-siphonage of the bowl's contents into the water system if the water pressure should drop while there is a stoppage in the water closet bowl. The diaphragm flush valve has two chambers separated by a relief valve mounted on a rubber diaphragm. The upper chamber is directly connected to the main water supply by a small bypass, and the lower (or flushing) chamber is connected to the 1 inch water supply line. Pressure applied to the handle releases the water under pressure on the inlet side of the relief valve. The unequal pressure which results lifts the diaphragm and lets water flow into the water closet bowl. Within about 10 seconds the water forces itself around the bypass and equalizes the pressure on the two sides, forcing the diaphragm down on its seat and shutting off the flow of water. Only one water closet can be served by a single flushometer. Flushometer-type valves are also used to supply water to urinals.

Tank Installation

Closet tanks are installed after the closet bowl has been set in place. Methods of installing the two types of low tanks differ greatly. Both tanks arrive from the manufacturer with all interior parts assembled.

Closed Coupled Tanks

1) Place the tank gasket over the portion of the flush valve body that extends through to the underside of the tank. Coat the outside of the gasket with pipe compound.

2) Place the tank carefully on the water closet bowl. The tank gasket will fit the opening of the inlet to the closet bowl.

3) Place rubber sealing washers on the tank bolts. Insert the tank bolts from inside the tank through the preformed holes in the tank so that they will pass through the preformed holes in the closet bowl. Install the washers and nuts and tighten them until the closet tank is drawn down firmly on the gasket against the closet bowl. Use a level, as the nuts should be tightened so that the tank is horizontal. Do not overtighten the nuts, as this can crack either the closet tank or the closet bowl.

4) Connect the water supply tube to the inlet side of the ballcock valve that extends through the underside of the tank. The water supply tube is assembled and connected as described previously for lavatories.

5) Fill the tank with water and adjust the float rod for the correct water level. Flush the water closet and check all connections for leaks.

Wall-Hung Closet Tanks

1) A mounting board for this type of tank should have been installed during rough carpentry work. The mounting board is usually cut from a 2 x 4 stud and should finish flush with the face of the partition studs. When the closet tank is ready to be installed, two wood screws approximately 1½ inches long are screwed through preformed openings at the rear of the tank. The screws anchor the tank firmly to the mounting board. This places the weight of the tank and its contents on the mounting board and not on the closet elbow (flush ell), which could cause these joints to leak.

2) Place one end of the closet elbow (flush ell) into the inlet opening on the back of the water closet bowl and the other end into the opening in the closet tank. Coat both slip joint washers with pipe compound and tighten both slip joint nuts until they form watertight joints.

3) Connect the water supply tube to the inlet side of the ballcock valve, as described above for the close coupled tank.

4) Fill the tank with water and adjust the float rod for the correct water level. Flush the water closet and check for leaks.

See Figure 19-26 for a detail view of a wall-hung closet tank and bowl installed in place.

Seats for water closets must be constructed of smooth nonabsorbent materials and must fit the water closet bowl. For example, do not install a round front seat on an elongated bowl.

Installation of Urinals

The two general designs for urinals that meet most code standards today are the wall-hung type and the floor mounted stall type. They are constructed of vitreous china and operate with flushometer valves.

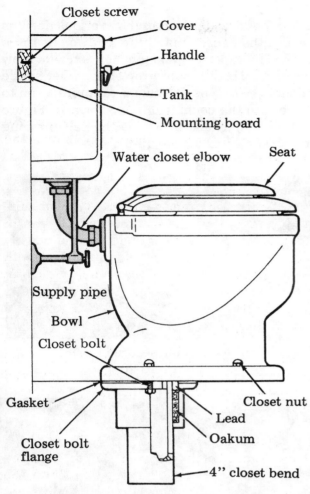

- Closet screw
- Cover
- Handle
- Tank
- Mounting board
- Water closet elbow
- Seat
- Supply pipe
- Bowl
- Closet bolt
- Gasket
- Closet nut
- Lead
- Oakum
- Closet bolt flange
- 4" closet bend

Detail of water closet and wall-hung
tank in place
Figure 19-26

Wall-Hung Urinals

A wall-hung urinal must be rigidly supported by a concealed metal carrier or other approved backing so that no strain is transmitted to the pipe connection. The joint between the urinal and the finished wall surface must be grouted with white cement or other suitable material that can provide a watertight seal.

Standard mounting requires that the waste piping be a minimum of 2 inches in diameter and the opening 21 inches from finished floor to the center of the waste pipe.

As with the water closet, most wall-hung urinals are designed with an integral trap.

Floor-Mounted Stall Urinals

A floor-mounted stall urinal must be recessed slightly below the finished floor to provide drainage. Its weight is transmitted to

the subflooring and does not require any other support. The back is recessed slightly into the finished wall and therefore must be grouted to provide a watertight joint.

The waste piping must be no less than 2 inches in diameter, and the opening should be 8½ inches from the finished wall to the center of the waste pipe.

A 2-inch trap must be installed in the waste line directly beneath the bottom of the urinal. The waste opening should be provided with a beehive-type strainer.

Flushometer-type valves are used with both types of urinals. These valves generally require a minimum ¾-inch water line to supply their water needs. (Some types may use a ½-inch water supply opening, but this size is not accepted by some codes.) The flushometer must complete the normal flushing cycle automatically after it is manually activated. It should deliver water at a rate that will flush all surfaces of the urinal. The valve must open fully and close tightly at normal water pressure. The flushometer must also have some means of regulating water flow. Only one urinal can be served by a single flushometer. See Chapter 18 for the roughing-in dimensions of wall-hung and floor-mounted urinals.

Laundry Sinks

Laundry sinks or wash tubs are usually placed in the utility area or room near the clothes washing machine. This is both convenient and economical because both fixtures can use the same hot and cold water and waste pipe.

Laundry sinks may be made of concrete, enameled pressed steel, enameled cast-iron, or vitreous china. They may be of the cabinet type, but most often are supported by a metal frame. They are made in single and double-compartment styles. Figure 19-27 shows a countertop

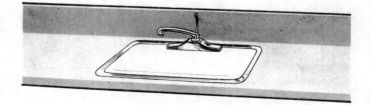

Laundry sink supported by countertop
Figure 19-27

installation. Figure 19-28 illustrates two laundry sinks supported by a metal stand.

The waste pipe for laundry sinks or tubs must be a minimum of 1½ inches in diameter. Traps, tailpieces, and continuous waste pipe must be a minimum of 1½ inches o.d. Each compartment in a laundry sink should have a waste outlet with a suitable stopper for retaining water.

A double-compartment sink with two adjacent laundry sinks can use waste outlet piping connected by a continuous waste (see Figure 19-28). This provides a single outlet and trap to connect to the drainage system. The faucet is usually of the swing spout type and may or may not have threads for attaching a hose.

steel. Sinks are manufactured in various designs and lengths. There may be one, two, or three bowls. The most popular kitchen sink for new construction is the countertop sink, either single or double bowl type. A sink may require a rim to secure it to the countertop, as shown in Figure 19-29. Another sink design is the self-rimming type as shown in Figure 19-30.

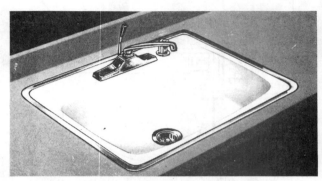

Countertop sink secured by rim
Figure 19-29

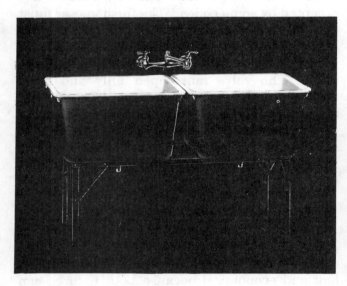

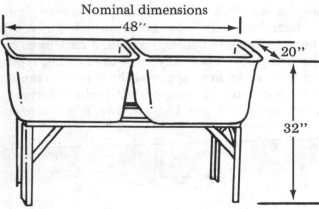

Nominal dimensions

Double laundry sink supported
by cast iron stand
Figure 19-28

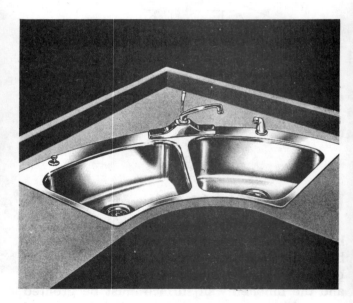

Nominal dimensions

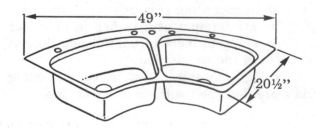

Corner stainless steel double bowl
self-rimming countertop sink
Figure 19-30

Kitchen Sinks

Kitchen sinks are usually made of enameled cast iron, enameled pressed steel, and stainless

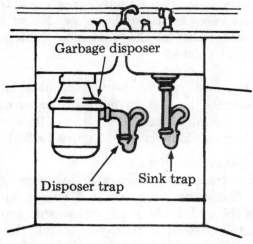

Food waste disposer installation
Figure 19-31

The waste pipe for any sink must be a minimum of 1½ inches in diameter. Traps, tail pieces, and continuous waste pipe must be a minimum of 1½ inches o.d. A double-compartment sink (as in Figure 19-30) may have a continuous waste connecting the two bowls together. The continuous waste may be of the end outlet or the center outlet type with a single trap. See Figure 7-6. The faucet may or may not have a spray for rinsing purposes.

Kitchen sinks are installed and secured to the countertop in the same way as countertop lavatories described previously.

Domestic sinks need a waste opening at least three and one half inches in diameter. This is necessary to fit the standard size sink basket strainer as well as to provide an opening large enough for a waste disposal unit to be installed if one is desired.

Food Waste [Garbage] Disposers

Food waste disposers are generally installed in all new living units constructed today unless prohibited by local authority. Contrary to belief, food waste disposers will not clog up the drainage system if manufacturers' instructions are followed. Instead, they help keep it unclogged because of the high pressure under which they force food waste into the sink waste line.

Some codes require that in two-compartment sinks the disposer discharge waste through a separate trap and waste line. In such cases two traps and two waste lines must be installed, as shown in Figure 19-31. Some codes permit the

disposer and second sink compartment to use one trap and one waste line. Where two traps and waste lines are required, use a hi-lo fitting 2 inches in diameter with two 1½ inch double vertical tappings not more than 6 inches apart. See Figure 19-32.

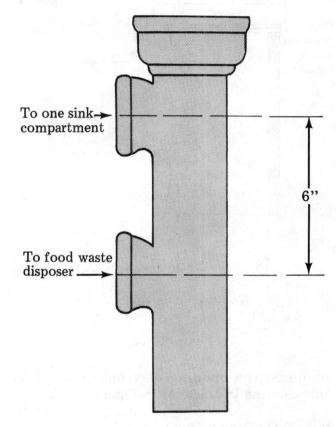

Figure 19-32

Food waste disposers may be installed on single compartment sinks using one trap and waste line. There are no special requirements to consider.

A food waste disposer can be installed on a double compartment sink in an existing home if a second waste opening is not available. But a special directional tee must be used to channel the flushed garbage away from the other sink compartment. See the directional tee in Figure 19-33. A 1½-inch trap is required for a food waste disposer.

Under-Counter Dishwasher

Under-counter dishwashers can be installed with either side against the wall at the end section of a kitchen cabinet (right or left), but never directly under a sink. In new construction the cabinet opening for a dishwasher is provided when kitchen cabinets are installed. For existing

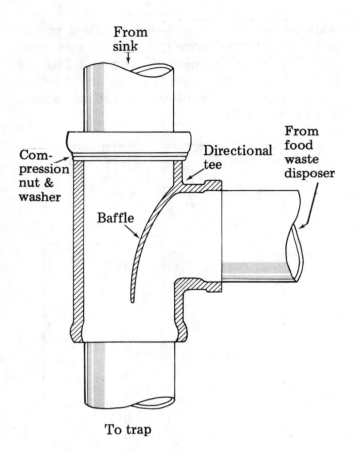

From sink

Compression nut & washer

Directional tee

From food waste disposer

Baffle

To trap

Figure 19-33

cabinets, the opening may be installed by following the instructions in Figure 19-34.

Hot Water Supply Line

The hot water supply line to the inlet valve of the dishwasher may be of any approved water piping material. However, hard or soft copper with brass or copper fittings are generally preferred because they are easy to work with. Do not use rubber hose or connectors. The constant pressure and high temperature will in time rupture this material. The drain hose supplied with the unit must not be used as a water supply or coupler.

Follow the guidelines below when installing the hot water supply line:

- An accessible hand shut-off valve must be installed in the adjacent sink cabinet.

- The water supply line must not be smaller than 3/8 inch I.D.

- A compression fitting or union must be

used to make the final connection of the inlet valve to the dishwasher. Never use a solder fitting; the heat may damage the inlet valve.

- For best results the water heater should supply the dishwasher with water heated to approximately 150°. (See "Hot Water Supply Connection," Figure 19-35.)

Dishwasher Installation

The drain hose from a dishwasher with a pump discharge must rise to a height equal to that of the underside of the dishwasher top. It is absolutely essential that you install the high loop fitting provided with each dishwasher drain hose. This prevents back-ups into the dishwasher if the sink should become clogged. Install the drain hose so that it is not kinked. Securely clamp the hose to the underside of the countertop.

Some plumbing codes require that an air gap fitting be installed in the drain line. When this is the case, the air gap assembly is mounted on the sink or countertop as shown in Figure 19-36. The air gap assembly, when required by code, must be mounted above the sink overflow rim. One hole 1¼ inches wide must be drilled in the sink or countertop and the drain gap assembly secured through it with the collar and nut at the proper height so that the cover can snap into position over the assembly.

The dishwasher pump and the ½-inch outside diameter inlet tube of the air gap assembly should be connected with the hose and clamp that comes with the dishwasher. Install the drain air gap assembly as close to the dishwasher as possible so that the dishwasher hose can be used. Do not locate a dishwasher more than 5 feet from the sink waste connection. (This distance may vary with some codes).

If a food disposal unit is installed in a sink, the waste from the dishwasher must connect to the opening provided in the body of the food disposer. See Figure 19-37. If a food disposal unit is not used, the waste from the dishwasher must connect to a fitting called a "dishwasher Y branch" as shown in Figure 19-37 and Figure 19-38.

Water Heaters

There are two types of water heaters commonly used in single family or small dwelling units: 1) the electric water heater and

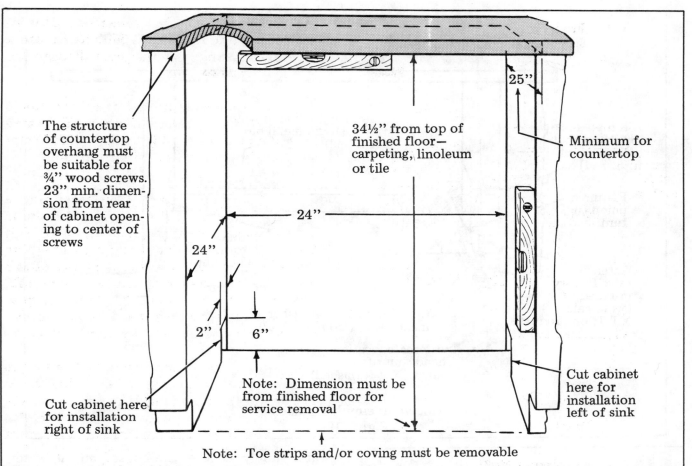

The structure of countertop overhang must be suitable for ¾" wood screws. 23" min. dimension from rear of cabinet opening to center of screws

34½" from top of finished floor— carpeting, linoleum or tile

Minimum for countertop

24"

24"

25"

2"

6"

Cut cabinet here for installation right of sink

Cut cabinet here for installation left of sink

Note: Dimension must be from finished floor for service removal

Note: Toe strips and/or coving must be removable

Note: The cabinet opening must be square, level, and proper size. The dishwasher must not be installed or secured with the frame distorted or twisted.

This dishwasher is not provided with complete enclosure. Enclosures for sides and back to be provided at time of installation.

Cabinet opening
Figure 19-34

2) the gas-fired water heater. The electric heater is the more popular and more often used.

There are generally three designs of electric water heater: 1) the upright model (Figure 19-39), 2) the utility model (Figure 19-40), and 3) the undercounter or table-top model (Figure 19-41).

Water Heater Installation

Select a level location accessible to water lines and the power supply. Do not locate the heater where water lines could be subject to freezing. To prevent heat loss through the pipes, try to install the heater in the center of the area with the greatest hot water use. Locate the heater so that access panels and drain valves are accessible.

Before starting the installation, close the main water supply valve, open a water faucet to

relieve the house pressure, and then close the faucet.

When the code requires a check valve in the cold water inlet line to the heater, a pressure relief valve must be installed between the check valve and the heater. However, installing a pressure relief valve in this location does not eliminate the requirement for a temperature-pressure relief valve in one of the locations shown in Figure 14-3.

The relief pipe should terminate near the floor drain, where permitted by code, or other suitable location not subject to blocking or freezing. (See Chapter 14.) Do not thread, plug, cap, or use the relief pipe for any other purpose. Leave an air gap of approximately 6 inches between the end of the relief pipe and the floor drain or other approved surfaces.

After installing water lines, open the main

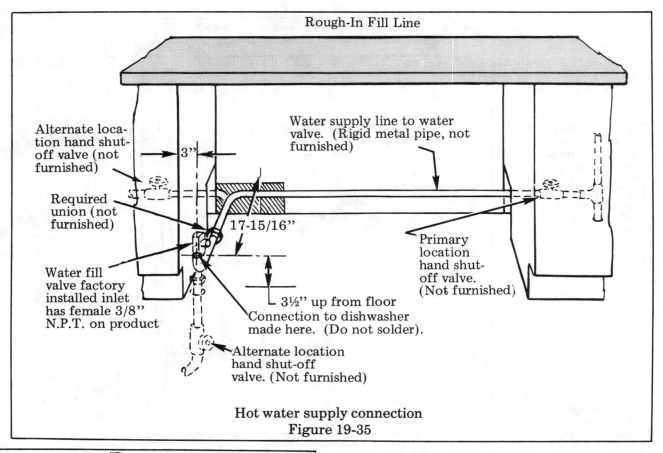

Hot water supply connection
Figure 19-35

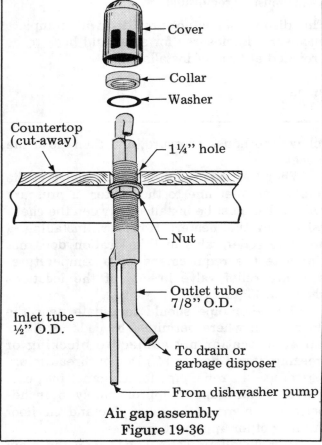

Air gap assembly
Figure 19-36

water supply valve to the house and the water heater control valve and fill the heater. Open several hot water faucets to allow trapped air in the heater tank to escape from the system while the heater is filling. Do not turn on the electrical current (or gas supply) to the water heater until the tank has been completely filled with water. The heating elements are damaged if heated when not immersed in water. When water passes through the faucets, close them and check for possible leaks in the system.

Electrical Connections

Before any electrical connections are made, or before the gas is turned on and lighted, be certain that the heater is full of water and that the valve in the cold water supply line is open. It is recommended that an electrical contractor make the final electrical connections. Otherwise, the warranty for the water heater may be in jeopardy. For a gas heater, a gas supplier must light the heater and make final adjustments.

See Figures 19-39, 19-40, 19-41, and 19-42 as well as Chapter 14 for further information on piping, electric and gas water heaters.

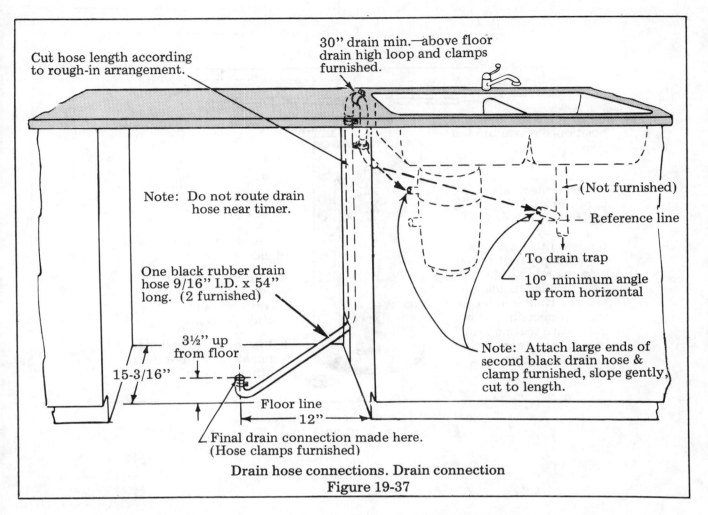

Cut hose length according
to rough-in arrangement.

30" drain min.—above floor
drain high loop and clamps
furnished.

Note: Do not route drain
hose near timer.

One black rubber drain
hose 9/16" I.D. x 54"
long. (2 furnished)

(Not furnished)

— Reference line

To drain trap

10° minimum angle
up from horizontal

3½" up
from floor

15-3/16"

Note: Attach large ends of
second black drain hose &
clamp furnished, slope gently,
cut to length.

Floor line
12"

Final drain connection made here.
(Hose clamps furnished)

Drain hose connections. Drain connection
Figure 19-37

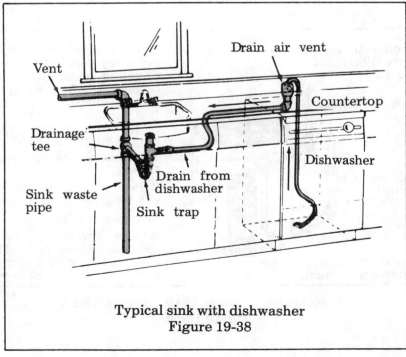

Vent

Drain air vent

Countertop

Drainage
tee

Dishwasher

Sink waste
pipe

Drain from
dishwasher

Sink trap

Typical sink with dishwasher
Figure 19-38

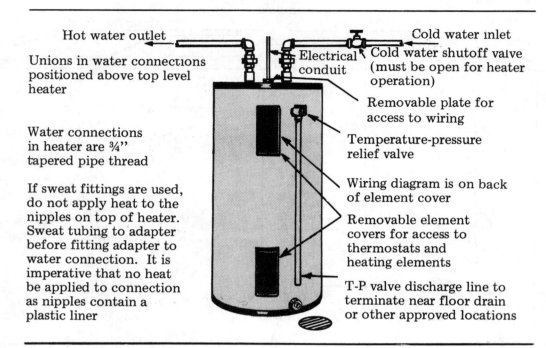

Hot water outlet

Unions in water connections positioned above top level heater

Water connections in heater are ¾" tapered pipe thread

If sweat fittings are used, do not apply heat to the nipples on top of heater. Sweat tubing to adapter before fitting adapter to water connection. It is imperative that no heat be applied to connection as nipples contain a plastic liner

Cold water inlet

Electrical conduit

Cold water shutoff valve (must be open for heater operation)

Removable plate for access to wiring

Temperature-pressure relief valve

Wiring diagram is on back of element cover

Removable element covers for access to thermostats and heating elements

T-P valve discharge line to terminate near floor drain or other approved locations

Typical electric water heater connections
Figure 19-39

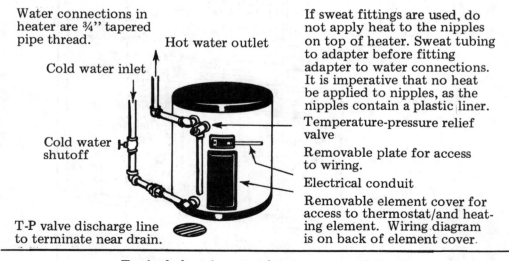

Water connections in heater are ¾" tapered pipe thread.

Cold water inlet

Hot water outlet

Cold water shutoff

T-P valve discharge line to terminate near drain.

If sweat fittings are used, do not apply heat to the nipples on top of heater. Sweat tubing to adapter before fitting adapter to water connections. It is imperative that no heat be applied to nipples, as the nipples contain a plastic liner.

Temperature-pressure relief valve

Removable plate for access to wiring.

Electrical conduit

Removable element cover for access to thermostat/and heating element. Wiring diagram is on back of element cover.

Typical electric water heater connections
Figure 19-40

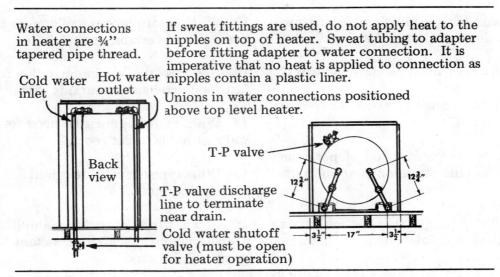

Water connections in heater are ¾" tapered pipe thread.

If sweat fittings are used, do not apply heat to the nipples on top of heater. Sweat tubing to adapter before fitting adapter to water connection. It is imperative that no heat is applied to connection as nipples contain a plastic liner.

Cold water inlet   Hot water outlet

Unions in water connections positioned above top level heater.

Back view

T-P valve

T-P valve discharge line to terminate near drain.

Cold water shutoff valve (must be open for heater operation)

$12\frac{3}{4}$" $12\frac{3}{4}$"

$3\frac{1}{2}$" 17" $3\frac{1}{2}$"

Typical electric water heater connections
Figure 19-41

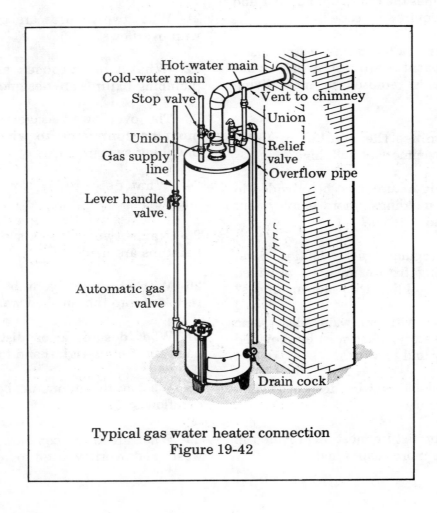

Hot-water main
Cold-water main
Stop valve
Union
Gas supply line
Lever handle valve
Automatic gas valve
Vent to chimney
Union
Relief valve
Overflow pipe
Drain cock

Typical gas water heater connection
Figure 19-42

Questions

1- Why have well-recognized plumbing fixture standards been developed over the years?

2- Plumbing fixtures must be free of what?

3- Plumbing fixtures must be manufactured of what type of material?

4- When fixtures are constructed of pervious materials, what specific condition is not permitted?

5- What potentially dangerous situation exists when bathrooms do not have adequate lighting or ventilation?

6- What are the minimum fixture requirements for a single family residence?

7- What determines the minimum number and type of fixtures required by code?

8- Besides the usual kitchen sink, water closet, lavatory, and bathtub (or shower), what additional fixture is required in most living units?

9- What determines the number of toilet facilities in light commercial buildings?

10- What plumbing fixture must be provided in light commercial buildings where employment does not exceed 45 persons?

11- Besides the regular toilet facilities, what additional plumbing fixture must be provided for a small office building with eleven offices?

12- In a toilet room with one water closet, one urinal, and one lavatory, what must be provided to ensure decency and privacy?

13- How must toilet bowls be designed when they are for public use?

14- What plumbing fixture must be provided in a place employing more than 5 males?

15- What determines the minimum toilet facilities in small restaurants?

16- How are the minimum toilet facilities for fast food drive-in restaurants calculated?

17- What material must be used for floors and walls of public toilet rooms?

18- What type of seat is required for toilet bowls serving the public?

19- In what type of eating establishment must a dishwashing machine or suitable 3-compartment sink be used?

20- What must be provided for employees' use in establishments where food is prepared and served to the public?

21- What two fixtures are generally provided with overflows?

22- What two purposes are served when plumbing fixtures are designed with overflows?

23- The overflow passageway from a fixture must be connected to what portion of the plumbing system?

24- What determines the size of a fixture strainer?

25- Name two materials of which modern bathtubs are made.

26- With what must bathtubs be provided when recessed into the finished walls?

27- What design factor distinguishes a left-hand tub from a right-hand tub?

28- What is the minimum bathtub waste and overflow size?

29- What are three common types of bathtub waste and overflow used today?

30- Describe the operation of a trip waste and overflow.

31- What must be replaced when the stopper or cylinder of a waste and overflow no longer remains in an open or closed position?

32- What type of shower is found in most homes?

33- How is water from a tub transferred to the shower head when there is a diverter spout?

34- What is the main advantage of having a shower stall?

35- What are the two general types of shower stalls used in today's plumbing?

36- Why is accuracy very important in roughing-in the waste opening for a shower stall?

37- Give two advantages of a tiled shower over a stall shower.

38- What is the most important requirement for a shower?

39- What is the minimum size waste outlet for a shower compartment?

40- What is the minimum floor space required for any shower compartment?

41- What is the minimum weight per square foot required by most codes for lead shower pans?

42- What must be provided by the carpenter before a shower pan can be installed?

43- What are the design requirements for shower strainers?

44- What protection must be given shower pans installed on concrete floors?

45- In securing the shower pan, what is the maximum distance from the top of it that nails or screws may be used?

46- How high should the sides of a shower pan extend above the finished curb?

47- When during construction of the building should a shower pan be prepared for inspection?

48- When may shower pans be omitted?

49- Shower rods are generally installed how high above the finished floor?

50- Shower doors are generally manufactured from what materials?

51- What two common materials are used in manufacturing lavatories?

52- What two types of lavatories are most often installed?

53- What is the standard height from the finished floor to the overflow rim of a lavatory?

54- How does the design of the lavatory determine the type of faucet to be used?

55- What is the disadvantage of having a single hot and cold water faucet on a lavatory?

56- Why are self-closing faucets sometimes used on commercial lavatories?

57- What are the two types of drain assemblies used on lavatories?

58- Why should lavatory faucets and waste assemblies be installed before the lavatory is in place?

59- How should wall-hung lavatories be finished at the wall contact point?

60- Name three types of flushing action for today's water closets.

61- What part of a combination water closet is installed first?

62- What type of closet flange is used in a plastic drainage system?

63- What is used to make a tight joint between a closet bowl outlet and the building waste pipe opening?

64- Why should a level be used when installing a closet bowl and tank?

65- Why are tank type water closets seldom used in public toilet rooms?

66- What are the two types of low closet tanks?

67- What does a refill tube accomplish?

68- What is the purpose of an overflow tube in a water closet tank?

69- Why are flushometers generally required in toilet rooms serving large numbers of people?

70- With which other commercially used fixture is the flushometer type valve used?

71- What should be done after a water closet is completely installed?

72- Why is it necessary to adjust the float rod in a water closet tank?

73- Toilet seats must be made of what kind of material?

74- What are the two most common urinal designs?

75- What is the minimum size waste pipe for urinals?

76- Why do most wall-hung urinals *not* require a trap?

77- Why must stall urinals be recessed slightly below the finished floor?

78- At what location in the waste pipe must a trap be installed to serve a stall urinal?

79- Why must a flushometer for a urinal deliver water at a certain rate?

80- How may the two compartments of a laundry sink be connected so that one outlet and one trap may be used?

81- What size trap must be used on a kitchen sink waste pipe?

82- What feature of a sink faucet is optional?

83- What is the minimum size waste opening for a domestic sink?

84- How is the drainage system affected when a food waste disposer is used according to the manufacturer's instructions?

85- What do some codes require when a waste disposer is newly installed in a two-compartment sink?

86- What fitting must be used when a waste disposer is installed on an existing two-compartment sink?

87- What size trap must be used to connect the waste pipe to a food disposer?

88- On which side of the kitchen sink may an undercounter dishwasher be installed?

89- Is a dishwasher served by both hot and cold water? Why or why not?

90- Why should a rubber hose *not* be used to connect the water supply to a dishwasher?

91- What water piping materials are generally used to supply water to a dishwasher?

92- Why should you not use a solder fitting on the inlet valve of a dishwasher?

93- Why is it essential to install the high loop fitting on the drain hose?

94- Where must the high loop fitting be installed?

95- When a plumbing code requires that an air gap fitting be used, where must it be installed?

96- By most code standards, at what maximum distance may a dishwasher be installed from the sink waste connection?

97- Where a food disposal unit is installed in a sink, to what must the waste from the dishwasher be connected?

98- What are the two most common types of water heaters installed in residential units?

99- Name three important criteria used in selecting the location for a water heater.

Maintenance Of Plumbing Systems

20

Clogged Drains

The major plumbing maintenance problem confronting most homeowners is clogged drains. Although there may be an occasional defective drainage system, the most common cause of clogged drains is foreign objects in the drainage system. Foreign matter which finds its way into the drainage system usually enters through the kitchen sink or the bathroom water closet.

The Kitchen Sink Some of the materials most detrimental to the kitchen sink drain pipe are grease, cooking fats, butter, gravy, and coffee grounds. These should *not* be disposed of through the kitchen drain but should be discarded along with the regular household garbage.

The Bathroom Some of the more common objects that cause stoppages affecting the entire building drainage system are combs, pencils, toys, baby diapers, paper tissues, sanitary napkins, overrim-type bowl deodorants, bath oils, and hair.

Stoppages occur, of course, even with normal use. But if the plumbing system is not abused it can be expected to give many years of carefree service.

Identifying Types and Locations of Stoppages

Figure 20-1 shows a typical two-bath floor plan with kitchen and utility room. In this plan the points in the drainage system where stoppages are most likely to occur are circled in the drawings in the following figures. It is safe to say that stoppages in the plumbing system very seldom occur in straight horizontal runs of piping, vertical drops from fixtures, and fixture traps (except those in water closets). Stoppages usually occur where two pipes are joined together with a fitting for a change of direction. Under the code, all fixture traps must be self

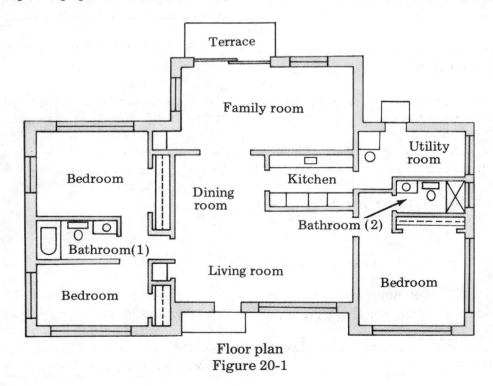

Floor plan
Figure 20-1

cleaning. (Interceptor traps are the only traps which are *not* self cleaning.)

When a complete stoppage exists in a drainage system connected to a public sewage collection system, it is wise to check first if the total system is functioning properly. A system may become *surcharged* (overloaded) at peak periods, lift station pumps may break down, or there may be a power failure. If any of these malfunctions occurs, the public sewage system can back up into the private sewers.

The first step in locating the source of trouble is to check if the next-door neighbor is also experiencing any drainage difficulties. If not, assume that there is a complete stoppage rather than a malfunction of the total system.

In Figure 20-2, a cleanout is located on the building sewer immediately outside the building wall. Most codes require that an accessible two-way cleanout be located within 5 feet of the building and another at the property line. The minimum size of this cleanout is usually 4 inches. Use a 14-inch pipe wrench to remove the cleanout plug. This plug may be "frozen" if it has not been used previously. If it is, place as much pressure as possible on the wrench and rap the handle hard with a hammer to loosen the plug. A soldering torch may be used on a *brass* plug if it still does not release from the cleanout body. The heat will probably unfreeze the threads. The last resort is to cut the plug out of the cleanout body with a hammer and sharp chisel.

When replacing the cleanout plug use pipe compound on the threads and tighten the plug snugly into the cleanout body.

Using Cables to Free Stoppages

If an electric sewer cable is not available to release the stoppage, one of the two types of cable should do the job adequately. The first is a flat 100-foot steel tape approximately ¾-inch wide with a ball on one end. The second is a ½-inch flexible 75 or 100 foot steel spring cable in a cylinder. The cable can be turned by using a handle attached to the cylinder.

Where the sewer pipe is installed fairly straight and there is plenty of space to work in, use the flat steel tape cable. The strength of the flat steel tape should easily break up the stoppage so that it can be flushed away.

Another type of complete stoppage, illustrated in Figure 20-2, may occur when a building's drainage system is connected to a septic tank. To help determine its location, use the flexible spring cable. If the stoppage is in one of the offset fittings, this cable should be strong enough to dislodge the obstruction. If the stoppage is in the inlet tee of the septic tank, the cable's flexibility allows it to make the 90° turn in the tank inlet tee. Insert the cable far enough to go through the inlet tee, then remove it. If the stoppage still exists, assume that the problem is in the septic tank or the drainfield. The tank may be in need of cleaning or the drainfield may need replacing, or both. If this is true, call a septic tank specialist.

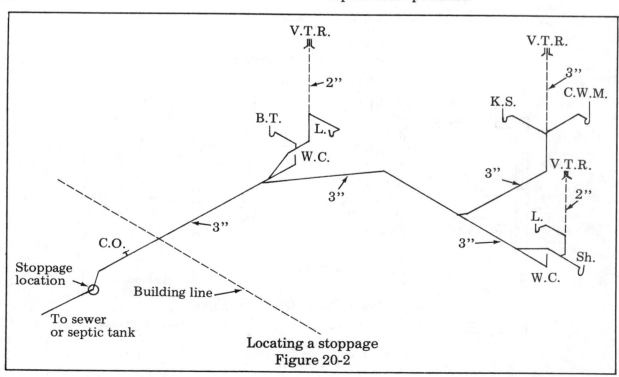

Locating a stoppage
Figure 20-2

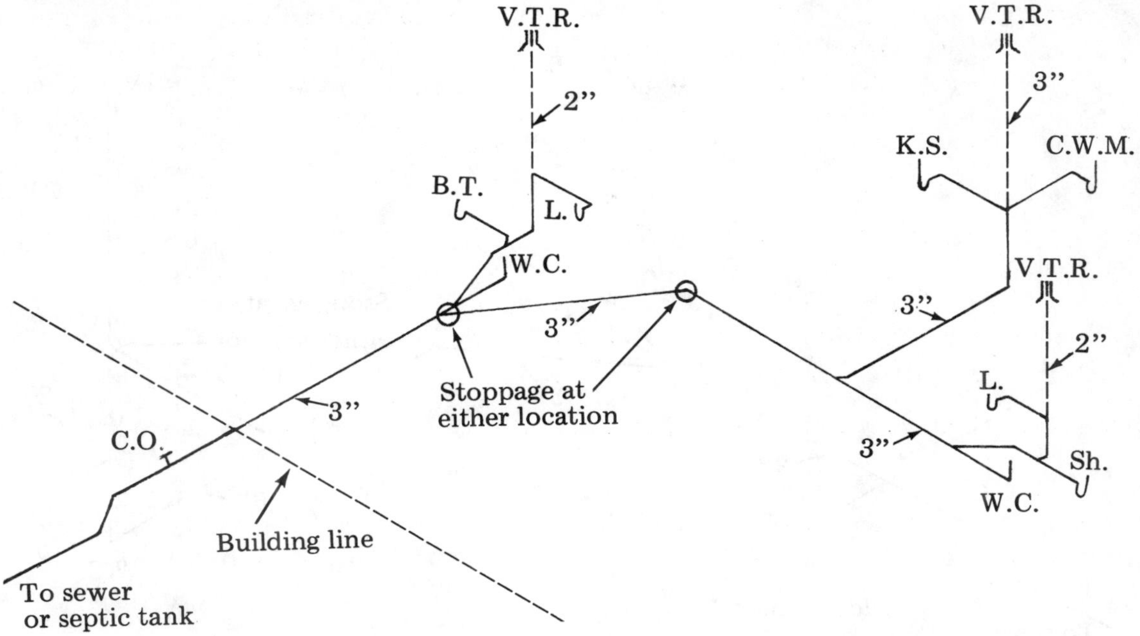

Locating a stoppage
Figure 20-3

Figure 20-3 shows a partial building drain stoppage. Bathroom 1 is not affected. Bathroom 2, the kitchen sink, and the utility room are affected. The location of the stoppage would have to be either in the fitting (1/8 bend) for a change in direction or where the building drain pipe connects with a wye (Y) fitting into the main drain pipe below bathroom 1. Some codes do not require that a cleanout be extended if it is accessible from the outside. The vent pipes extending above the roof are considered adequate for the entry of a cleaning cable. If the cable is to go through the vent pipes (see Figure 20-3), the vent serving the sink and washing machine would be the right entry point.

Here, the cable must pass through two 90° offsets and possibly two 45° offsets. An electric sewer cable would be ideal for this type of stoppage. But if one is not available, use a flexible cable in a cylinder. The cable should move fairly easily through the pipe and fittings until it reaches the blockage.

When you make contact with the obstruction, tighten the thumb screw on the cylinder outlet to grip the cable firmly. Leave about a foot of excess cable above the vent pipe. Do the same if a cleanout fitting is available for use. Keep firm pressure on the cable as the cylinder is turned very forcefully. This should break up the blockage.

Before retrieving the cable fill the sink with hot water and flush the line thoroughly. Do this

several times. If the water drains away freely remove the cable.

Figure 20-4 shows another common partial building waste pipe stoppage. Here the kitchen sink and clothes washing machine are affected. The blockage may be located in the sweep at the base of the stack or in the horizontal waste pipe where it connects with a combination into the building drain.

Figure 20-5 illustrates still another type of blockage that affects the shower and lavatory in the number 2 bathroom. If the ground-level cleanout is not available, use the vent pipe opening above the roof to locate both types of stoppages. Again, first preference would be an electric sewer cable. If one is not available, use a cylinder sewer cable. In the example illustrated in Figure 20-5 the cable should be 3/8-inch in diameter, as the pipes are smaller and the blockage is of a lighter substance. The procedure is the same as described above for cylinder-type cables. When the obstruction is broken up, flush the waste pipes with hot water before retrieving the cable. If the water flows away freely, retrieve the cable.

Clogged Sinks

One type of stoppage that often occurs in the fixture branch (drain) is shown in Figure 20-6 and 20-7. The sink will simply no longer drain. The stoppage usually occurs at the junction of the fixture drain pipe and the waste and vent pipe.

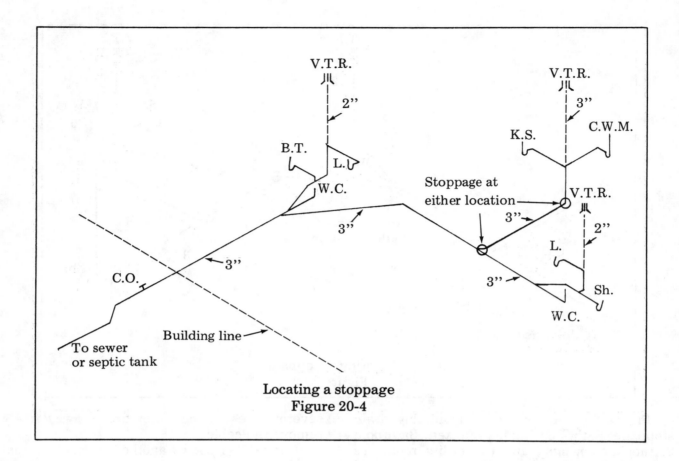

Locating a stoppage
Figure 20-4

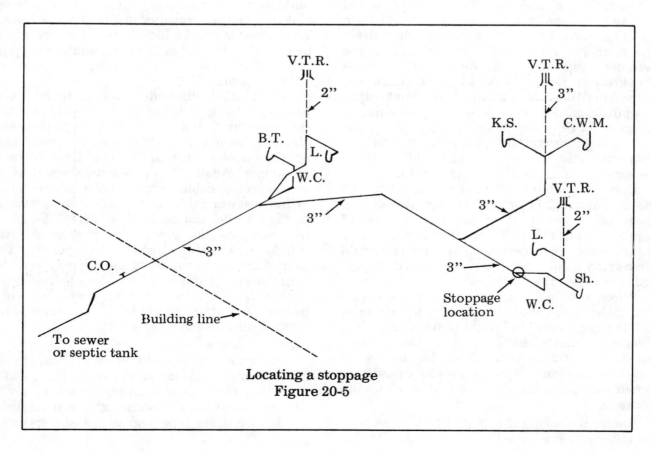

Locating a stoppage
Figure 20-5

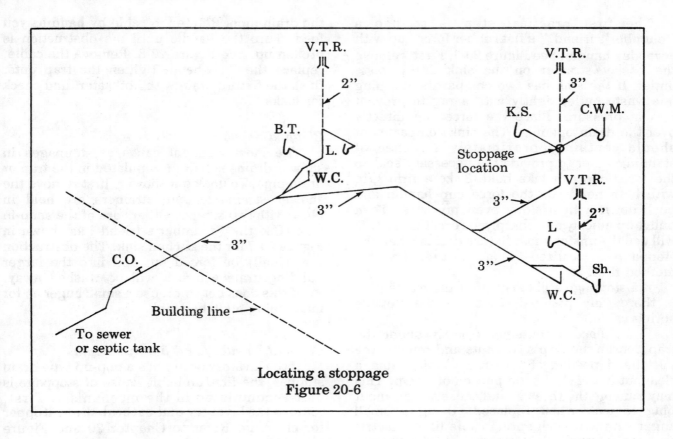

Locating a stoppage
Figure 20-6

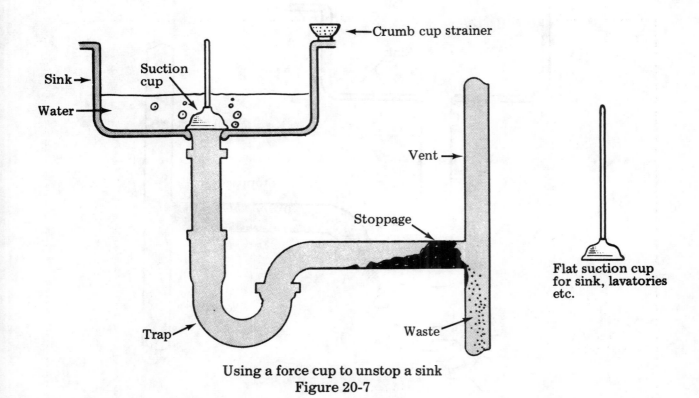

Using a force cup to unstop a sink
Figure 20-7

The first practical step is to use a "plumber's friend," a flat rubber force cup with a wooden handle. See Figure 20-7. First, remove the basket strainer on the sink drain waste outlet. If the sink has two compartments, plug one waste outlet tightly with a rag to prevent loss of pressure. Place the force cup directly over the drain opening. The sink compartment should contain approximately 4 inches of standing water to provide the necessary seal so the force cup can take hold. Take a firm grip with both hands on the force cup handle and push down with a slow, even pressure. Then pull it up quickly and sharply several times. This will usually unstop most fixture drains when the stoppage is localized to one fixture. Try this method first.

If a stoppage still exists after the waste pipe is thoroughly plunged, then try the flexible spring cable.

First, place a container directly under the trap, loosen the trap's two nuts and remove the "J" bend portion of the trap. Avoid using a cleanout in the "J" bend portion of a trap. This may damage the threads and cause the cleanout plug to leak when replaced. Using a small auger with a ¼-inch spring cable (if an electric auger is not available), feed the cable slowly into the drain pipe. Rotate the cable by hand as you feed. Turn the handle until the obstruction is broken up. See Figure 20-8. Remove the cable, replace the "J" bend, tighten the trap nuts, flush the fixture drain with hot water, and check for leaks.

Clogged Showers

The most frequent cause of stoppages in shower drains is hair accumulated in the trap or drain pipe. To unclog a shower, first remove the shower strainer. Some strainers are held in place with two screws. Others are of the snap-in type. Use the "plumber's friend" as shown in Figure 20-7 for the kitchen sink. The obstruction can usually be forced out and into the larger building drain where it will be washed away.

If this does not work, use a small auger as for unstopping a kitchen sink.

Clogged Lavatory and Bathtub Drains

If the lavatory or tub has a pop-up type drain stopper, the first probable cause of stoppage is hair accumulated in the mechanism. First, remove the lavatory and bathtub drain stopper for cleaning. Refer to Chapter 20 and Figure 20-9.

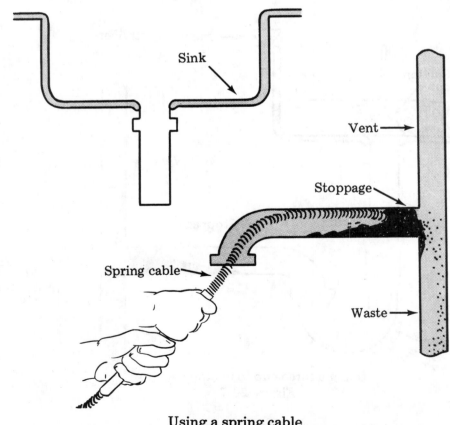

Using a spring cable
Figure 20-8

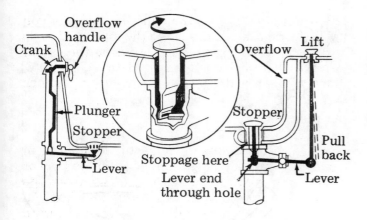

Typical tub and lavatory stoppers
Figure 20-9

When the drain stopper is not at fault, the next most probable cause is a stoppage in the trap or waste line. Use a "plumber's friend" as shown in Figure 20-7. This should be effective in forcing the obstruction out. Be sure to tightly seal the overflow of the lavatory or tub with a rag before using the force cup. When the pumping of the force cup reduces the water level in the bathtub, add more water to keep the force cup covered until the flow through the drain becomes normal.

In tubs with a drum trap, the entire cover screws out of the body. A fine penetrating oil applied all around the cover edge will help loosen frozen threads. You may have to use a cold chisel and hammer to free the cover. See Figure 20-10. If the trap is not clogged, use a small auger as for unstopping a kitchen sink drain line.

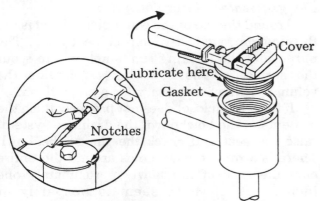

Removing a drum trap cover
Figure 20-10

Clogged Water Closets

The water closet also has its share of stoppages. The water closet trap is a single unit with the bowl. The passageway of the trap is designed to pass "acceptable" materials no

larger than two inches. The only way a stoppage can occur is if someone puts something in the fixture that is larger than the trap is designed to accept. Find the cause of this type of stoppage before you try to unclog it.

There are two tools normally used to unstop a water closet bowl: the closet auger (Figure 20-11) and the force cup especially designed for

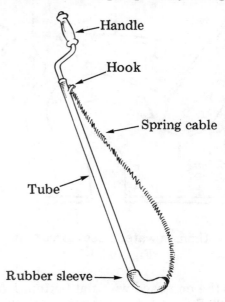

Closet auger
Figure 20-11

a closet bowl (Figure 20-12). Determine whether the cause of the blockage is something that will result in additional stoppage if it is forced into the drainage system. If not, use the closet force cup as shown in Figure 20-12. But if the blockage is caused by cloth or some other material that should be retrieved and not forced into the drainage system, use the closet auger as shown in Figure 20-13. If the blockage is caused by a comb, pencil or other object that cannot be retrieved, try forcing it through the trap into the drainage system using the force cup and the closet auger alternately. It may flow with the sewage to the treatment plant and cause no further problem. Should these methods fail, the only solution remaining is to remove the water closet bowl and remove the object from the underside.

To be certain the blockage has been removed, flush several fairly heavy loads of toilet tissue through the bowl. If it flushes clean and there is no further evidence that other fixtures are affected, you can assume the blockage has been broken up and washed away.

Water Closet Force Cup

The closet force cup does not have a flat base

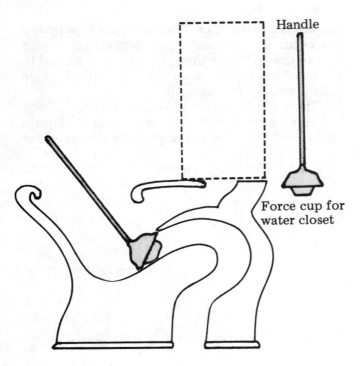

Using a water closet force cup
Figure 20-12

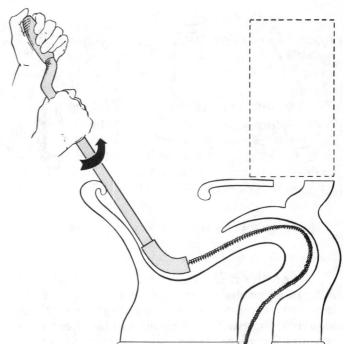

Using a closet auger on a water closet stoppage
Figure 20-13

such as the one used with flat bottom fixtures in Figure 20-7. Instead, it has a specially designed end that fits the opening of the water closet trap discharge side. Refer again to Figure 20-12. This type of force cup works almost the same way as the flat-based cup.

The Closet Auger

The closet auger consists of a hollow metal tube approximately three feet long and curved at one end. A solid metal rod passes through this tube and extends to the curved end. The auger has a handle on one end for turning and a spring steel cable with an attached hook on the other end. See Figure 20-11. The curved end of the tube has a rubber sleeve or other soft protective covering to prevent the metal from scratching the bowl's surface. To prepare the auger for use, pull the rod all the way up through the tube until the hook reaches the curved portion of the tube. To use the auger, insert the curved end of the tube into the trap opening and force the spring cable back through the tube until the obstruction is reached. Turn the handle, pressing down on the spring cable as shown in Figure 20-13. Try to hook and retrieve the object trapped in the bowl.

Hard Water Problems

"Hard" water has high amounts of iron and

other dissolved minerals. These minerals collect and settle to the bottom of the water heater and water closet tanks. They remain soft like wet mud as long as they are covered by water. But when they dry out—sometimes at intervals—a residue of tiny hard pieces is formed that can clog up small openings. Some of these problem areas and how to correct them are described below.

Clogged Water Closet Flush Rim

Around the top of the water closet bowl is the flushing rim, against which the seat rests. The purpose of the flushing rim is not only to scour the surface of the bowl but to provide the volume of water necessary for a clean flush.

If the water closet does not properly flush the contents of the bowl down the drainage system, raise the seat and check the rim of the bowl. There is a row of small holes around the entire circumference of the bowl approximately one inch apart. If these seem to be partly or completely closed, they need to be cleaned. A six to eight inch length of clotheshanger wire makes an ideal tool for this. Push one end into and through each hole. Then flush the water closet to expel any small particles. You may have to repeat this several times.

Clogged Shower Heads

Another victim of this type of clogging is the

shower head. If some of the holes in the shower head do not allow water to pass through, or if small low volume sprays at a right angle to the main spray are found, the head may need cleaning.

The head is easily removed and taken apart for cleaning. The portion of the head from which the water sprays is called the face plate. It is held in place by a small knob or a screw in its center. Remove the face plate and soak it in vinegar. The vinegar loosens the mineral deposits. Scrub it thoroughly with a brush and replace it.

Water Heater Maintenance

Whenever a water heater must be drained, the electric power or the gas supply source must be shut off first. *Only then should the water supply be turned off.* Two general precautions should be taken to ensure efficient operation and long tank life:

1) Drain the heater through the drain valve until the water runs clear. Do this at least once a year and more often if the sediment content warrants it. Failure to do this may result in noisy operation and lime and sediment buildup in the bottom of the tank. At the same time, check the temperature-pressure relief valve by raising the test lever at the top of this valve to make certain that the waterways are clear.

2) If the building is left unoccupied during cold weather months, drain the heater completely to prevent the tank from freezing.

Causes of Reduced or Lost Water Pressure

Clogged Aerator Nearly all combination lavatory and sink faucets today have an aerator at the end of the spout. See Figure 20-14. The aerator blends air with the water, thus conserving water and preventing it from splattering. The aerator has several fine screens inside it and these may become clogged. Remove the aerator to clean the screens. First wrap the outside with adhesive tape to keep from marring the chrome finish. Then turn the aerator counterclockwise with pliers to remove it. Take out the screens and thoroughly clean them until all debris is removed. Then reassemble exactly as before. If the aerator is the female type, place a small amount of pipe compound on the outside threads of the spout only. If it is the male type, place some pipe compound on the outside threads of the aerator only. Tighten the aerator enough to prevent water from seeping out around the threads. Then remove the tape.

Clogged Screens The hot and cold water hoses that supply water to clothes washers are equipped with filter screens. Each hose has a cone shaped screen to prevent water-formed scale and other sediment from entering the washer tub.

When water enters the washing machine tub under reduced pressure, the screens are probably partially clogged and need to be cleaned. The screens can be removed with your fingers after disconnecting each hose from its faucet. Clean each screen thoroughly and replace as before.

Water is usually left "on" at all times in a clothes washing machine. When the house is to be vacant for any length of time, the faucets should be turned off. This prevents flooding if one of the hoses should rupture.

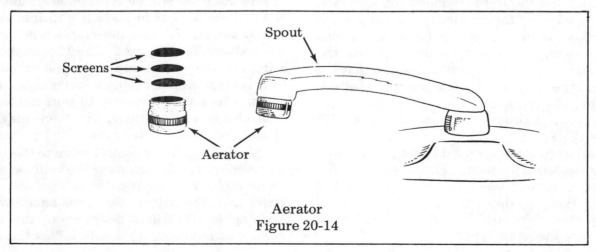

Aerator
Figure 20-14

Complete Loss of Hot Water

When investigating this problem, first find out if there is a loss of electric power or gas. If this is not the case, it may be that someone (possibly a child) has been "playing plumber." The shutoff valve controlling the flow of cold water to the heater may have been closed. Open the valve by turning the handle counterclockwise.

Complete Loss of Hot and Cold Water

There are three common causes of total water loss to a building.

1) The house control valve may have been closed by someone, since this valve is accessible. Open the valve by turning it counterclockwise.

2) The utility company supplying water to the community may have shut off the water to make emergency repairs. A telephone call to the utility company can verify whether or not this is the cause and the approximate length of time the water will be off.

3) In colder climates frozen water can cause a blockage in the main service pipe. If this is the problem, see the procedure for thawing frozen water pipes later in this chapter.

Reduced or Lost Water Pressure

Gradual reduction in water pressure from a compression-type faucet (hot water side only) may be caused by the wrong type of washer installed in the faucet. Heat expands washers that are not designed for hot water use, and the flow through the faucet is reduced considerably. To check whether the washer is the cause, turn on the hot water side of the faucet about one quarter turn. If water pressure falls as the temperature of the water rises, assume that the wrong type of washer has been installed. Replace the washer with a proper one. (Replacement procedures are discussed in Chapter 21.)

Another cause of reduced water pressure on a compression-type faucet may be that a flat washer has been replaced with a beveled washer. Worn washers which have given good service should always be replaced with new washers of the same type.

Pressure Fluctuation

There are two probable causes of pressure fluctuation on either the hot or cold water side of a compression-type faucet.

1) When the faucet handle is turned to an open position there may be an alternating strong and weak water flow accompanied by a chattering or whistling noise. This is usually caused by a loose screw that holds the washer tightly in place at the bottom of the faucet stem. See Figure 20-15. The washer thus floats up and

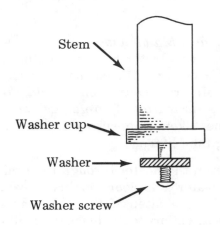

Loose washer
Figure 20-15

down on the loose screw, causing the pressure to fluctuate and, in many cases, emit an irritating chatter.

To correct this, remove the stem as shown in Chapter 21 and tighten the screw.

2) The second cause of pressure fluctuation is rare but must still be considered. If the screw holding the washer in place is not tightened, it can loosen and fall into the supply tube beneath the fixture. See Figure 20-16. The washer can adhere to the washer cup at the end of the stem and remain firmly in place even without the screw. The faucet continues to work normally to control the flow of water, but there may be a fluctuation in the water flow.

Shut off the water control valve to the fixture and remove the faucet stem assembly as if you were replacing the washer. Close the waste opening on the fixture. Cup your hand over the opening in the faucet body where the faucet stem was removed. This will deflect the water

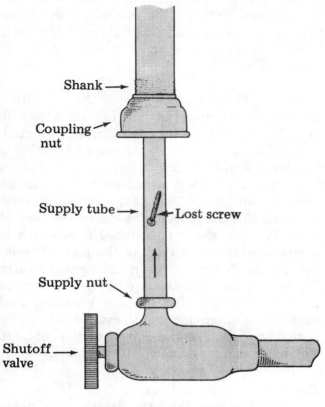

Lost bibb screw
Figure 20-16

down into the fixture bowl. Have someone open the control valve and then close it immediately. Your hand will deflect the surge of water downward and should cause the screw to fall into the fixture bowl as it flows out with the water.

If this is unsuccessful, disconnect the supply tube beneath the fixture. At each end of the supply tube there is a nut. Loosen the coupling nut with a basin wrench; loosen the compression nut with a small smooth-jawed adjustable wrench. Remove the supply tube. The screw should then fall out. If the screw is good, reinstall it. If not, replace it with a new one. Replace the supply tube, covering the threads with a small amount of pipe compound. Tighten the nuts snugly, turn the water on, and check for leaks.

Water Hammer

Water hammer is a banging noise emitted when a faucet is closed quickly and the flow of water is suddenly stopped. The cause of water hammer is air lost from the air chambers. It may affect the entire water system, but it is generally isolated to either the hot or the cold water line or

to a particular bathroom. To correct water hammer, follow in order the three steps outlined below.

1) Shut off the main house valve and open one or more outside faucets. Then open all faucets, hot and cold, inside the building for a few minutes. This allows the water in the pipes to drain from the system and permits air to fill the pipes. After the water has stopped draining through the outside faucets, close them and all inside faucets. Then open the house valve. As water refills the pipes, the air will be pushed back into the air chambers again. This works very well when the piping system is installed high enough above the outside faucets to permit complete draining.

2) If the water system pipes are installed in such a way as *not* to permit complete drainage of the system (such as beneath the floor), try another method:
Shut off the main house valve and open one or more outside faucets as before. Then open the faucets on one fixture one faucet at a time, hot and cold. Remove the aerator if the faucet has one. Place your mouth over the faucet spout opening, take a deep breath, and blow with all your might. This usually blows the trapped water out and through the hose faucet.
Repeat with the other hot and cold water faucets in each fixture. This procedure forces sufficient air into the pipes to replace the lost air in the air chambers and should eliminate the banging noise.

3) If the two methods above are not successful, install exposed air chambers in both hot and cold water pipes. The easiest and most likely place for these would be near the water heater. An air chamber should be one pipe size larger than the pipe to which it connects, and a minimum of 12 inches long. (See #3, Figure 15-4 and Figure 16-7.)

Thawing Frozen Pipes

In cold climates, water pipes can freeze in the winter. The old saying, "an ounce of prevention is worth a pound of cure," applies here. All exposed piping should be insulated well and should be checked before winter so that any unsound insulation can be replaced.

But if pipes do freeze, only heating the pipes

can thaw them. Heat may be applied in several ways.

1) If there is no danger of fire, use a blowtorch or a burning twist of newspaper. To thaw a pipe using an open flame, open the nearest faucet to indicate when the flow of water starts. Apply heat at one end of the pipe and work the flame along the entire length of the pipe. Continue heating the pipe until the water flows from the faucet.

2) If there is danger of fire, wrap the affected section of pipe with rags and pour boiling water over the rags. This also gives satisfactory results. Leave a faucet open to indicate when water is flowing again.

3) Frozen water supply pipes located where heat cannot be readily applied may be thawed by the following procedure. Close the house valve and remove the fitting that closes the exposed end of the pipe. Then insert a smaller pipe or rubber hose with a funnel into the pipe until it contacts the ice blockage. See Figure 20-17. Then pour boiling water into the funnel.

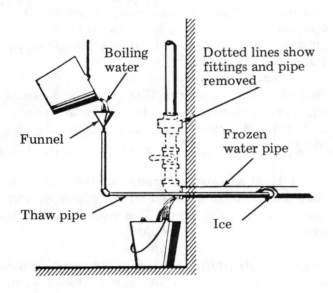

Thawing an underground or
otherwise inaccessible pipe
Figure 20-17

The hot water circulates to the ice blockage through the smaller pipe. It then comes back out the opening where the smaller pipe enters. Catch this spillage in a pail if water damage is likely. As the ice melts, push the "thaw pipe"

deeper into the building pipe. Then withdraw the pipe quickly when the flow starts and allow the flow to continue until the thaw pipe has been completely withdrawn and the building pipe is cleared of ice.

4) Inaccessible pipes may be thawed with a low voltage electrical thawing unit which may be attached to any alternating current electric outlet. The unit has a transformer that takes a relatively small amount of current and in turn sends a large amount of current at low voltage through the pipes. Two heavy cables may be clamped to each side of the frozen section of pipe. The transformer produces heat which is transmitted to all portions of the pipe between the clamps. To thaw a frozen underground water service, connect one clamp where the service pipe is accessible within the building and connect the other clamp to the street valve. The heat transmitted to the frozen section of pipe usually gives quick, satisfactory results.

Cleaning Discolored Fixtures
No matter how dingy the fixtures look, never attempt to remove stains with scouring powder or pads. These destroy the glaze. Instead, clean fixtures with any fine-grained cleaning powder or with a newer liquid cleaner called "soft scrub."

Stained tubs or sinks may be whitened by filling the fixture with warm water and adding some chlorine bleach. For best results let this stand for several hours.

If the water source is a well, a water softener should be used. It improves the water's taste and color and removes most of the staining agents from the water, especially iron. Iron stains porcelain fixtures a deep reddish brown, which is practically impossible to remove without damage to the fixture finish.

Vinegar is usually effective on stubborn mineral stains. Soak a pad in vinegar and place it directly over the stain for an hour or longer. Rust stains sometimes respond if you rub the area with a fresh slice of lemon. If stains do not respond to these efforts and heavy-duty cleaners must be used, follow the instructions on the label carefully to avoid damaging the porcelain finish.

Cleaning Toilet Seats, Chrome, and Tile
It is recommended that only a soft cloth and

mild soap be used for cleaning toilet seats, chrome, or tile. To prevent mildew from forming on tile walls, shower unit enclosures, and bathtubs, always wipe excess water from the tile or bath enclosure with a towel after use. Also wipe the chrome shower or tub trim dry at the same time. Both will remain bright and new-looking for years if properly cared for.

Questions

1- What is the major cause of clogged drains?

2- In what portions of a drainage system do stoppages rarely occur?

3- At what point is a stoppage most likely to occur in a drainage system?

4- Besides a local blockage on private property, what else might be the cause of a complete stoppage of a drainage system?

5- In order to make stoppages accessible for unclogging, what do most codes require on a private sewer?

6- If a brass cleanout plug is frozen and will not loosen, what other procedure should be tried?

7- What are some of the factors determining the right kind of sewer cable to select for a particular type of stoppage?

8- When an accessible cleanout is not available, what portion of a sanitary system is considered adequate for cleaning purposes?

9- Which plumbing fixture is most subject to clogging?

10- What is a "plumber's friend"?

11- How should a two-compartment sink be blocked up to prevent loss of pressure?

12- What is the most likely cause of a clogged shower drain?

13- What tool should be tried first to unstop a shower drain?

14- When the drain stopper of a lavatory or bathtub is not causing the stoppage, what tool should be tried first to clear the trap or drain line?

15- What is the passageway of a water closet trap designed to do?

16- Why is it important to determine the cause of a stoppage in a water closet trap before proceeding to unclog it?

17- What two tools are usually used to unstop a water closet bowl?

18- What tool should be used to clear a closet bowl trap in a stoppage caused by cloth?

19- How should you check to be certain a closet bowl blockage has been removed?

20- What substances are contained in "hard" water?

21- Where is the flushing rim of a closet bowl?

22- What is one of the first checks you should make when a water closet is not flushing properly?

23- What procedure should be used to clear mineral deposits from a shower head's face plate?

24- What should be the first step before shutting off the water supply to an improperly-working water heater?

25- What yearly check should be made on a water heater to find out whether it is working properly?

26- What two things are accomplished when an aerator is used on the end of a faucet spout?

27- If an aerator has male threads, where

should the pipe compound be applied when reassembling?

28- What is the main cause of reduced water pressure to a clothes washing machine?

29- What should be done to a washing machine to protect the property from possible water damage when the house is to be vacant for any length of time?

30- What is the most likely cause of a complete loss of hot water in a building?

31- What would be the most probable cause of a complete loss of water to a residence in Chicago during January?

32- What would be the most probable cause of a complete loss of water to a residence in Key West, Florida, in January?

33- What is the most probable cause of a gradual pressure reduction in a compression faucet on the hot water side?

34- What are other possible causes for reduced pressure on a compression faucet?

35- What usually causes a chattering or whistling noise in a compression type faucet when the water is turned on?

36- How do you recover a lost bibb screw that has dropped into the supply tube?

37- What is the major cause of water hammer?

38- What three steps should be tried to eliminate water hammer in a water distribution system?

39- "An ounce of prevention is worth a pound of cure": How does this apply to protecting a water piping system from freezing?

40- What is used to thaw water frozen in pipes?

41- Where there is no danger of fire, what procedure should be used to thaw a frozen section of pipe?

42- Where danger of fire exists, what procedure should be used to thaw a section of frozen water pipe?

43- What method is used to thaw pipe that is inaccessible?

44- What should you *not* use to rid plumbing fixtures of heavy stains?

45- How may bathtub or sink stains be removed safely?

46- How can fixture stains be avoided when the source of water is a well?

47- What substance may be used to remove stubborn mineral stains without damage to the fixture finish?

48- To avoid damaging the finish, what is recommended for cleaning toilet seats, tile, and chrome?

21
General Plumbing Repairs

Leaks account for much of the water loss in any system. For example, a drop of water that takes 2½ seconds to form and drip to the sink wastes about 365 gallons of water per year. If the drip is from the hot water side, the leak wastes both 30 gallons of water a month and 65 kilowatt hours of electricity costing about $2.50. General plumbing repair work includes changing worn-out washers and packings as well as repairing other types of leaks in the system. Every plumber should be able to find, diagnose, and eliminate leaks.

Valves

Gate and globe valves are the most common types of shutoff valves used to control water flow to fixtures and to appliances in home water systems. Figure 21-1 shows that they are very

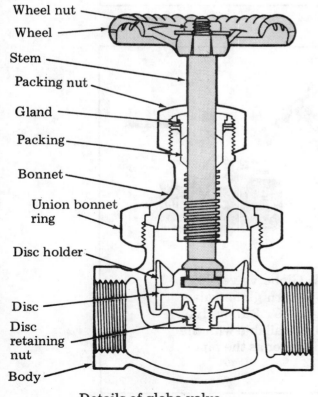

Wheel nut
Wheel
Stem
Packing nut
Gland
Packing
Bonnet
Union bonnet ring
Disc holder
Disc
Disc retaining nut
Body

Details of globe valve
Figure 21-1

similar in construction. The repair instructions below apply to both types.

Valves come in many styles and may differ somewhat in general design from those shown in Figures 21-1, 21-2, and 21-3. Because of the larger number of valves used in plumbing systems, it is essential that a plumber know how to maintain and repair them. Valve leaks are generally limited to leaking washers or leaking packing nuts.

Globe Valve Repair

When a globe valve (also known as a *compression valve*) no longer controls the flow of water, the washer on the end of the stem is probably worn and needs to be replaced. Shut off the water at the building main and drain the line. Then disassemble the valve in this order: remove the handle or wheel, the packing nut, the packing, and the stem. Fit the handle or wheel back on the stem and use it to unscrew and remove the stem from the valve body. Do *not* use a wrench to remove the stem. See Figures 21-1 and 21-2 for valve parts detail.

Remove the screw and the worn washer from the stem. Clean the washer cup and install a new washer of the proper size and type. Reassemble the parts in reverse order: the stem, the packing, the packing nut, and then the handle.

If the washer in a globe or gate valve requires frequent replacement, the replacements may be the wrong type for the seat. Or the seat may be pitted and rough, scoring and wearing away the washer prematurely. Most globe valves have seats which may be replaced when they become worn or pitted. Replaceable seats have either square or hexagonal water passages, and the seat removal tool must be either square or hexagonal to fit the passage. Seat dressing tools are available for both replaceable and non-replaceable seats.

If water leaks around the stem, either the packing nut has loosened and needs tightening

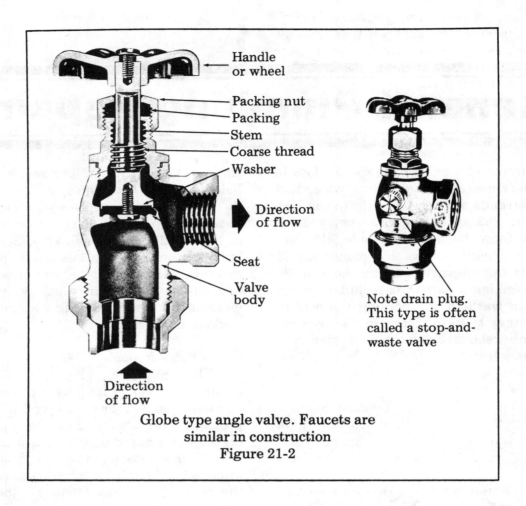

Handle or wheel
Packing nut
Packing
Stem
Coarse thread
Washer
Direction of flow
Seat
Valve body
Direction of flow

Note drain plug. This type is often called a stop-and-waste valve

Globe type angle valve. Faucets are similar in construction
Figure 21-2

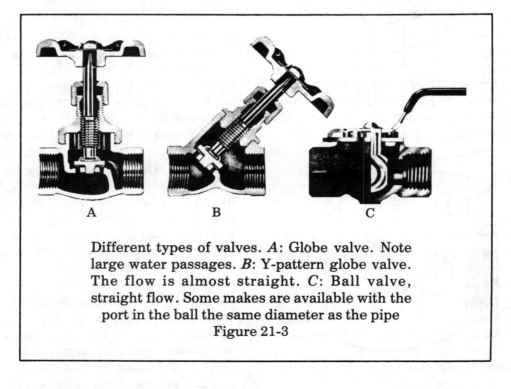

A   B   C

Different types of valves. A: Globe valve. Note large water passages. B: Y-pattern globe valve. The flow is almost straight. C: Ball valve, straight flow. Some makes are available with the port in the ball the same diameter as the pipe
Figure 21-3

or the packing is worn and needs replacing. Tighten the packing nut slightly with a suitable wrench such as an adjustable crescent-type wrench. If the leak continues, remove the handle and then the packing nut. Wrap a few strands of graphite packing string or lamp wick string dipped in pipe joint compound around the stem. Slide the packing nut down the stem and tighten it with a wrench as before.

Gate Valve Repair

Gate valve repair usually requires only replacing the stem packing. The procedure is the same as for globe valves.

Compression Faucet Repair

Most plumbing fixtures today have ordinary compression type faucets—two separate valve and water supply units with a common spout. The washer in each side of the faucet body (one for hot and one for cold) eventually wears down and must be replaced. Each unit must be repaired independently.

Dripping faucets are the most common plumbing problem you may be called upon to correct. Normally, a new washer in the faucet is all that is required.

To replace the washer in a compression faucet, first shut off the water supply to the faucet. Most newer buildings have shutoff valves directly below the fixture, one for the hot water supply and one for the cold. Turn the valve handle clockwise to shut off the water. In older buildings where no such valve is available, shut off the main house control valve. This valve is usually located on the water service pipe close to the point of entry to the building (see Figure 12-1) or in some other accessible place.

Figure 21-4 lists the stem assembly parts of a typical compression faucet. The appearance of these parts varies from brand to brand, but their function remains the same regardless of the manufacturer, the type of faucet, and its particular use. Here are some of the most common problems plumbers face in servicing compression faucets:

1) When water leaks from under the faucet handle, the stem packing (packing washer) almost certainly needs replacing. The original packing washer is preformed to fit the stem

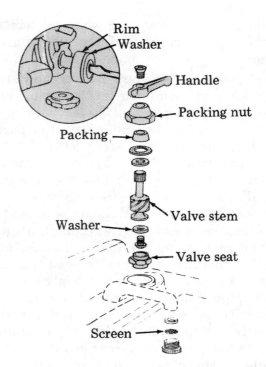

Typical faucet assembly
Figure 21-4

firmly, and the packing nut holds the washer in place. With thousands of turns the stem wears away the packing or the packing nut loosens, allowing water to seep out around the stem.

This is a very simple repair. Remove the handle screw and handle first. (Some handle screws are located beneath a snap-on button.) Use a smooth-jawed adjustable wrench. Adjust the wrench to fit the packing nut and tighten the nut slightly. Often this adjustment takes care of the leak. If the leak continues, remove the packing nut completely. Remove the old packing washer and replace it with a new packing washer of the same size if one is available. If one is not available, wrap a few strands of graphite packing string or lamp wick string dipped in pipe joint compound around the stem. Slide the packing nut down the stem and tighten it.

Some newer compression faucets have one or two rubber "o" rings secured to the stem in shallow grooves instead of a packing washer. The "o" rings serve the same purpose as the packing washers and prevent water from seeping out around the stem. Here the replacement procedure is to completely remove the stem from the faucet body and replace the old "o" rings with new ones of the same size (not thicker or thinner).

2) When water drips from a faucet spout, a

tub spout, or a shower head when the handle is fully closed, the cause is almost always a worn washer.

To replace the washer, shut off the water to the faucet and then open the faucet to relieve some of the water pressure. Remove the handle screw and the handle and loosen the packing nut. Then unscrew the stem from the valve body. The bibb screw holding the damaged washer in place may be brittle due to age and difficult to remove with a screwdriver. If so, grip the head of the screw with pliers and remove it by turning counterclockwise. Clean the washer cup and replace the old washer and the screw with the same size and type. Remember that flat washers are used on seats with a crown or a round ridge. Tapered or rounded washers are used with tapered seats. A perfect fit is essential; a near fit is *not* good enough. Reassemble the faucet in reverse order; turn the handle to the closed position, tighten the packing nut, and reset the handle to the proper position before replacing the handle screw.

If a compression washer requires frequent replacement, it may be that the washer has become scored by a pitted or rough seat. Use a seat dressing tool to smooth the surface of the seat if the roughness is slight. If the seat is badly pitted, remove the seat and replace it with one of the same type and size.

On old faucets which have not had their washers replaced regularly, the handle may have frozen to the stem over the years. If you can not remove the handle, try a tool called a *handle puller*. If this does not work, rap the underside of the faucet handle with the plastic end of a screwdriver. This usually loosens the handle without damaging its chrome finish.

3) Occasionally a compression faucet becomes very noisy when water is flowing through it. This may be caused by a loose washer or by worn threads on the stem and worn receiver threads in the faucet body. Either condition allows the stem to vibrate or chatter. Determine the exact cause by pressing down on the handle. If the stem vibration stops, the source of trouble is worn threads and not a loose washer. If the stem vibration does not stop, the washer is probably loose.

Replacement stems are available. However, if the receiving threads are worn excessively in the faucet body, a new stem would not eliminate

the problem completely. In some faucets it is possible to replace the stem receiver, the stem, and the seat, thus restoring all normal wearing parts in the faucet. If the faucet is old and the finish is worn, replace the entire faucet.

There are several new faucet designs available which are aimed at easier operation and a long, drip-free service life. The American Standard Aquaseal faucet shown in Figure 21-5

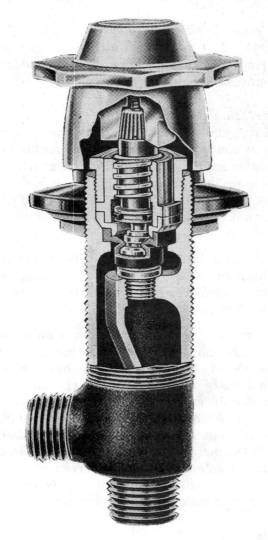

American Standard Aquaseal faucet
Figure 21-5

is such a faucet. Instead of the conventional "washer-grinding" principle, the Aquaseal works on the diaphragm principle of water control. All moving parts above the Aquaseal are outside the flow area, lubrication on the stem threads is effective for the life of the fitting, and there is no seat washer wear to

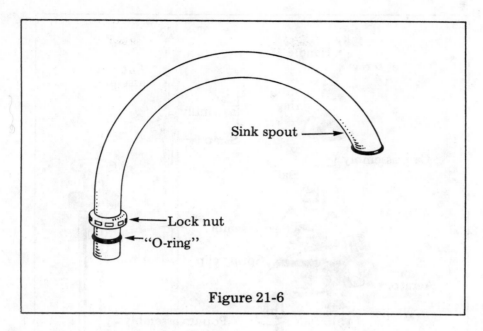

Figure 21-6

cause leaks and dripping as in ordinary fixture faucets.

4) A problem common to all compression-type sink faucets is a leak around the lock nut that secures the swing spout to the faucet body. Continuous side-to-side movement of the spout wears down the rubber "o" ring or rings located at the base of the spout. See Figure 21-6.

To repair this leak, first loosen the lock nut at the spout base. Some lock nuts are designed with square corners so a smooth-jawed wrench can be used. Others are round, requiring the use of channel lock pliers or a similar tool. Wrap round nuts with adhesive tape to avoid marring the chrome finish. Once the lock nut is loosened. the spout lifts out. Replace the "o" ring washer with a new one of the same size.

Some types of older faucets may have a split graphite washer. Replace this with a washer of like type or use an adequate amount of graphite string. Reassemble the faucet and tighten the lock nut. Remove the adhesive tape and test for leaks. Note that the water to the faucet does not have to be shut off for this repair.

Single Handle Mixing Faucets

Many years of maintenance-free use can be expected before repairs are necessary on single handle mixing faucets. This is because the moving parts can control the flow of water without grinding the washer against a metal seat. Today there is increased demand for these mixing faucets on sinks, lavatories, bathtubs,

and showers. Every plumbing professional should know how to maintain and service them.

The parts for the Delta and Moen mixing faucets and valves are shown in Figures 21-7 through 21-12. Other manufacturers' mixing faucets and valves may differ somewhat, but these illustrations will be quite helpful for disassembly and reassembly of other makes.*

Repairing the Delta Faucet

Delta faucets and valves are designed to control water temperature and flow with a single ball that rests against a seat assembly. Figure 21-7 shows a Delta lavatory faucet, Figure 21-8 a Delta deck-type sink faucet, and Figure 21-9 a Delta tub valve. The routine maintenance instructions for each type of faucet are shown in Figure 21-7A, 21-8A, and 21-9A. Each of these faucets has identical parts to control the water flow. When you learn to repair one Delta faucet you can repair all faucets in the Delta line. Follow these steps to make proper repairs on a Delta faucet. Remember that because the Delta sink faucet has a swing spout, *two* "o" ring washers must be used instead of one. This is the only difference between repairing this faucet and repairing the other Delta faucets.

1) If the faucet leaks from under the handle, loosen the set screw and pull off the handle.

*The author gratefully acknowledges the permission granted him for the use of the Delta illustrations by the Delta Faucet Company, Greensburg, Indiana, and for the use of the Moen illustrations by Stanadyne, Elyria, Ohio.

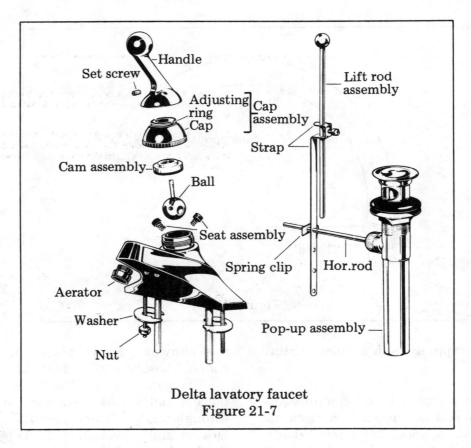

Delta lavatory faucet
Figure 21-7

Tighten the adjusting ring until no water leaks around the stem when the faucet is on and pressure is exerted on the handle to force the ball into the socket.

2) If the faucet drips at the spout outlet, the tub spout, or the shower head, shut off both hot and cold water. Then loosen the set screw and pull off the handle. Unscrew and remove the cap assembly. Pull up on the ball stem to remove the cam and ball assembly. Replace the rubber seats and springs. Check the ball and replace it if there is a sharp edge or any roughness around either of the two small holes. Reassemble the faucet in the reverse order. Make sure that the *slot* inside the ball and the *lug* on the side of the cam is *inserted into the slot* on the side of the body. Screw the cap down until it is *tight*.

3) If the faucet leaks from either the bottom or the top of the spout, shut off both hot and cold water, loosen the set screw, and pull off the handle. Unscrew and remove the cap assembly. Turn the spout slowly and pull it up to remove. Replace the two body "o-rings." Reassemble in the reverse order. Tighten the cap until it is *tight*.

Repairing the Moen Faucet

Moen faucets and valves are designed to control water temperature and flow with a single cartridge. Figure 21-10 shows a Moen lavatory faucet, Figure 21-11 a Moen deck-type sink faucet, and Figure 21-12 a Moen tub valve. Each of these has identical parts to control the water flow. As with Delta, the swing spout on the sink faucet requires that *two* "o" rings be installed instead of one. The following steps explain how to make repairs on Moen faucets:

1) If the faucet drips at the spout outlet, tub spout, or shower head, first shut off both hot and cold water. Then remove the handle cover, handle screw, handle, and stop tube. Lift the retainer clip and pull the cartridge out by the stem from the faucet or valve body.

To reassemble, reinsert the replacement cartridge by pushing it all the way into the faucet or valve body *until the front of the ears on the cartridge shell are flush and aligned with the body*. See Figures 21-10 and 21-12. Replace the retainer clip so that its legs straddle the cartridge ears and slide down into the bottom slot in the body. This prevents the cartridge from rotating and locks it into the body.

Delta faucets have earned the reputation of superiority in design, engineering, performace and durability in millions of home installations. Routine faucet maintenance to compensate for foreign materials in varying water conditions, will restore like new performance and extend the life of the faucet. Ease and simplicity of service and maintenance is another advantage of having Delta faucets throughout the home.

A. If you should have a leak under handle—tighten adjusting ring following steps 1 and 9. Reassemble as in step 10.

B. If you should have a leak from spout—shut off water supply, and follow steps 1, 2, 3, 4 and 5. Reassemble as in steps 6, 7 and 8. Set adjusting ring as in 9. Replace handle as in 10.

1. Loosen set screw and lift off handle.

2. Unscrew cap assembly and lift off.

3. Remove cam assembly and ball by lifting up on ball stem.

4. Remove seats.

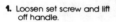

5. Place new seats over springs and insert into sockets in body.

6. Place ball into body over seats.

7. Place cam assembly over stem of ball and engage tab with slot in body. Push down.

8. Partially unscrew adjusting ring and then place cap assembly over ball stem and screw down tight onto body.

9. Tighten adjusting ring until no water will leak around stem when faucet is on and pressure is exerted on handle to force ball into socket.

10. Replace handle. Tighten handle screw—tight.

Use only genuine Delta replacement parts.
—All Delta Lever, Single Handle Valves.
Kit 3614—"O" Rings (2)
        Cam Assembly
        Seat Assembly (2)
        Wrench

DELTA ® Delta Faucet Company
P.O. Box 31 · Greensburg, Indiana 47240
A Division of Masco Corporation of Indiana

Routine maintenance instructions for Delta series 500, 510, 520, and 530 single control lavatory faucets
Figure 21-7A

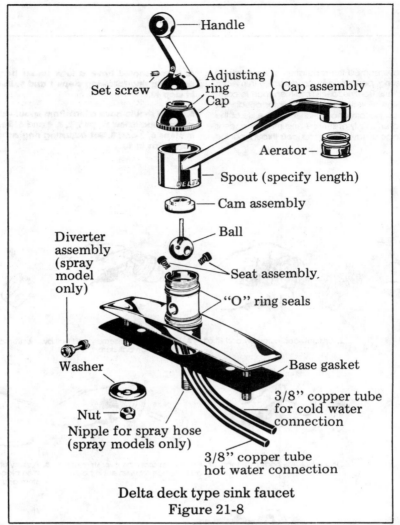

Delta deck type sink faucet
Figure 21-8

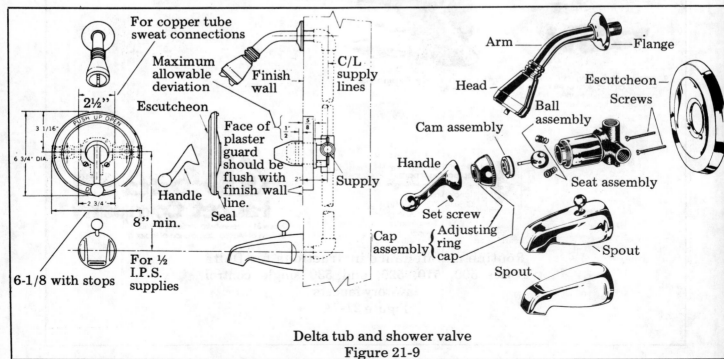

Delta tub and shower valve
Figure 21-9

Delta faucets have earned the reputation of superiority in design, engineering, performance and durability in millions of home installations. Routine faucet maintenance to compensate for foreign materials in varying water conditions, will restore like new performance and extend the life of the faucet. Ease and simplicity of service and maintenance is another advantage of having Delta faucets throughout the home.

A. If you should have a leak under handle—tighten adjusting ring by following steps 1, 14, 15 and 16.

B. If you should have a leak from spout—replace seats and springs by following steps 1, 2, 3, 4 and 10. Then reassemble with steps 11, 12, 13, 14, 15 and 16.

C. If you should have a leak from around spout collar—replace "O" rings by following steps 1, 2, 5, 6 and 9. Then reassemble by following steps 13, 14, 15 and 16.

D. If spray does not work properly—check hose for wear or cuts. Clean or replace diverter part No. 320 by following steps 1, 2, 5, 7, 8 and 9A. Then reassemble, following steps 13, 14, 15 and 16.

1. Loosen set screw and lift off handle.

2. Turn water supply off and unscrew cap assembly and lift off.

3. Remove cam assembly and ball by lifting up on ball stem.

4. Remove seats.

5. Rotate spout gently and lift off.

6. Cut "O" rings and remove from body.

7. To remove diverter assembly, pull straight out with fingers.

8. Place diverter assembly into cavity inside of body as far as possible.

9. Stretch "O" rings and snap into grooves on body.
9A. Push spout straight down over body gently, and rotate until it rests on plastic slip ring.

10. Place **new seats** over springs and insert into sockets in body

11. Place ball into body over seats, making certain body pin is in ball slot.

12. Place cam assembly over stem of ball and engage tab with slot in body. Push down.

13. Partially unscrew adjusting ring and then place cap assembly over stem and screw down tight onto body.

14. Turn on water supply. Tighten adjusting ring until no water will leak around stem when faucet is on and pressure is exerted on handle to force ball into socket.

15. Replace handle and tighten set screw tight.

16. Important. Remove aerator, clean and flush faucet. Then replace aerator.

Use only genuine Delta replacement parts.
Kit 3614—(4) "O" Rings, (1) Cam Assembly
(2) Seats and Springs
(1) Wrench
Kit 4993—(2) Seats and Springs

Delta Faucet Company
P.O. Box 47, Greensburg, Indiana 47240
A Division of Masco Corporation of Indiana

Routine maintenance instructions for Delta series 100, 300, 350, 400, and 450 single lever kitchen faucets
Figure 21-8A

Delta Faucets have earned the reputation of superiority in design, engineering, performance and durability in millions of home installations. Routine faucet maintenance to compensate for foreign materials in varying water conditions can restore like new and extend the life and performance of the faucet. And, ease and simplicity of service and maintenance is another advantage of having Delta faucets throughout the home.

A. If you should have leak under handle—tighten adjusting ring by following steps 1 and 9 leaving water supply and faucet turned on. Replace handle as in step 14.

B. If you should have leak from spout—replace seats and springs by following steps 1, 2, 3, and 4. Reassemble following steps 5, 6, 7 and 8. Turn water supply on and follow steps 9 and 14.

C. If diverter fails to operate property follow steps 1, 10, 11, and 12. Reassemble with steps 13 and 14.

Shut off water supply

1. Loosen set screw and lift off handle.

2. Unscrew cap assembly and lift off.

3. Remove cam assembly and ball by lifting up on ball stem.

4. Remove seats and Springs

5. Place new seats over springs and insert into sockets in body.

6. Place ball into body over seats.

7. Place cam assembly over stem of ball and engage tab with slot in body. Push down.

8. Partially unscrew adjusting ring and then place cap assembly over ball stem and screw down tight onto body.

Turn on water supply

9. Tighten ring until no water will leak around stem when faucet is on and pressure is exerted on handle to force ball into socket.

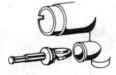

10. Remove escutcheon.

11. Unscrew diverter. Check for sediment, clean off, and flush cavity. If flapper is nicked, or plunger sticks, replace with new diverter.

12. Screw diverter back into body.

13. Replace escutcheon.

14. Replace handle. Tighten handle screw—tight.

PRESSURE BALANCED VALVES

If unable to maintain constant temperature, clean balancing spool as follows:

1. Remove handle and escutcheon.

2. Close stops.

3. Unscrew balancing spool assembly (part No. 574) and remove—slowly—so as not to damage "O" Ring Seats. NOTE: On shower only installations water trapped in shower riser will drain out when balancing spool assembly is removed.

4. Examine valve hole for chips or any other foreign matter.

5. Remove any chips from spool sleeve before attempting to remove spool from inside of sleeve.

6. Remove spool from sleeve and clean all deposits from both sleeve and spool.

7. When spool will slide freely in sleeve, replace assembly in original position.

8. Open stops and check cap of spool assembly and stems of check stops for leaks before reinstalling escutcheon.

Delta Faucet Company

Use only genuine Delta replacement parts.
Kit 3614—"O" Rings (4)
            Cam Assembly (1)
            Seats (2)
            Wrench (1)
Kit 4993—Seats and Springs (2)

P.O. Box 47 • Greensburg, Indiana 47240 • A Division of Masco Corporation of Indiana

Routine maintenance instructions for Delta series 600 single lever bath valves, with or without pressure balance
Figure 21-9A

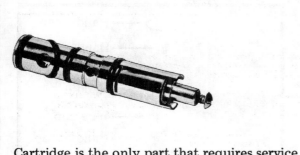

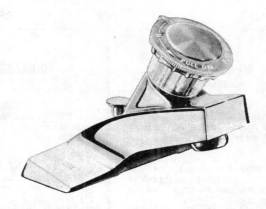

Cartridge is the only part that requires service

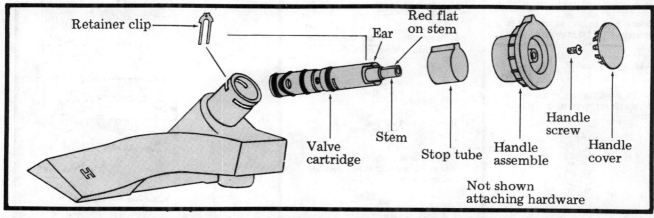

Retainer clip

Ear

Red flat on stem

Valve cartridge

Stem

Stop tube

Handle assemble

Handle screw

Handle cover

Not shown attaching hardware

To Disassemble: Remove handle cover, handle, and stop tube. Lift out retainer clip and pull the cartridge out of the body by the stem.

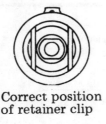

Correct position of retainer clip

To Re-Assemble: Reinsert cartridge by pushing it all the way into the body *until the front of the ears on the cartridge shell are flush and aligned with the body*. Replace the retainer clip so that the legs straddle the cartridge ears and slide down into the bottom slot in the body. This prevents the cartridge from rotating and locks it in the body. Reinstall stop tube, handle, and handle cover. The red flat on the stem must point up when mounting the handle.

Moen lavatory faucet
Figure 21-10

Reinstall the stop tube, handle, handle screw, and handle cover. When mounting the handle be sure the *red* flat on the stem is pointing *up*.

2) If the sink faucet drips at the spout outlet or leaks from either the top or the bottom of the spout, follow the step-by-step outline in Figure 21-11 to replace the cartridge or the two body "o" rings.

Other Types of Washers and Packings
Other parts of the plumbing system contain washers and packings that may eventually need replacing.

Trap Washers
The fixture trap design most commonly used for such installations as lavatories, kitchen sinks, and laundry trays is called a *P-trap*. P-traps are usually constructed of chromed brass but may be of plastic where plastic systems are permitted. The trap is two-piece and is held together by two nuts and two

TO DISASSEMBLE
(Need Pliers, Screwdriver, Flat-Jawed Wrench)

1. Turn off both hot and cold water supplies. Pry off handle cap. Remove handle screw and pull handle off.

2. Remove wire clip by hooking one end of wire with finger, pulling it out of groove on slotted stop guide, and lifting it over cartridge stem.

3. Remove handle stop. Unscrew and remove retainer nut. Lift off slotted stop guide.

4. Lift and twist spout off.

5. With screwdriver, pry up both legs of cartridge clip and remove it. See illustration below.

6. Grasping stem of cartridge with pliers, pull out cartridge as shown below.

If necessary to Flush Supply Lines, Turn On Both Hot and Cold Water Supplies Slowly.

TO REASSEMBLE

7. With stem of cartridge pulled out, put in cartridge by pushing down on its ears. See illustration below.

8. Turn cartridge so its ears face front and back of faucet as shown below.

EARS

9. Replace cartridge clip by pushing down on both sides so it locks on both sides. See illustration below.

10. Turn red (notched) flat of cartridge stem so it faces rear of faucet. Replace spout, pushing down and twisting until it nearly touches escutcheon.

NOTE: For cross piping installations where supply piping is reversed, red (notched) flat faces front of faucet.

11. Replace slotted stop guide so it straddles cartridge ears and lies flat against cartridge clip. Screw on retainer nut with fingers. Do not cross thread. Tighten with wrench.

12. Put on white handle stop. Push cartridge stem all the way down. Replace wire clip making sure both sides catch in slots of stop guide. See illustration at right.

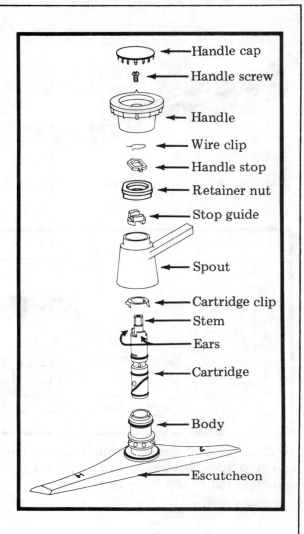

13. Push handle onto cartridge stem with handle's pointer facing rear of faucet.

14. Replace handle screw. Tighten securely. Replace handle cap.

NOTE: If hot and cold water reversed, red (notched) flat of cartridge stem is just opposite of correct position. (See Step 10.)

NOTE: For proper water flow, aerator must be free of foreign particles. If flow is weak or irregular, unscrew aerator, clean and replace.

Moen deck type sink faucet
Figure 21-11

Cartridge is the only part that requires service

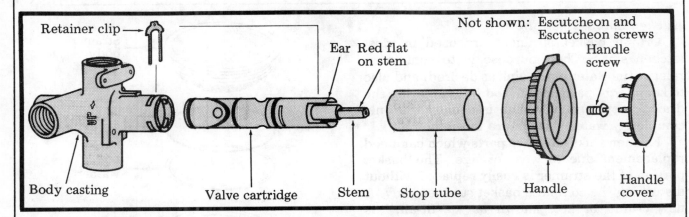

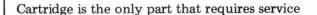

Retainer clip

Not shown: Escutcheon and Escutcheon screws

Handle screw

Ear   Red flat on stem

Body casting   Valve cartridge   Stem   Stop tube   Handle   Handle cover

To Disassemble: Remove handle cover, handle, and stop tube. Lift out retainer clip and pull the cartridge out of the body by the stem.

Correct position of retainer clip

To Re-Assemble: Reinsert cartridge by pushing it all the way into the body *until the front of the ears on the cartridge shell are flush and aligned with the body.* Replace the retainer clip so that the legs straddle the cartridge ears and slide down into the bottom slot in the body. This prevents the cartridge from rotating and locks it in the body. Reinstall stop tube, handle, and handle cover. The red flat on the stem must point up when mounting the handle.

Moen tub and shower valve
Figure 21-12

washers called *square cut washers.* They must be replaced when they begin to deteriorate. Square cut washers for kitchen sinks and laundry trays are 1½ inches in diameter, while washers for lavatories are 1¼ inches in diameter.

To replace these washers, use a ten-inch pipe wrench to loosen the two nuts. The "J" bend portion of the trap will contain water, so place a container directly beneath the trap. See Figure 21-13. After loosening the nuts, remove the "J" bend portion of the trap and the old washers. Replace them with new washers of the right size. Spread a small amount of pipe joint compound around and over each washer and tighten the nuts snugly. Fill the fixture with water and run water through the trap to check for leaks.

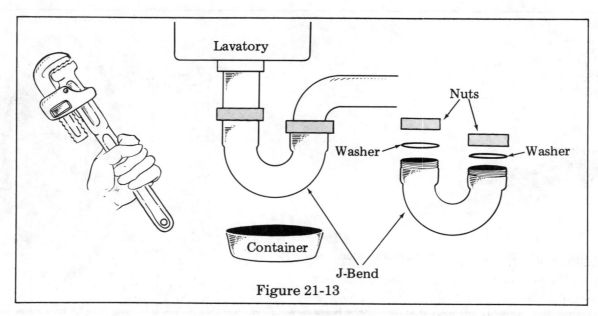

Figure 21-13

Kitchen Sink Basket Strainer

Crumb basket strainers are used in most kitchen sinks. Their purpose is to contain or release the water in the sink as desired, and also to keep large particles of food from entering the drainage system. The sink is usually the only completely watertight fixture in the home.

Kitchen sinks have two parts which can need replacement due to wear or age. The basket portion of the strainer is easily replaced without disturbing the complete basket strainer body. A new crumb basket, known as a "fit-all," is available for replacement. It fits very well in most types of strainer bodies. The entire strainer body must be removed to replace the rubber washer. See Figure 21-14. This would be the ideal time, of course, to replace the old strainer body with a new unit if it is marred or worn from age.

To remove the strainer body or install a new washer or a complete unit, first loosen the 1½-inch tail piece nut shown in Figure 21-15. If the proper tool (a *combination lock nut wrench*) is not available to loosen the strainer lock nut, a 14-inch pipe wrench may be used. The jaws should open wide enough to grip the lock nut, but because of the difficulty of keeping a grip on the nut with a pipe wrench a second person's help will probably be needed.

Whichever tool is used, place the handle end of a pair of pliers between the crossbars in the strainer body to keep the body portion of the strainer from turning. Use a medium size screwdriver as in Figure 21-16 for additional leverage. Then unscrew the lock nut.

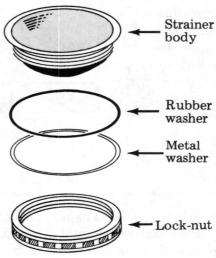

Figure 21-14

After the complete unit has been removed (and if the old unit is to be reused), remove the old putty with a knife or the blade of a screwdriver. Whether the old strainer unit is to be reused or a new unit installed, place a generous amount of putty around the flange portion of the strainer. Reassemble the drain and tighten it into the original position. Fill the sink with water and check for leaks.

Sink Tail Piece Washer

The tail piece for a kitchen sink, shown in Figures 21-15 and 21-17, fits between the strainer unit and the upper (first) slip joint nut on the trap. The top of the tail piece has a flat surface that rests against the base of the strainer unit. It requires a flat 1½-inch washer.

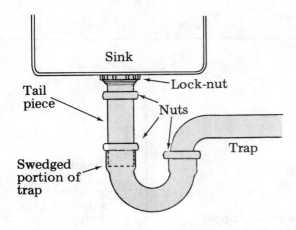

Figure 21-15

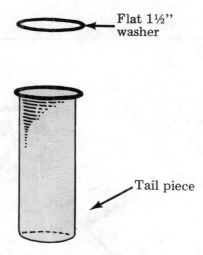

Figure 21-17

To replace this washer, loosen the nut on the strainer unit and also the top nut on the trap. The tail piece should drop downward into the swedged portion of the trap a quarter of an inch or so. This should be sufficient to permit removal of the worn washer and the replacement of a new one without disturbing the rest of the trap. Coat both sides of the washer with pipe compound, reassemble, and tighten both nuts firmly. Fill the sink with water and release it through the drain to check for leaks.

Lavatory Pop-Up Assembly and Washers
The lavatory waste assembly has two

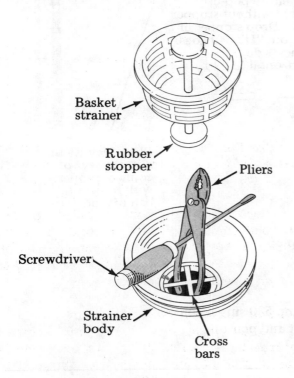

Figure 21-16

washers. The *mack washer* seals the opening on the bottom side of the lavatory. The second washer, or *seal*, prevents leakage around the pop-up rod pivot ball. The parts of a Delta lavatory pop-up assembly are shown in Figure 21-18. A pop-up assembly is also shown in Figure 19-18.

If water drips from the bottom of the lavatory, the mack washer is not holding and should be replaced. (Remember to first place a container beneath the lavatory trap to catch the water in the trap.) Remove the "J" bend portion of the trap. Loosen the compression nut on the pop-up rod and slide the rod free from the pop-up body. Unscrew the pop-up body from the drain assembly flange. Lift the flange out and remove the old putty. Place a small ring of new putty on the underside of the flange and press it firmly down into its original position. Place a new mack washer, beveled side up, over the threaded part of the pop-up body. Tighten the pop-up body into the flange. Use a pipe compound on the threaded parts and on each side of the mack washer. Lift the metal washer to the bottom side of the mack washer and tighten the lock-nut. Replace the "J" bend of the trap. Then connect the pop-up rod to the pop-up body as before and tighten the compression nut. Fill the lavatory bowl with water and test for leaks.

The second washer or seal is located on the pop-up rod assembly. This seal is not difficult to replace. The only tool necessary is a pair of pliers. Most lift rods are secured to the pop-up rod with either a set screw or a spring clip. The complete pop-up assembly can be taken out by

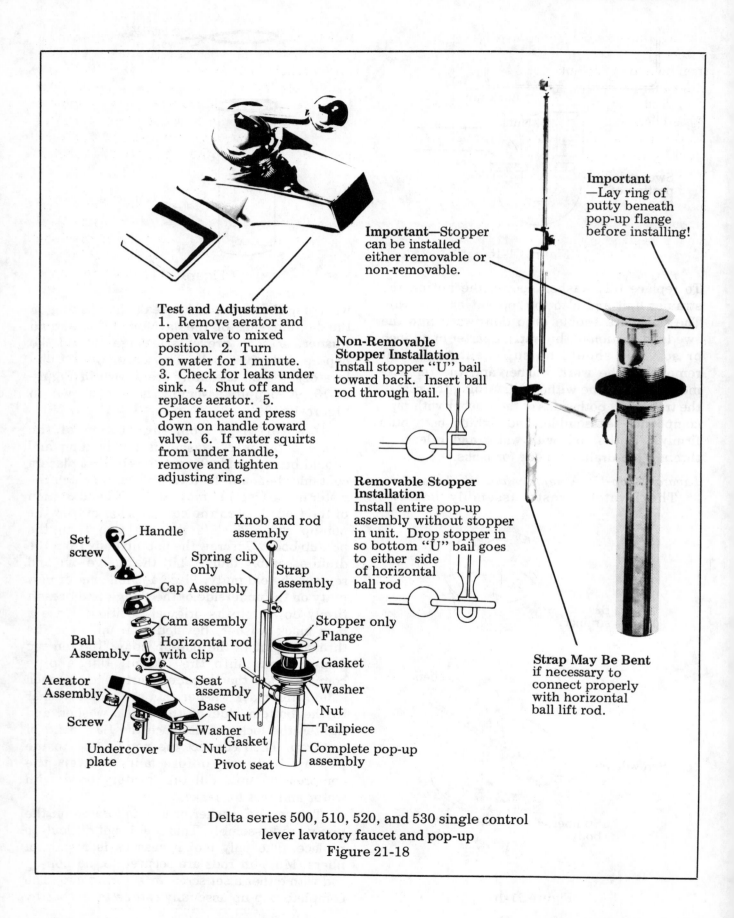

Important—Stopper can be installed either removable or non-removable.

Important —Lay ring of putty beneath pop-up flange before installing!

Test and Adjustment
1. Remove aerator and open valve to mixed position. 2. Turn on water for 1 minute. 3. Check for leaks under sink. 4. Shut off and replace aerator. 5. Open faucet and press down on handle toward valve. 6. If water squirts from under handle, remove and tighten adjusting ring.

Non-Removable Stopper Installation
Install stopper "U" bail toward back. Insert ball rod through bail.

Removable Stopper Installation
Install entire pop-up assembly without stopper in unit. Drop stopper in so bottom "U" bail goes to either side of horizontal ball rod

Strap May Be Bent if necessary to connect properly with horizontal ball lift rod.

Set screw
Handle
Knob and rod assembly
Spring clip only
Strap assembly
Cap Assembly
Cam assembly
Stopper only
Flange
Ball Assembly
Horizontal rod with clip
Gasket
Seat assembly
Washer
Aerator Assembly
Base
Nut
Nut
Screw
Washer
Tailpiece
Undercover plate
Nut
Gasket
Complete pop-up assembly
Pivot seat

Delta series 500, 510, 520, and 530 single control lever lavatory faucet and pop-up
Figure 21-18

freeing the lift rod from the pop-up rod and unscrewing the compression nut. A new seal can then be added. Replace the assembly in reverse order and check for leaks.

Water Closet Tank Washers

Each close coupled closet tank is held securely to the closet bowl by two bolts with washers. If too much pressure is placed against the tank in everyday use, the bolt washers may loosen and create a leak. If water drips from one or both tank bolts, try tightening the bolts. In most cases this stops the leak. The top of the bolt has a slot for a screwdriver head. Use a screwdriver large enough so the head fits snugly into the slot. Tighten the nut beneath the tank with a pair of pliers or a socket wrench. Do not overtighten, as this may crack the tank. If the leak continues, first shut off the water and drain the tank. Then remove the tank bolt and install new washers.

Wall-Mounted Tank Leaks

A wall-mounted water closet tank supplies water to the bowl through a two-inch flush elbow. See Figures 21-19 and 19-26. Two slip nuts and rubber washers secure the flush ell to the closet tank and bowl. If water drips from one or both nuts, the washers should be replaced. If a spud wrench is not available, use a 14-inch pipe wrench to loosen and remove the nuts. Install new two-inch square cut washers. Coat both sides of the washers with pipe compound and retighten the nuts firmly. (See detail of water closet elbow parts, Figure 21-20.) It is not necessary to shut off the water for this repair.

Water Closet Spud Washer

The spud washer fits over the end of the flush valve. It provides a watertight seal between the closet tank and bowl of close coupled water closet combinations. If a leak occurs and water seeps from beneath the tank, the spud washer should be replaced.

Shut off the water to the tank and drain the tank of all water. Disconnect the water supply from the tank. Remove the tank bolts and lift the tank up from the bowl. The spud washer is now accessible for replacement. Reassemble the water closet, fill the tank with water, and check for leaks. (See "Tank Installation," Chapter 19.)

Water Closet Leaks at Floor Connection

When water seeps out from beneath the base of a water closet bowl, the setting seal is not

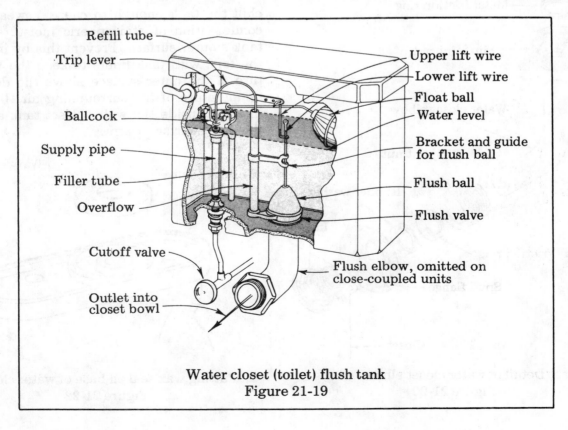

Water closet (toilet) flush tank
Figure 21-19

holding against the heavy water discharge pressure of each flush. This could be the result of building settlement.

To replace the seal, the tank and bowl must be removed completely. Remove the tank as described above for replacing the spud washer. Then remove the closet bolt nuts and lift the bowl up from the floor flange. See Figure 21-21. Remove the old seal and install a new wax seal, as in Figure 21-22. Reinstall both the bowl and the tank as before. (See ''Bowl and Tank Installation,'' Chapter 19.) Fill the tank with water and flush several times to check for leaks.

''Sweating'' Water Closet Tanks

Cold water entering a water closet tank may

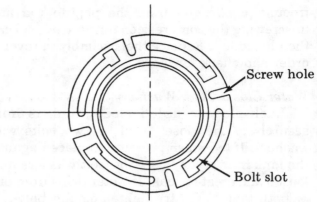

Plan Of Floor Flange

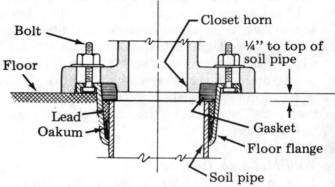

Connection of water closet to floor and soil pipe
Figure 21-21

chill the tank enough to cause ''sweating'' or condensation of atmospheric moisture on the tank's outer surface. Prevent this by insulating the tank with tank liners to keep the temperature of the outer surface above the dew point temperature of the surrounding air. Insulating liners fit inside the water closet tank and keep the outer surface warm.

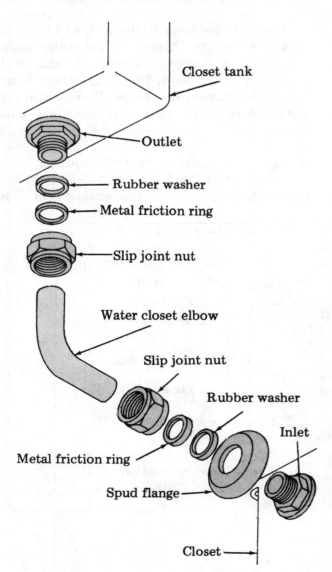

Detail of water closet elbow
Figure 21-20

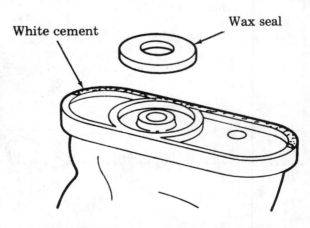

Putting wax seal on base of water closet
Figure 21-22

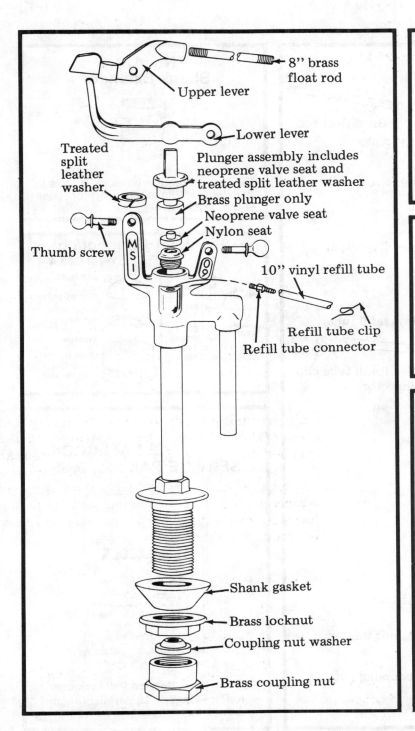

8" brass float rod

Upper lever

Lower lever

Treated split leather washer

Plunger assembly includes neoprene valve seat and treated split leather washer

Brass plunger only

Neoprene valve seat

Nylon seat

Thumb screw

10" vinyl refill tube

Refill tube clip

Refill tube connector

Shank gasket

Brass locknut

Coupling nut washer

Brass coupling nut

Mansfield 09 ballcock
Figure 21-23

No. 422 Shank Extender

Extends all regular ballcock shanks to 2¾" length. Eliminates need for repair shank ballcocks.

Slip Joint Connection

⅝"	West Coast
447-0371 ⅝" Rubber Cone Gasket	447-0281 ½" Rubber Cone Gasket
936-0370 ⅝" Brass Friction Washer	936-0065 ½" Brass Friction Washer
	609-1165 Brass Coupling Nut. ½" Inlet Hole

03-09-XQ16 BALLCOCK SERVICE PAK No. 7066

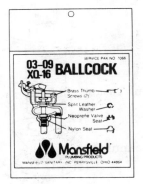

Service Pak includes:
Thumb screw (2) • Nylon seat
Neoprene valve seat • Split washer

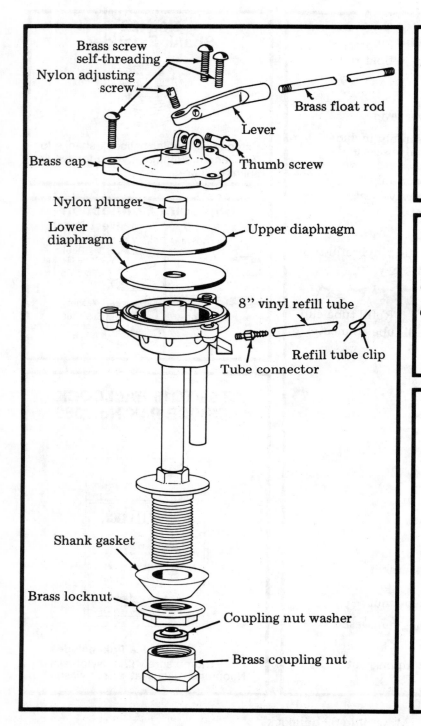

Brass screw self-threading

Nylon adjusting screw

Brass float rod

Lever

Brass cap

Thumb screw

Nylon plunger

Lower diaphragm

Upper diaphragm

8" vinyl refill tube

Tube connector

Refill tube clip

Shank gasket

Brass locknut

Coupling nut washer

Brass coupling nut

No. 422
Shank Extender

Extends all regular ballcock shanks to 2¾" length. Eliminates need for repair shank ballcocks.

Slip Joint Connection

⅝"	West Coast
447-0371 ⅝" Rubber Cone Gasket	447-0281 ½" Rubber Cone Gasket
936-0370 ⅝" Brass Friction Washer	936-0065 ½" Brass Friction Washer
	609-1165 Brass Coupling Nut. ½" Inlet Hole

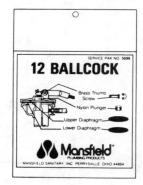

12 BALLCOCK
SERVICE PAK No. 5699

Service Pak includes:
Nylon plunger ● Upper diaphragm
Lower diaphragm ● Thumb screw

Mansfield 12 ballcock
Figure 21-24

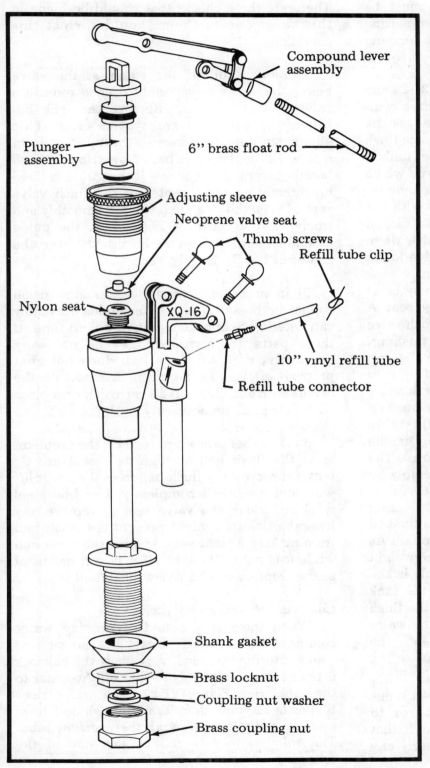

Mansfield XQ-16 ballcock
Figure 21-25

No. 422 Shank Extender

For extending all regular ballcock shanks to 2¾" length. Eliminates need for repair shank ballcocks.

Slip Joint Connection

5⁄8"	West Coast
447-0371 5⁄8" Rubber Cone Gasket	447-0281 ½" Rubber Cone Gasket
936-0370 5⁄8" Brass Friction Washer	936-0065 ½" Brass Friction Washer
	609-1165 Brass Coupling Nut. ½" Inlet Hole

03-09-XQ-16 BALLCOCK SERVICE PAK NO. 7066

Service Pak includes:

Thumb screw (2) • Nylon seat
Neoprene valve seat • Split washer

Repairing Water Closet Tanks

The typical water closet tank in Figure 21-19 contains many working parts which must be kept in good repair. Inspecting and repairing these internal parts ensures efficient flushing action and prevents loss of water through leaking valves. A valve called a *ballcock* is the main working part. See Figures 21-23, 21-24 and 21-25, for components and parts of three commonly used ballcocks. The ballcock causes the closet tank to refill automatically after each flush and shuts off the water when the closet tank is full. The ballcock is attached to a *float rod* which is in turn attached to a *float ball*. As the tank refills after a flush, the float ball rises with the water level in the closet tank. The rod attached to the float ball closes the ballcock valve when the proper level is reached. A flush valve forms the outlet connection between the closet tank and the water closet bowl. This valve may be of brass, copper, or plastic with a machined seat. A rubber *flush ball* or *flapper* keeps the flush valve opening closed except during actual flushing. The rubber flush ball is connected to the *trip lever* by a lower and upper *lift wire*. This wire raises the rubber flush ball to begin the flushing cycle when the operating handle is pushed down. The *flapper type* uses an adjustable beaded chain attached to the operating handle in lieu of the lower and upper lift wires. The flapper does not require a guide assembly as does the rubber flush ball. The float ball is full of air and remains in the raised position, floating on the water's surface until the toilet is flushed and the water level in the tank begins to fall. As it descends within the draining tank, the weight of the float ball opens the valve assembly in the ballcock. Fresh water begins to enter the tank through the valve, and gravity causes the flush ball to return to its seat as the last of the water passes through the flush valve and into the closet bowl.

Flushing mechanisms and water closet tanks vary in design from one manufacturer to another. But they are all similar enough that these general repair instructions fit nearly any design.

Water Dripping Into Bowl

When a trickle of water continues to flow from the tank into the bowl after the flushing cycle is complete, the rubber flush ball is probably not seating properly. See Figure 21-26. There are three checks that should be made in this case; one of them should correct this condition.

1) First, shut off the water at the valve below the water closet tank. (Codes require a valve at this location.) Remove the tank lid. Flush the toilet to empty the tank. This eliminates any resistance to the up and down movement of the flush ball. Now trip the flush handle several times to see if the guide is lined up directly over the center of the flush valve seat. If it is, the flush ball will fall smoothly and freely onto its seat. If not, adjust the guide assembly until the guide is directly over the center of the flush valve seat.

2) In areas of corrosive water, the usual copper or brass lift wires and guide assembly can deteriorate in a comparatively short time. If these parts are corroded, they do not work smoothly, and the flush ball does not seat properly. If this is the case, disassemble the mechanism and clean off all corrosion or replace with new parts as necessary.

3) If neither procedure corrects the problem, raise the flush ball as high as it will go. Or better, unscrew the flush ball from the lower lift wire and remove it completely. Use fine steel wool and clean the valve seat to remove any irregularities that might prevent the flush ball from making a tight seal. Inspect the flush ball while it is out of the tank. If it is soft or out of shape, replace it with a new flush ball.

Continual Running of Water

When there is a sound of running water coming from the water closet but no sign of water entering the bowl, very likely the ballcock is not closing completely to shut off the water to the tank. See Figure 21-27. The water rises higher in the tank than its normal shutoff level and leaves the tank through the overflow tube. The solution to this can be any one of the following:

1) If the float ball is riding high on the surface of the water, bend the float rod down slightly. Use both hands and work carefully to avoid placing too much strain on the ballcock

assembly. Then flush the tank. The water should not rise as high this time. (The normal water level is approximately one inch below the overflow tube.)

2) If the float ball rides low in the tank water, it has probably lost its buoyancy (become water logged) and should be replaced. Unscrew the float ball from the float rod and replace it with one of copper, plastic, or foam. One material works as well as another.

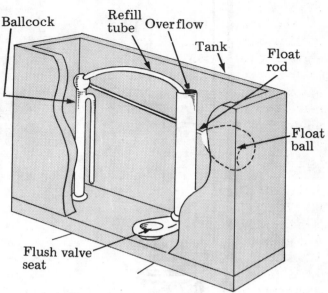

Continual running water
Figure 21-27

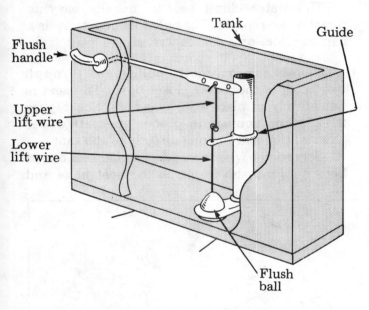

Water dripping into bowl
Figure 21-26

3) If the float ball is not the cause, the ballcock should be repaired or replaced.

The valve of the ballcock is made of brass, copper, or plastic and has a machined plunger. The plunger has two washers. One is a split round leather or rubber washer which encircles the plunger to keep water from squirting out around the top of the ballcock. The other washer is a round washer in the bottom of the plunger. This washer closes off the flow of water to the closet tank when the float ball reaches the proper water level. These two washers are the likely cause of the running water sound.

To replace the washers, shut off the water and drain the tank. Two thumbscrews hold the plunger in place. Unscrew both of these and lift out the plunger. Unscrew the cup on the bottom and insert a new washer of rubber or leather. Remove the split leather washer and replace it

with a new one. If the ballcock assembly is badly corroded or is old, install a new unit rather than replace the washers.

To replace the ballcock, first shut off the water to the closet tank. Flush the tank. This leaves about an inch of water in the bottom of the tank that cannot drain away. Place a shallow pan directly beneath the shutoff valve to catch this water. Unscrew the float ball, the float rod, and refill tube and remove these from the tank. Use a pair of channel lock pliers to loosen the supply nut and ballcock shank nut. Lift the ballcock up and out. Replace it with a new ballcock. Be sure the beveled part of the rubber washer is facing down toward the bottom of the tank. Coat the washer and the threads with pipe compound. Retighten the ballcock shank nut snugly. Do not overtighten, as this may crack the tank. Reconnect the supply nut and replace the float rod, the float, and the refill tube. Open the shutoff valve to refill the closet tank. Check the supply and ballcock shank nuts for leaks.

Flush Valve

The flush valve (Figure 21-28) consists of a machined seat and an overflow tube. It is seldom the cause of a leakage problem. The machined seat is not subject to wear since the rubber flush ball which closes the opening is soft. The valve is made of brass, copper, or

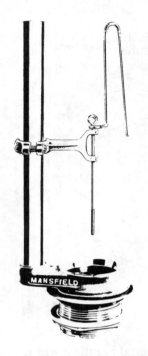

Mansfield Douglas-type flush valve
Figure 21-28

the overflow. Insert the lower lift wire through the ring at the bottom of the upper lift wire and the opening in the guide holder. Screw the threaded portion of the lower lift wire into the rubber flush ball. Center the flush ball over the flush valve opening. Adjust the loose guide holder forward or backward until the ball seats properly. Securely tighten the guide holder to the overflow tube while it is in that position. See Figure 21-28.

Water Closet Seats

The water closet seat is usually easy to remove and replace. The only tool necessary is a pair of pliers or a socket wrench.

Two brass or plastic bolts pass through two holes in the water closet bowl. The seat is held firmly in place with two nuts. See Figure 21-29. For some one-piece (special) water closets, the seat fastens directly to the tank.

Sometimes when removing an old seat one or both nuts may be frozen to the seat bolts and

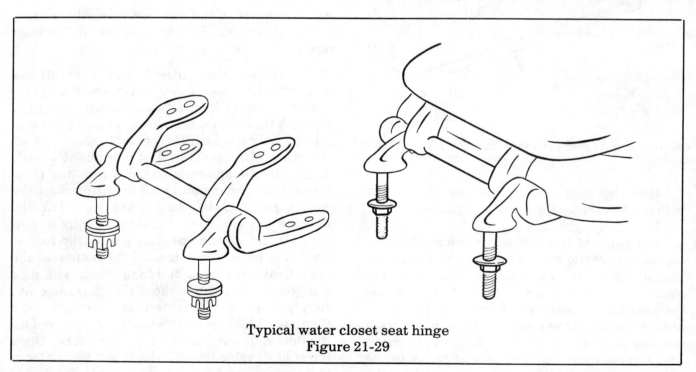

Typical water closet seat hinge
Figure 21-29

plastic. It is not usually subject to ordinary corrosion and usually lasts for the life of the closet tank. If repair is necessary, the flush valve must be replaced as a unit.

Flush Ball Guide

To install this unit, swing the refill tube to one side and loosely place the guide holder over

will not loosen. If time is not critical, place some penetrating oil around each bolt. If this does not loosen the nuts in half an hour or so, cut through the bolts with a new fine tooth hacksaw blade without the frame. Insert the blade between the top of the bowl and the hinge portion of the seat. Push the blade slowly and easily back and forth until you cut through the bolt. Then replace the seat. Tighten the nuts snugly so the seat will not wobble.

Note: Never sit or stand on a seat cover (lid). This places a strain on the hinge and may cause the seat cover to split. The hinge may break or the hinge screws may pull out.

Use only a mild soap, a soft cloth and warm water when cleaning a seat. Abrasive cleaners destroy the seat's finish.

Using Valves to Locate the Leak

Valves may be used to determine the location of a leak in a water distribution system. Follow the steps below to isolate the leak to a given section of piping. This method can save considerable time and can help you decide whether to repair the leak or to repipe with a new material.

1) Close the house valve. Remove the cover of the water meter and watch for movement of the meter's hands. If the hands are moving, a leak has developed in the building water service pipe between the meter and the house valve.

The pipe would then have to be uncovered until the leak is found. Pipes that are split because of freezing or that are weakened by corrosion must of course be replaced. In an emergency a small hole in a pipe can often be repaired with a pipe clamp. (A pipe clamp consists of a rubber patch and a hinged frame that is pulled together tightly by two bolts. The rubber patch seals the leak.) If a pipe clamp is not available, a sharpened wooden stick can be forced tightly into the hole. As the wood swells, it seals the leak without materially affecting the flow of water through the pipe. Remember, these measures are for emergency repair *only* and should be followed by permanent repair as soon as practicable.

If a small section of piping is split or corroded through, cut out the bad section with a hacksaw. Then join the two sections of piping with a *dresser coupling* to make a watertight connection. (See Figure 21-30.) A dresser coupling can provide a permanent repair if the rest of the piping is sound.

Sometimes a leak occurs at a threaded connection. This can often be stopped by unscrewing the fitting and applying pipe

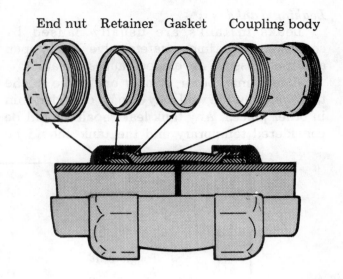

End nut   Retainer   Gasket   Coupling body

Dresser couplings
Figure 21-30

compound to the threads. This seals the joint when the fitting is screwed onto the pipe again.

If you are sure that the building water service piping is not leaking, then try step two.

2) Open the house valve and close the control valve to the water heater. Recheck the water meter hands for movement. If the hands are moving, the leak is located in the cold water distributing pipes beneath the floor. Rule out a slow leak in the cold water distributing pipes by placing an ear against one of the inside fixture faucets. If no sound of running water is detected, the leak is in the hot water distributing pipes. Leaks in cold water distribution systems under concrete floors are very difficult to locate. When a leak is traced to the cold water system, usually it is to the advantage of the customer to simply abandon the old pipes underneath and install new ones overhead or around the outside of the building.

Leaks in hot water piping under a concrete slab are much easier to locate so repairs can be made. Locate hot spots by removing your shoes and walking over the general area where hot water pipes are installed. After breaking a hole in the floor and repairing the leak, let the hole remain open for a few days to make certain the repair has been successful.

If the hot water pipe shows signs of considerable deterioration, recommend that new piping be installed overhead.

Leaking Tanks

Leaks in tanks are usually caused by corrosion. Sometimes a safety valve fails to open and the pressure developed causes a leak.

A leak may occur at only one place in the tank wall, but the wall may also be corroded thin in other places. Any tank leak repair should be considered temporary and the tank should be replaced as soon as possible.

A leak can be temporarily repaired with a toggle bolt, a rubber gasket, and a metal washer as shown in Figure 21-31. You may have to drill the hole larger to insert the toggle bolt. Draw the bolt up tight to compress the rubber gasket against the tank wall to contain and seal the leak.

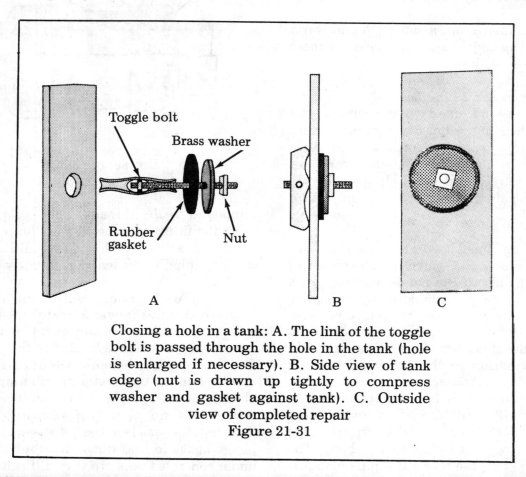

Closing a hole in a tank: A. The link of the toggle bolt is passed through the hole in the tank (hole is enlarged if necessary). B. Side view of tank edge (nut is drawn up tightly to compress washer and gasket against tank). C. Outside view of completed repair
Figure 21-31

Questions

1- What portion of a plumbing system is generally considered to be repair work?

2- What types of valves are normally used to control the water to a building?

3- Valve repairs are generally limited to _____ and _____.

4- When a globe valve does not control the water, what must be done to correct it?

5- What is the *first* step in disassembling a control valve?

6- What may be the problem when a washer requires frequent replacement?

7- How are replaceable valve seats identified?

8- What is the name of the tool used to remove pits or rough surfaces from valve or faucet seats?

9- What should be done first to repair a leak around the stem of a control valve?

10- What type of faucet is now most often used on plumbing fixtures?

11- What is the most common repair required on fixture faucets?

12- In which direction must a valve handle be turned to shut off the water?

13- Where is a house control valve usually located?

14- When water leaks from under the handle, what is usually needed to repair it?

15- What do newer type compression faucets use to prevent leaks from under the handle?

16- What repair is needed when water drips from the spout?

17- What is the procedure for removing a frozen faucet handle when a "handle puller" is not available?

18- When water flowing through a faucet is "noisy," what (other than a loose washer) may be the cause?

19- What must be replaced when water leaks around the swing spout of a faucet?

20- What precautions should be taken when pliers are needed to loosen a chrome swing spout lock nut?

21- Why do single handle mixing faucets last longer than other types before repairs are needed?

22- How do the Delta and Moen faucets differ in controlling water temperature and flow?

23- Describe and size the two washers in a sink trap.

24- What is the U-shaped portion of a P-trap called?

25- Name the two parts of a sink basket strainer that usually must be replaced because of age.

26- What distinguishes a washer designed for a sink tailpiece from one designed for a P-trap?

27- What is the name of the washer that seals the bottom of a lavatory bowl?

28- What should be tried first to stop a leak at the water closet tank bolt?

29- Why is it important not to overtighten water closet tank bolts?

30- What is the device that provides the waterway between a wall-mounted tank and a closet bowl?

31- Where is a water closet spud washer located?

32- What would be the probable cause of a water closet leak at the floor connection?

33- Why does a water closet tank sometimes "sweat"?

34- What may be used to stop tanks from "sweating"?

35- What is the valve called that controls the water to a water closet tank?

36- Describe the three checks that should be made when a trickle of water continues to flow from the tank into the bowl.

37- Describe the three checks that should be made when there is the continual sound of running water coming from a water closet but no visible water entering the bowl.

38- What must one be familiar with to repair a ballcock?

39- What should be done first to repair or replace a ballcock?

40- What does the guide in a water closet tank do?

41- When the nuts are frozen to the bolts of a water closet seat and will not come loose, how can you remove the old seat?

Simple Gas Installations

Matter occurs in three physical states: solid, liquid, and gas. Although gas is much ligher per cubic measure than the other two forms of matter, it does have weight. It can be forced through very small spaces, a valuable characteristic. Gas has neither a fixed shape nor a fixed volume. It is made of constantly moving atoms. When these atoms are forced into a container, they take on the container's shape but occupy only about one thousandth of the container's interior space.

Gas liquifies when it is cooled to below its boiling point. When this temperature is reached, the gas particles are pulled together to form a liquid. This principle is used in making liquid oxygen. Since natural gas is clean, dry, and has no odor, a gas leak in a pipeline or in pipes within a building might go undetected until an explosion occurred. Therefore, a chemical odorant is added to gas before it enters the pipelines. This odor warns anyone in the area of escaping gas before the concentration can reach danger level.

Kinds of Gas

Natural gas or methane, contains chemical impurities which are valuable for uses other than as a fuel. These impurities are removed before the gas is piped to the customer. The natural gas which millions of consumers use as fuel in homes and industries is known as *dry* or *sweet gas*. Natural gas is not poisonous but can cause suffocation in a closed space. It is also explosive under certain conditions.

Manufactured gas is produced chiefly from coal. It burns with a blue flame and is generally added to other fuels to increase its heating capacity. Manufactured gas is also used by consumers as fuel in homes and industries. It can be poisonous since it contains carbon monoxide. It is explosive under certain conditions.

Liquified petroleum gas is also known as LP or bottled gas. It is produced in plants that process natural gas. LP gas consists primarily of butane or propane, or a mixture of both. LP gas liquifies under moderate pressure, and this makes it easy to transport and store in special tanks. When an LPG tank supplies a building's gas piping system, the liquid is allowed to drop to normal atmospheric pressure and temperature and return to its original gaseous state.

LP gas is heavier than air, colorless, and nonpoisonous. Since it is easily containerized and transported, it is convenient to use as fuel for homes and businesses in remote areas.

Gas Piping

The *gas service pipe* is sized and installed under the direct control of the gas supplier. This pipe extends from its connection at the gas supplier's pipe (located on public property) to the gas meter (usually located outside the building). The installer is responsible for sizing and installing the gas supply piping only within the building.

Plumbing codes seldom if ever govern the sizing or installation of gas supply systems. The plumbing code usually refers the installer to the local gas code. The gas code is a separate book but is similar to the plumbing code in complexity. Trying to learn and comply with these codes can be frustrating and discouraging to anyone wishing to do simple gas installations.

Unfortunately, information relevant to the installer is scattered throughout the local gas code. This chapter and the next are intended to give plumbers making simple gas installations a working knowledge of the essential requirements of the gas code. The information here does not replace the gas code but should meet most code requirements for residential or simple installations. Commercial installations and less common materials used with gas systems are examined in the more advanced *Plumber's Handbook* by this author.

Sizing Gas Systems

The building gas main and branch lines can be sized once you know the maximum gas demand at each appliance outlet and the length of piping required to reach the most remote outlet. Other sizing factors such as pressure loss, specific gravity and diversity are already accounted for in the tables in the gas code and in Table 22-1.

Gas appliance manufacturers always attach a metal data plate to each appliance in a visible location. This plate shows the maximum input rate in Btu's. (Btu is the abbreviation for "British thermal unit." One Btu is the quantity of heat required to raise the temperature of one pound of water one degree Fahrenheit.)

The tables in the code prescribe the sizing of gas piping in cubic feet of gas rather than in Btu's. Thus, each Btu input rating has to be converted to cubic feet of gas before sizing the distribution piping.

You can assume that each cubic foot of natural gas releases 1,000 Btu's per hour. The Btu rating of some gas varies from this figure, but using 1,000 Btu's per cubic foot is a safe assumption.

Assume you are sizing pipe for a cooking range with a maximum demand of 68,000 Btu's per hour. Divide the value in Btu's by 1,000 to find the demand in cubic feet per hour. Thus, 68,000 Btu's ÷ 1,000 = 68 cubic feet per hour.

On used appliances the Btu rating on the data plate may not be legible or the plate itself may be missing. In such a case, it is a safe practice to make the appliance inlet pipe no smaller than the supply pipe serving the appliance. A larger size supply pipe could be installed without violating the code, but the appliance would not function any better. The supply pipe should never be smaller than the appliance's inlet pipe, and under no circumstances can it be smaller than ½ inch.

Table 22-1 can be used with a sizing method that is quick and easy. *Use it only to help you understand how to use the table in your local gas code.* The low pressure gas table, as here, is the one most commonly used by professionals. Low pressure gas is used in millions of home and business installations.

Length In Feet	Nominal Iron Pipe Size in Inches*			
	½	¾	1	
10	176	361	681	
20	121	251	466	
30	98	201	376	
40	83	171	321	
50	74	152	286	
60	67	139	261	Residential
70	62	126	241	
80	58	119	221	
90	54	111	206	
Column One	Two	Three	Four	

*More complete sizes are found in your local gas code and in *Plumber's Handbook* by this author.

Maximum capacity of pipe in cubic feet of gas per hour (natural gas @ 1,000 Btu/cubic foot)

Table 22-1

Figure 22-2 shows a simple gas piping system similar to what you might find in most single family residences. Note that each section of piping must be sized to serve the Btu input rating of the appropriate appliance outlet. Study this illustration and the explanation in the rest of this chapter until you can size each section of pipe correctly.

To size the gas piping system in Figure 22-2, use Table 22-1 and follow the procedure outlined below. Assume that the total developed length of gas piping in Figure 22-2 is 60 feet from the meter to outlet A. Find that distance in Table 22-1. Underline all figures in each vertical demand column opposite 60. (This row is shaded in Table 22-1.) The pipe size listed at the top of each demand column is the size that carries the volumes listed in that column.

You are now ready to size each section of pipe shown in Figure 22-2. The maximum gas demand of outlet A is 40,000 Btu's per hour, or 40 cubic feet per hour (c.f.h.). The correct pipe size is ½ inch (Column Two). The maximum demand of outlet B is 4,000 Btu's or 4 c.f.h., and the correct pipe size is ½ inch (Column Two). The combined maximum demand for A and B is 44 c.f.h. The correct pipe size for section C is ½ inch (Column Two). Maximum demand of outlet D is 68,000 Btu per hour, or 68 c.f.h., and the correct pipe size is ¾ inch (Column Three). The combined maximum gas demand for A, B, C, and D is 112 c.f.h. The correct pipe size for section E is ¾ inch (Column Three). The maximum demand of outlet F is 140,000 Btu per hour, or 140 c.f.h., and the correct pipe size is 1 inch (Column Four). The combined maximum gas demand for A, B, C, D, E, and F is 252 c.f.h. This is the maximum gas demand for the entire residence. The correct pipe size for section G is 1 inch (Column Four).

When you master the sizing procedure in Figure 22-2, size each section of piping in Figures 22-3 and 22-4 and have someone knowledgeable in plumbing check your work. The more practice you get, the more proficient you will become.

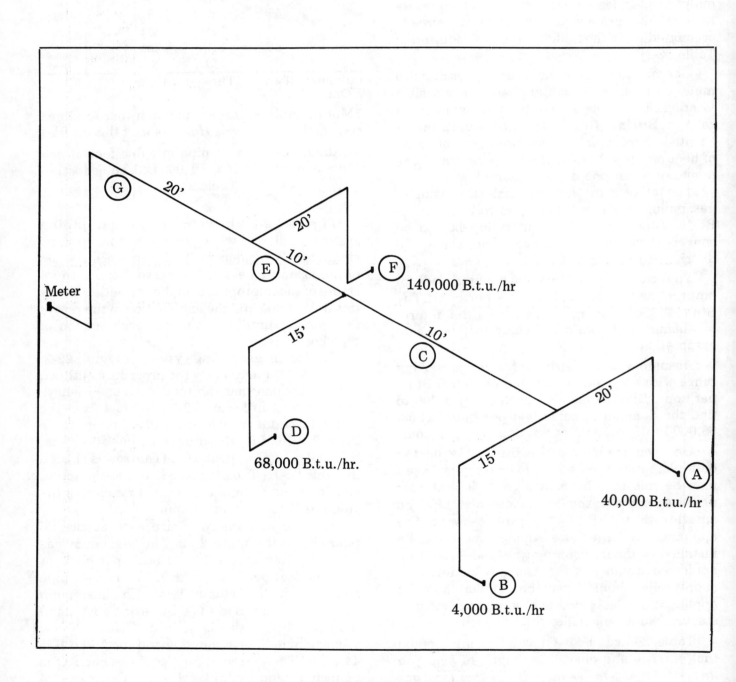

Natural gas residential installation
Figure 22-2

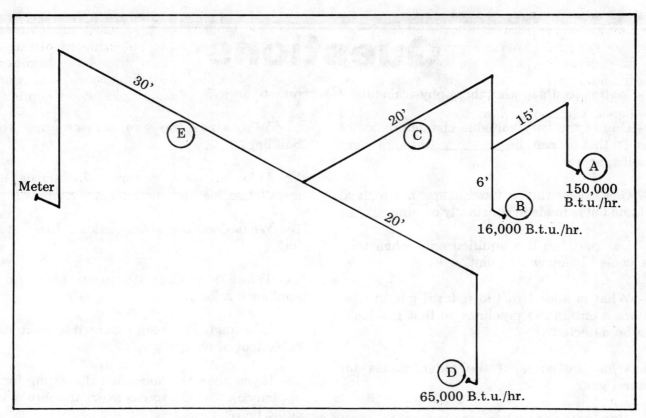

Figure 22-3

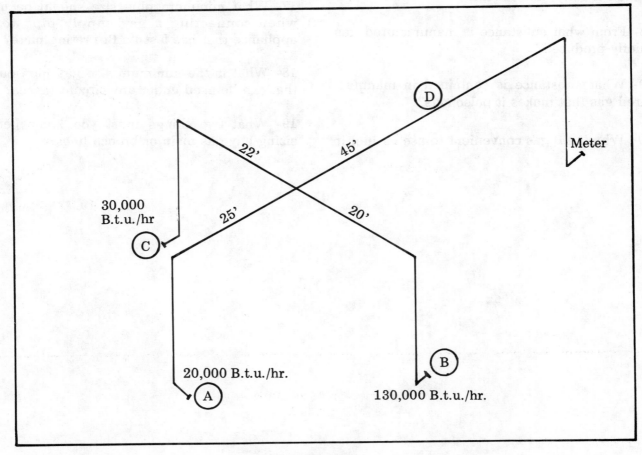

Figure 22-4

Questions

1- Matter occurs in what three physical states?

2- One of the very valuable characteristics of gas is that it can be _____ through very small spaces.

3- Gas has neither a fixed shape nor a fixed volume but is made of constantly moving _____.

4- Gas particles are liquified only when they are cooled below what point?

5- What is added to the natural gas system before it enters the pipelines so that gas leaks can be detected?

6- What are some of the other names for natural gas?

7- How can natural gas cause death even though it is not poisonous?

8- From what substance is manufactured gas chiefly produced?

9- What substance is contained in manufactured gas that makes it poisonous?

10- Why is LP gas convenient to use as fuel in remote areas?

11- Who sizes the gas service pipe to a building?

12- Who or what governs the sizing and installation methods of gas supply piping?

13- What does the abbreviation ''Btu'' stand for?

14- What does the Btu input rate for an appliance reflect?

15- How many Btu's are assumed to be in each cubic foot of natural gas?

16- If you know the maximum Btu rating for an appliance, how do you convert the Btu's into cubic feet?

17- What minimum pipe size should be used when connecting a gas supply pipe to an appliance that has lost its Btu rating plate?

18- What is the minimum size gas pipe outlet that can be used under any circumstances?

19- What two things must you know before sizing any gas main or branch lines?

23
Materials For Simple Gas Installations

Your local gas code regulates the materials and installation methods for gas systems. Certain materials are limited in use and others are prohibited. This chapter reviews the two more commonly used piping materials for residential and simple gas installations.

When selecting the materials for gas supply pipes or fittings, consider the characteristics of your particular gas supply and its effect on the inside of the pipe. For example, gases in certain areas are classed as corrosive. These gases contain an average of 0.3 or more grains of hydrogen sulfide per 100 cubic feet. If your community gas supply is corrosive, certain types of materials commonly used for gas piping will not be accepted by your local gas supplier. If the corrosive character of the gas to be used is not known, call your gas supplier before the gas system is installed.

Materials

Galvanized steel pipe is the most common gas piping material in use. Because of its versatility and strength, it can be installed equally well underground and above ground. Galvanized steel pipe can also be installed where there is corrosive gas. Galvanized steel pipe in simple gas installations is threaded. Fittings should be of the same material as the pipe. Threaded joints must be sealed tight with an approved pipe compound. Galvanized steel pipe can not be installed under a concrete slab.

Copper seamless pipe or copper tubing, type K or L, can be used for interior gas piping if first approved by the local authority or the area gas supplier. Copper pipe or copper tubing can not be used if the gas is corrosive. Copper pipe or copper tubing can be used outside underground but can not be installed under a concrete slab. Fittings must be of the same material as the pipe or tubing. Where copper pipe is selected, joints between fittings and pipe should be made with a hard solder, usually a silver solder.

Joints between fittings and pipe can also be brazed. This usually means that a filler of brass is used to join the metals. This type of gas piping generally requires special equipment and a knowledgeable plumbing professional and should not be attempted by an inexperience plumber.

Copper tubing may be used with approved gas flare fittings. Copper tubing may be concealed within partitions if the tubing is continuous and of one piece. Flare fittings used for gas openings or for joining two sections of pipe must not be concealed. Copper tubing may be used to advantage where the system can be exposed or is accessible, as in a building crawl area or basement.

Installation of Gas Piping

Place underground gas piping deep enough to protect the pipe from damage from sharp tools. A six-inch depth is a good minimum, though this is not governed by the code. Pipe installed in corrosive soils should be protected with an approved wrapping or one or two coats of asphaltum paint. Where freezing temperatures occur, the trench bottom should be below the frost line to prevent freezing and rupturing of the pipe. Insulate pipe where it enters a building above ground, in crawl spaces, and anywhere it is not protected from the cold. This is necessary because many gases contain moisture which can freeze and block the flow of gas through the pipe.

Lay underground gas piping in open trenches on a firm bed of earth. The pipe should be securely supported to prevent sagging and excessive stress during backfill. Use only fine material for backfilling.

Occasionally gas piping must be installed underground under a building slab. When this is unavoidable, the code permits this type of installation only if the following three conditions are met: 1) The entire length of gas piping up and through the floor must be encased in conduit. 2) The termination of the conduit above the floor must be sealed to prevent the entrance of any gas into the building in case of a leak. 3) The termination of the conduit outside the building must be tightly sealed to prevent water from entering the conduit. A vent must be extended above grade and secured to the conduit. This vent conveys any leaking gas to the outside of the building. See Figure 23-1.

Installing gas appliances can be a problem when they are to be located on a concrete slab in the center of a room away from adjoining partitions. Installation can be rather unsightly when walls are not available to conceal the gas piping. In this case, the piping must be installed in an open channel cut into the concrete floor. The channel must be as deep as the outside diameter of the pipe. A removable grill or cover must be installed over the channel so there is access for repair or replacement of the gas piping. See Figure 23-2.

When gas piping must pass through

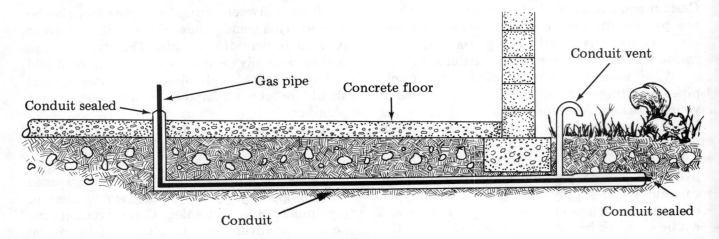

Figure 23-1

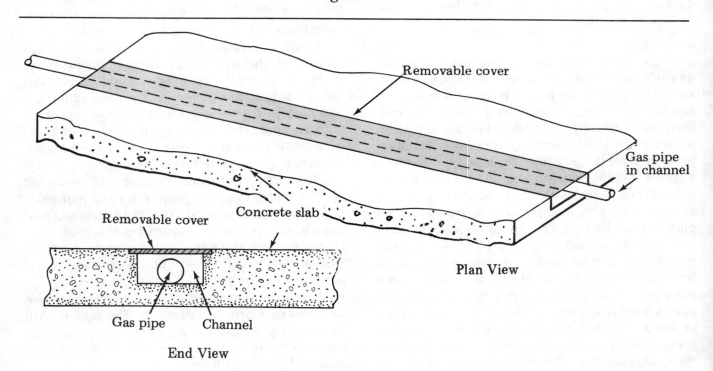

Plan View

End View

Figure 23-2

suspended concrete slabs (as in a basement ceiling) or through masonry walls, the pipe must be properly protected against corrosion by sleeving or painting with asphaltum paint. Where gas piping must be installed in vertical masonry walls, adequate chases must be provided to protect the pipe. See Figure 23-3.

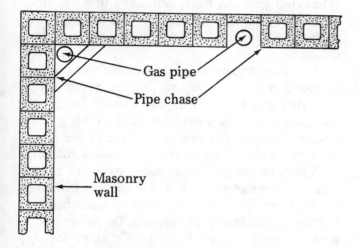

Gas piping installed in vertical masonry wall
Plan view
Figure 23-3

Horizontal and vertical supports for galvanized gas piping, copper pipe and tubing should be the same as those described for similar water piping materials in Chapter 16.

When installing galvanized gas piping, copper pipe and tubing horizontally in wood partition walls, give the following protection to the building structure and to the pipe or tubing: 1) Install short runs of horizontal gas piping or tubing which do not require additional joints through a hole drilled in the center of the partition stud. 2) Install longer runs of horizontal gas piping or tubing in notches cut deep enough into the wall stud to conceal the pipe or tubing. (To avoid weakening the stud, don't cut **deeper than 40% of its total width**). 3) Protect copper pipe or tubing in a notched partition with a metal stud guard to avoid penetration by lath nails. See Figure 16-6.

Metal stud partitions are replacing wood partitions in many new buildings. The metal studs are hollow rather than solid. The gas pipe or tubing is installed through manufactured openings provided in the center of the stud. The pipe or tubing must be wrapped with an

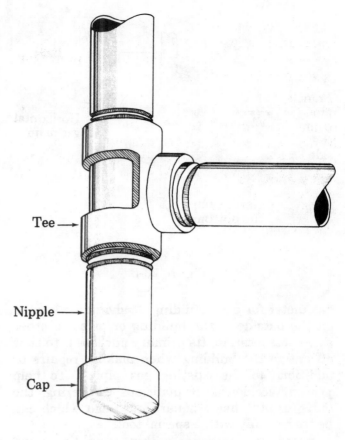

Gas drip pipe
Figure 23-4

approved material to prevent contact with the studs. Secure the pipe to the studs with tie wire.

Drip Pipe

The gas main must be installed to drain dry, and the pitch or grade must be toward the gas meter. Many gases contain moisture. Most codes require that drip pipes be provided to receive condensation that forms within the pipe. The drip pipe must also be accessible for emptying. It is usually assembled from a tee, nipple, and cap as shown in Figure 23-4 and should not be smaller than the pipe it serves. The drip pipe, if exposed, should be protected from freezing in colder climate zones.

Connect gas branch pipes to other horizontal pipes at the *top* or the *side* of the feeder pipe and never from the bottom. This keeps condensate from filling and obstructing the flow of gas through the branch lines. See Figure 23-5.

Shutoff Valves

A shutoff valve must be provided near the

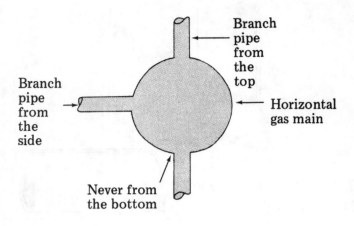

Figure 23-5

gas meter for each building. It should be located on the outside of the building or at some other accessible location. Its primary purpose is to shut off gas to the building when making repairs or additions to the existing gas piping. To help prevent accidental or malicious tampering, the valve usually has a square nut head which can be turned only with a special tool.

Each gas appliance within a building must have an accessible, manually operated shutoff valve. This valve must have a lever handle that does not require a tool, so that a homeowner may shut off the gas in case of emergency. This valve should be located as close as possible to the gas outlet pipe serving each appliance.

Shutoff valves are manufactured in two types: a straight pattern and an angle pattern. The shutoff valve must be installed not more than six feet from the appliance it serves.

Galvanized or copper pipe installed in a concealed location must not have unions, right or left couplings, running threads, bushings, or swing joints. Copper tubing must not have flare fittings.

Use only a ground joint union to make a new connection to an existing line in a concealed threaded gas piping system. Punch the center nut on the joint to prevent it from working loose under vibration. No new connection is permitted on a concealed copper tubing gas installation.

Threads for gas pipe must conform to the standards adopted by the American Standards Association. See Table 16-12 for the number and length of standard pipe threads. Do not use pipe with chipped or torn threads. Cutting, threading, and reaming gas pipe is done the same way as for water pipe in Chapter 16.

Do not conceal the completed gas installation until it is pressure tested. Cap each outlet and install a pressure gauge on one of the outlets. Pressure test at 10 to 20 pounds. The system must remain airtight (no loss of pressure) until it

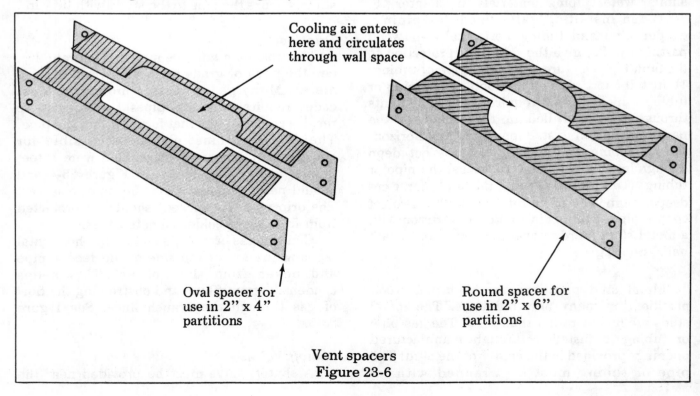

Vent spacers
Figure 23-6

is inspected. Leave the caps in place until the appliances are ready to be connected. If a leak occurs during the test of the rough piping, find its location by brushing a liquid soap around each joint. The leaking joint will blow bubbles and repairs can then be made.

It is the plumber's job to connect the gas from the wall outlet to the appliance. This connection can be made with either a rigid pipe or an approved flexible connector. When the appliances have been set in place and connected, the gas supplier purges the lines of air, checks for leaks at the joints that connect the

appliance to the gas system, lights the pilot lights, and adjusts each appliance.

Gas appliances such as water heaters and clothes dryers can be installed on the floor of a residental garage under certain circumstances: the floor of the garage must be higher than the driveway or adjacent ground and the combustion chamber must be a minimum of 18 inches above the floor or adjacent ground. The appliance may be installed in a separate room off the garage if the walls, ceiling, and door of the room have a one-hour fire rating. Ventilation must be provided through permanent

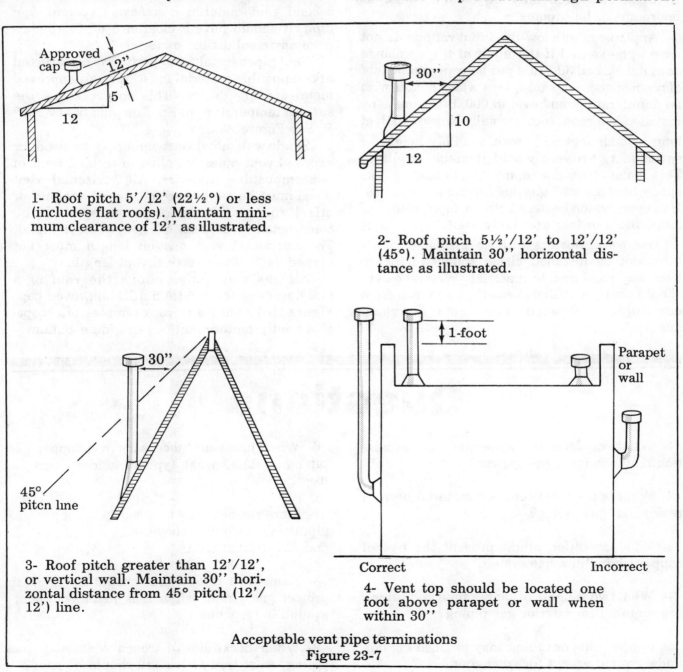

1- Roof pitch 5'/12' (22½°) or less (includes flat roofs). Maintain minimum clearance of 12'' as illustrated.

2- Roof pitch 5½'/12' to 12'/12' (45°). Maintain 30'' horizontal distance as illustrated.

3- Roof pitch greater than 12'/12', or vertical wall. Maintain 30'' horizontal distance from 45° pitch (12'/12') line.

4- Vent top should be located one foot above parapet or wall when within 30''

Acceptable vent pipe terminations
Figure 23-7

openings with a total free area of one square inch for each 5,000 Btu's per hour of input rating. One vent opening must be a minimum of 12 inches above the floor. The second opening should be a minimum of 12 inches below the ceiling. This permits air circulation for combustion and dilution of the flue gases. See Figure 14-1.

Gas appliances should be installed so there is access to the appliance for repairs, replacement, and cleaning, as well as for the intended use. Water heaters must not be installed in any living area which may be closed, such as bedrooms or bathrooms.

Appliances with low Btu input ratings do not have to be vented if the input of the appliance does not exceed 30 Btu's per hour per cubic foot of room space. For example, a water heater with an input rating not over 5,000 Btu's may be installed in a room (not normally closed) six feet long by four feet wide with a ceiling height of seven feet. No vent would then be required. This holds true for many free-standing gas space heaters and gas log burning fireplaces. However, water heaters with an input rating of 5,000 Btu's or less are rarely used.

Gas water heaters with an insulated jacket must not be installed closer than two inches from any combustible material. Water heaters should not be installed closer than one inch from enclosures constructed of one hour fire rated materials.

Appliances that must be vented should be installed as close to the vent pipe as possible. If a draft hood is required, the vent pipe should never be smaller than the opening of the draft hood. Gas appliances and their vents should be installed so there is sufficient clearance from combustible materials to avoid a fire hazard.

There are two acceptable types of concealed vent pipe and fittings. The most common is the *double wall metal pipe*. Stamped in the metal is the clearance distance specified by the manufacturer. A one-inch clearance from any combustible material is recommended. The second vent material is *asbestos cement flue pipe*. It should have a clearance of 1½ inches from any combustible material.

Vent pipes installed in partitions constructed of combustible material must have an approved metal spacing device. This spacer keeps the surface temperature of the flue pipe below 160° F. See Figure 23-6.

Single wall metal vent piping can be used for exposed vent pipes installed in a room built of noncombustible material. All horizontal vent pipes must be supported to prevent sagging or misalignment. Straps or hangers should be constructed of at least 20 gauge sheet metal. The horizontal vent section length must not exceed 75% of the vertical vent length.

All gas vent pipes above the roof of a building must terminate in a UL approved cap. Figure 23-7 shows several examples of acceptable vent pipe terminations outside a building.

Questions

1- What regulates the materials and installation methods for a gas system?

2- What is the most common material used in residential gas piping?

3- What condition might prevent the use of copper piping in a gas system?

4- What two weights of copper pipe and tubing are required for interior gas piping?

5- Copper pipe or tubing may be used outside underground except for what condition?

6- When joints are necessary in a copper gas piping system, what type of solder must be used?

7- Flare fittings may be used in a copper gas piping system only if they are_____ and_____.

8- Name two things that should be done to protect gas piping when it is installed underground in trenches.

9- When backfilling a trench containing gas piping, what type of backfill should be used?

10- Describe the three steps that must be taken when gas piping must be installed under a slab.

11- How should gas piping be installed to serve an appliance located in the center of a room?

12- What must be provided to protect gas piping installed in vertical masonry walls?

13- When might it be necessary to drill a hole in the center of the partition stud for horizontal gas piping?

14- Why must you not notch a partition deeper than ⅓ the width of the stud?

15- What is generally used to secure gas piping installed in metal stud partitions?

16- A gas main must be _____to drain dry.

17- What must be installed for certain gases containing moisture?

18- Why should gas branch pipes be taken only from the top or side of a gas feeder pipe?

19- What is the purpose of installing a shutoff valve near the gas meter?

20- What type of shutoff valve should each gas appliance in a building have?

21- What two types of shutoff valves are manufactured for appliances?

22-What is the maximum distance allowed from a shutoff valve to the appliance it serves?

23- When are left and right threaded fittings in a concealed gas piping system permitted?

24- What must be done to prevent a union in an existing concealed gas line from working loose?

25- When is a new connection in an existing concealed copper tubing gas line allowed?

26- To what standards must the threads for gas piping conform?

27- The procedure for preparing threads for gas piping is the same as for_____ _____.

28- What last major step must be taken before gas piping can be concealed?

29- What is the best and safest way to check gas piping for leaks?

30- What is the minimum height above the garage floor that the combustion chamber for a gas water heater may be set?

31- When a gas water heater is installed in a separate room, what type of ventilation openings must be provided?

32- Where must gas water heaters never be installed?

33- If a gas appliance input rating does not exceed 30 Btu's per hour per cubic foot of room space, what may be omitted?

34- What is the minimum separation between any combustible material and a gas water heater with an insulated jacket?

35- What size vent pipe would be required for a 30 gallon gas water heater with a four-inch draft hood?

36- What are the two types of vent piping materials acceptable for concealed installations?

37- What must be provided for vent pipes installed in partitions constructed of combustible material?

38- What gauge should metal straps or hangers be to support horizontal gas vent piping?

39- All gas vent pipes terminating above a roof must be equipped with what?

Private Swimming Pools

Approximately 75,000 new pools are installed in the United States each year. Most of these are considered permanent pools by plumbing codes because they are constructed in the ground and cannot be disassembled. Others are considered non-permanent because they can be disassembled and reassembled to their original condition. These are considered seasonal and for nonswimming purposes.

The only common type of swimming pool today is the *recirculating* type. This pool is equipped with a pump to recirculate water from the pool through a filter system. Some pool equipment for private use may have an automatic feeding device for adding chlorine or fluorine. Many private pool owners add their own chemicals, or they use a professional pool company to maintain the quality of the pool water.

The recirculating type swimming pool can accommodate heavy use and yet use a minimum of water. Fresh water is added as needed when pool water is lost by evaporation, splashing, or backwashing. Good filtration equipment assures pool users of water that is clear of organic matter and safe from harmful bacteria.

Water Supply

Water for the pool can come from a well or from the public water system. Many owners of private pools use the public water supply and attach a garden hose to a hose bibb to fill the pool. When this is done, the hose bibb must have a vacuum breaker to prevent a cross-connection. Pools may also have a direct connection to the public water system, if a fill spout with an air gap above the overflow rim of the pool is installed. See Figure 24-1.

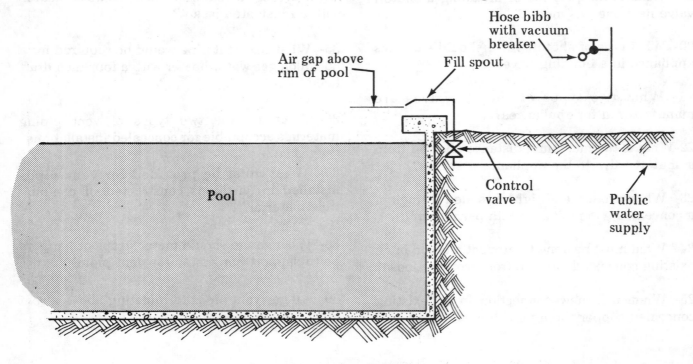

Water supply layout
Figure 24-1

If pool water comes from a well, the water supplied must be clean and meet the bacterial requirements for a domestic water supply. If it is not reasonably free of objectionable minerals, the filtration system must correct the deficiency.

Well water color should not exceed the standard of 100 and the iron content should not exceed 0.3 parts per million before filtration. If the raw well water does not meet these specifications, it must be given a preliminary treatment before it enters the pool.

Waste Water Disposal

Occasionally the pool must be emptied or the filter flushed. This requires disposal of waste water. Some of the most common methods of

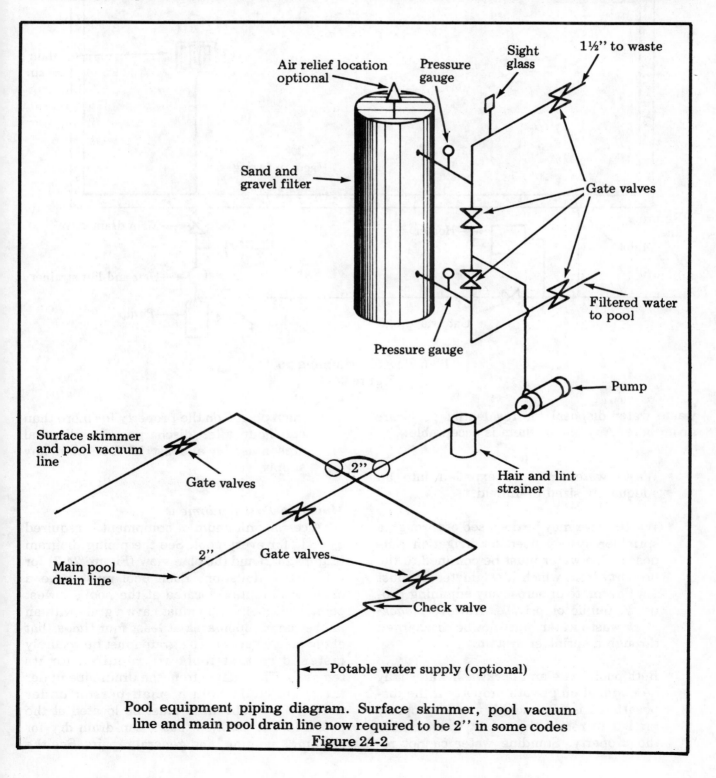

Pool equipment piping diagram. Surface skimmer, pool vacuum line and main pool drain line now required to be 2'' in some codes
Figure 24-2

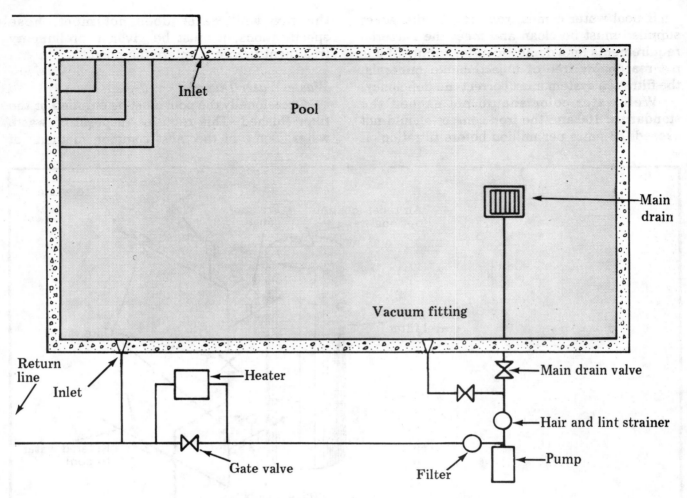

Plan view of swimming pool
Figure 24-3

waste water disposal for residential pools are given here. Any one of them is acceptable.

- Waste water may be expelled into an adequately sized drainfield.

- Waste water may be disposed of through a sprinkler system used for irrigation purposes. The water must be confined to the property from which it originates. It must not flow on to or across any adjoining property, public or private. However, *backwash* waste water must not be discharged through a sprinkler system.

- Both pool waste and backwash water may be puddled on private property if the disposal area is big enough and is properly graded to retain the waste water within the property. Standing water cannot re-

main pooled on the property for more than one hour after discharge. The disposal area must be at least 50 feet from any supply well.

Minimum Pool Equipment

A certain minimum of equipment is required by code for every pool. See the piping diagram (Figure 24-2) and the plan view (Figure 24-3) for a typical installation. Each pool must have a main drain (outlet) located at the pool's lowest point. This main drain must have a grate with an unobstructed open area at least four times that of the pipe it serves. The grate must be securely fastened so that tools are required for its removal. (The suction from the drain pipe under the grate could hold a small person under water.) The main drain must be located at the pool's lowest point so that it can drain dry for cleaning, painting, and general repairs. See the main pool detail in Figure 24-4 and the main drain installation in Figure 24-5.

A hair, lint, and sediment interceptor equipped with an easily removable screen must be installed in the suction line ahead of the pump. The screen must be sized to have a free area five times the cross sectional area of the suction pipe. See the interceptor detail in Figure 24-6.

Recirculation inlets (supply fittings) must be sized and spaced to produce uniform circulation of incoming water throughout the pool. One inlet is required for each 350 square feet of pool water surface or fraction thereof. See the supply

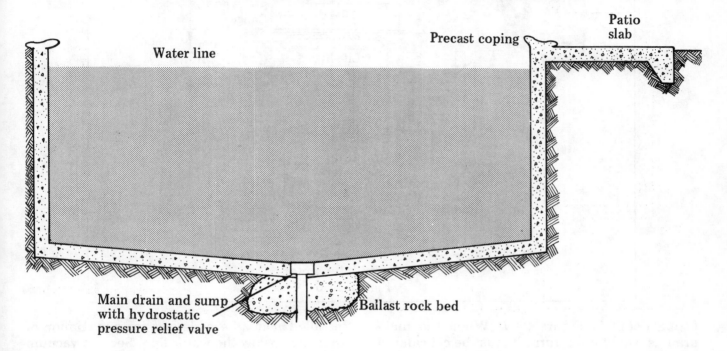

Figure 24-4

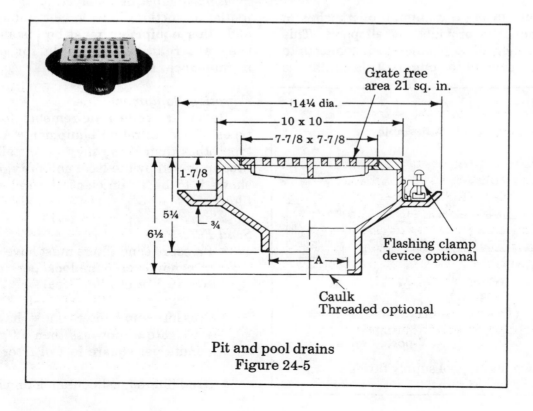

Pit and pool drains
Figure 24-5

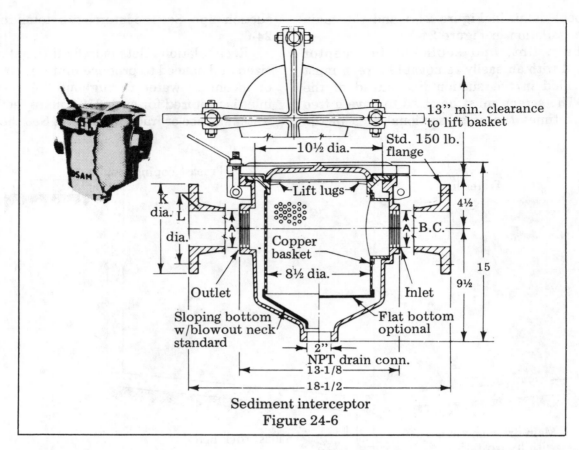

13" min. clearance to lift basket

Std. 150 lb. flange

10½ dia.

Lift lugs

Copper basket

8½ dia.

K dia.

L dia.

Outlet

A

A

B.C.

4½

15

9½

Inlet

Sloping bottom w/blowout neck standard

Flat bottom optional

2" NPT drain conn.

13-1/8

18-1/2

Sediment interceptor
Figure 24-6

fitting detail in Figure 24-7. When the main drain is used for a return, it must be considered an inlet (supply) and sized at a minimum of two inches.

A vacuum fitting a minimum of 2 inches in diameter must be provided on all pools. This fitting is connected to piping which connects to the suction side of a pump. This fitting is

installed in an accessible location a maximum of 18 inches below the water line. See the vacuum fitting detail in Figure 24-8.

A valve must be installed on the main drain (outlet or suction) line. Valves, pumps, filters, and other equipment must be installed so that they are readily accessible for operation, maintenance, and inspection.

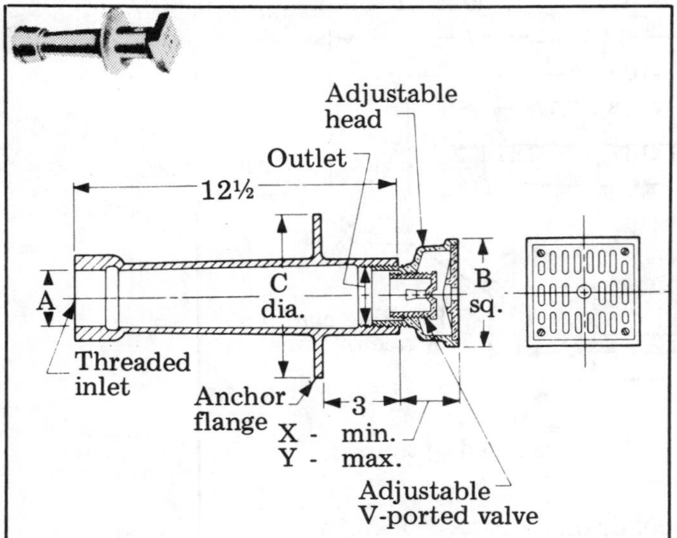

Filtration Equipment

Below are code requirements for common types of pool filtration equipment. A filter of a type other than these may be installed if it is first approved by the local authority, or if a test shows it to be as efficient as sand and gravel filters.

Sand Filters

- Pressure sand filters must have a filtration rate not over 5 gallons per minute per square foot of filter area.

- Pressure sand filters must have a backwash rate of not less than 12 gallons per minute per square foot of filter area.

- Sand filters must contain a minimum of 19

Adjustable head

Outlet

12½

C dia.

A

B sq.

Threaded inlet

Anchor flange

3

X - min.
Y - max.

Adjustable V-ported valve

Swimming pool supply fitting
Figure 24-7

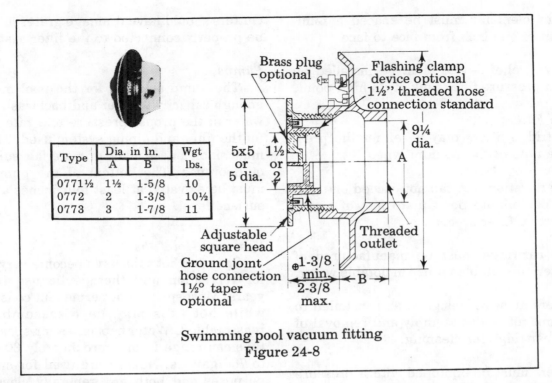

Type	Dia. in In.		Wgt
	A	B	lbs.
0771½	1½	1-5/8	10
0772	2	1-3/8	10½
0773	3	1-7/8	11

Swimming pool vacuum fitting
Figure 24-8

inches of suitable grades of screened, sharp silica sand properly supported on a graded silica gravel bed.

- A sufficient freeboard must be provided above the surface of the sand and below the overflow troughs or filter pipes to permit 50% expansion of the sand during the backwash cycle. There should be no loss of sand.

- The inflow and effluent lines must be provided with pressure gauges. See Figure 24-2.

- The backwash line must have a sight glass installed so that backwash water can be visibly checked for clarity. See Figure 24-2.

- Tanks longer than 24 inches must have an access hole measuring a minimum of 11 inches by 15 inches.

- All pressure filter tanks must be designed to remove air from the tank by an approved method or device. See Figure 24-2.

Diatomaceous Earth Filters
- Diatomaceous earth filters may be either the vacuum type or the pressure type.

- They must have a filtration rate of no more than two gallons per minute per square foot of effective filter area.

- A filter aid must be able to be introduced into the filter tank so that the filter septum or element is evenly precoated before it is put back in operation.

- The filter piping must be designed and installed during the precoating operation in such a way that the filter aid recirculates or discharges through the waste pipe and does not return to the pool water.

- Provision must be made for removing the caked diatomite by either backwash or disassembly.

- The filter must be designed and installed so that the filter elements can be removed easily.

- Pressure or vacuum gauges must be installed on these filters to determine the pressure differential across the filter and the need for cleaning.

- All pressure filter tanks must be tested to a minimum of 50 psi.

• Filter elements must be spaced a minimum of one inch from face to face.

• An air relief device must be installed on each pressure filter tank at its high point.

Cartridge Filters

• Cartridge filters may be either the pressure type or the vacuum type.

• The filtration rate can not exceed one gallon per minute per square foot of actual effective filter area.

• The cartridge must be manufactured of materials suitable for use in potable water.

• Filters must be designed and installed to permit ready disassembly and removal of the cartridge for cleaning.

• Filters must be equipped with a pressure or vacuum gauge. This gauge must be located to indicate when the filter needs cleaning.

• Filter tanks must be hydrostatically tested to a minimum pressure of 50 psi.

Surface Skimming

Swimming pools must have some form of surface skimming device to remove floating debris from the surface of a pool. One skimming device must be provided for each 600 square feet of pool surface or fraction thereof. The skimming device should have a volume control which will divert through the skimmer up to 50% of the filter capacity. Two types of skimming procedures are acceptable under most codes:

1) Skimmers built into the pool wall. These must have skimmer weirs that are automatically adjustable to variations in water level over a range of three inches. The skimmer weirs must be a minimum of five inches wide. A removable basket with a minimum surface of 75 cubic inches must be provided. The basket must be accessible for cleaning through an opening in the deck and must have a minimum diameter of five inches.

2) An overflow gutter at the end of the pool.

Gutters should have drainage grates and should be properly connected to the filter system.

Pumps

The pump selected for the pool must have enough capacity to filter and backwash the pool water at the proper pressure and rate required for the filter and piping system used. The pump must sit on a base made of materials suitable for outdoor use. The bottom of the pump motor must be elevated at least four inches above the surface level.

Spas and Hot Tubs

Spas and hot tubs have become very popular for recreation and therapeutic use in recent years. Most spas are of permanent construction, while hot tubs may be disassembled and reassembled. Water capacities for spas and hot tubs can range from approximately 300 gallons to 700 gallons. Neither are used for swimming purposes and both are generally classified as special pools.

Most codes require that spas and hot tubs be of the recirculating type. The water supply, waste water disposal, and minimum equipment requirements are the same as for swimming pools (see Figure 24-2). The pool heater must heat the water to a temperature high enough to serve the facility's intended purpose.

The national energy conservation law that became effective March 25, 1979, bans most outdoor pool heaters. But it does not prohibit use of heaters with spas and hot tubs because most codes do not define these installations as pools. Until there is further clarification, some authorities will issue permits for spas and hot tubs, other authorities will issue permits only if a letter from a physician is obtained stating that the facility is required for the patient's therapeutic use. Check with your local code authority before installing a spa or hot tub. A pool or spa without a heater may be undesirable.

Materials and Installation Methods

The piping materials and fittings of a swimming pool may be the same as those for potable water. This section reviews the two most commonly used piping materials for residential pools. Installation methods of other material types can be found in the more advanced *Plumber's Handbook* by this author.

Thermoplastic pipe must be continuously marked on opposite sides by the manufacturer with the size, type, schedule, U.S. Commercial Standard and National Sanitation Foundation seal of approval.

ABS or *PVC* plastic pipe and fittings must be schedule 40. The assembly and installation method is the same as for potable water piping of the same material.

Polyethylene plastic pipe is flexible and uses insert type fittings. Stainless steel clamps must be used to secure each connection. Polyethylene plastic pipe that is approved for a building water service line may be used for swimming pool piping on pressure lines only. The assembly and installation method is the same as for sprinkler piping of the same material. See Chapter 17.

Unless it is supported directly on existing ground, pool piping must be supported to the pool structure with pipe hangers or heavy duty strap iron. Hangers must be spaced a maximum of 4 feet apart, center to center. Backfill must be fine material free of rocks larger than ¾ inch in diameter. Ells installed on any suction line underground must be of the long radius type. Short radius 90° ells (generally used on water piping) are not to be used.

Dielectric fittings are required where dissimilar metals are installed in pool and filter piping.

Energy Conservation Law

The new energy conservation law, which went into effect nationally on March 15, 1979, virtually eliminates heated swimming pools. The only exception is for pools used for therapeutic purposes, and these must be approved by the local code authority.

Pools designed to use electric or fossil fuel heaters for pool water systems must meet the following energy conservation criteria before approval can be granted.

- The pool heater must have an automatic control to limit water temperature to no more than 80° F.

- An uncovered or unenclosed heated pool must have automatic controls to make the pool heater inoperative whenever the outdoor air temperature is below 60° F.

Pool Heaters

Gas-fired swimming pool heaters (after first being approved) must comply with A.G.A. and A.S.M.E. standards. Pool heating equipment must have a minimum of 70% thermal efficiency across the unit.

Pool heaters must be provided with a thermostatic or high temperature control which maintains a maximum temperature differential of 15° F, or with some other acceptable overheat protection device.

Pool water heating equipment installation is the same as for domestic water heaters. See Chapter 14.

A plumbing permit is required for the installation of pool piping and equipment. The entire pool pressure piping system, including the main drain, must be water-tested at 40 psi and proved tight before the installation is concealed.

Questions

1- A pool that is considered permanent can not be _____ .

2- What is the most common type of residential pool?

3- How does a good filtration system protect pool users?

4- What must be provided for a hose bibb when it is used for adding water to a swimming pool?

5- How is cross connection prevented when water is added to a swimming pool by means of a direct water supply connected to public water?

6- What are the special problems when well water is used to supply a swimming pool?

7- What are two acceptable ways of disposing of swimming pool waste water?

8- In what part of a pool is the main drain located?

9- Other than recirculating pool water, what other purpose does the main drain serve?

10- Why must the grate over a main pool drain be securely fastened so that tools are required for its removal?

11- Where must an interceptor be located in a pool piping system?

12- Why is it important to size and space pool inlets properly?

13- How many square feet of water surface may one pool inlet adequately serve?

14- What is the minimum pipe size connected to a main drain?

15- What is the minimum diameter size of a pool vacuum fitting?

16- What is the maximum depth below the water line to which a vacuum fitting may be installed?

17- When valves, pumps, filters, and other pool equipment are installed, what provision must be made for their continued operation and maintenance?

18- Name the types of filtration equipment accepted by most codes.

19- Describe the two methods commonly used for surface skimming a pool.

20- How high must the bottom of the motor for a swimming pool pump be mounted above the surface level?

21- How do most spas and hot tubs differ in the method of construction?

22- What is the general use of a spa?

23- What are the most commonly used materials for swimming pool piping?

24- If pool piping is not supported directly on the ground, how should it be supported?

25- What type of ell should be used underground on any swimming pool suction line?

26- Why must pool heaters have automatic controls to limit water temperature to no more than 80° F?

27- The installation of pool water heating equipment is the same as for what other type of heaters?

28- At what pressure must pool piping be water-tested to prove the system is tight?

29- What is required before installing and repairing pool piping?

25

Definitions, Abbreviations, And Symbols

Definitions

The terms included here are found in most plumbing codes. Some words in the code have become so descriptive and specialized that their meaning is different from what a dictionary might give.

Two words that appear repeatedly in every code are *shall* and *may*. To be able to comply with the code, you must clearly understand the specialized meanings of these words.

Shall means that compliance is mandatory and that the procedure or condition specified must be performed without deviation. For example, part of the requirements for waste disposal is that "sewage and liquid waste *shall* be treated and disposed of as hereinafter provided." *May* is a term of permission. When used in the code, it means "allowable" or "optional," but not required. For example, "Drinking fountains *may* be installed with indirect waste only for the purpose of resealing required traps of floor drains."

"Building drain" and "building sewer" are two terms often used improperly. Many professionals assume that both terms apply to the same part of the drainage system. However, note the code definition for each term: Building drain is the main horizontal collection system within the walls of a building which extends to five feet beyond the building line. (This distance may vary in some codes.) A building sewer is defined as that part of the horizontal drainage system outside the building line which connects to the building drain and conveys the liquid waste to a legal point of disposal.

Effective and constructive code interpretation is possible only when the words and terms used in the code are understood. Many definitions can be illustrated through isometric drawings. It is possible to have some variation and still remain within the intent of the code definitions. The alert professional will discover that most isometric drawings fit into these definitions. Isometric illustrations provide a better understanding of code definitions.

Definitions

Absorption Drainfield absorption area.

Air gap (in a water supply system) The unobstructed vertical distance through the free atmosphere between the lowest opening from any pipe or faucet supplying water to a tank, plumbing fixture, or other device, and the flood level rim of the receptacle.

Anaerobic Living without free oxygen. Anaerobic bacteria found in septic tanks are beneficial in digesting organic matter.

Approved Approved by the plumbing official or other authority given jurisdiction by the code.

Area drain A receptacle designed to collect surface or rain water from an open area.

Backfill That portion of the trench excavation up to the original earth line which is replaced after the sewer or other piping has been laid.

Backflow The flow of water or other liquids, mixtures, or substances into the distributing pipes of a potable supply of water and any other fixture or appliance, from any source, which is opposite to the intended direction of flow.

Backflow connection Any arrangement whereby backflow can occur.

Backflow preventer A device or means to prevent backflow into the potable water system.

Back Siphonage The flow of water or other liquids, mixtures or substances into the distributing pipes of a potable supply of water or any otner fixture, device, or appliance, from any source, which is opposite to the intended direction of flow due to negative pressure in the pipe.

Base The lowest point of any vertical pipe.

Battery of fixtures Any group of two or more similar adjacent plumbing fixtures which discharge into a common horizontal waste or soil branch.

Boiler blow-off An outlet on a boiler to permit emptying or discharging of water or sediment in the boiler.

Branch Any part of the piping system other than a main, riser or stack.

Branch interval A length of soil or waste stack (vertical pipe) generally one story in height (approximately nine feet, but not less than eight feet) into which the horizontal branches from one floor or story of a building are connected to the stack.

Branch vent The vent that connects one or more individual vents with a vent stack or stack vent.

Building drain The main horizontal sanitary collection system, inside the wall line of the building, which conveys sewage to the building sewer beginning five feet (more or less in some codes) outside the building wall. The building drain excludes the waste and vent stacks which receive the discharge from soil, waste and other drainage pipes, including storm water.

Building sewer That part of the horizontal piping of a drainage system which connects to the end of the building drain and conveys the contents to a public sewer, private sewer, or individual sewage disposal system.

Building storm drain A drain used to receive and convey rain water, surface water, ground water, subsurface water and other clear water waste, and discharge these waste products into a building storm sewer or a combined building sewer beginning five feet outside the building wall.

Building storm sewer Connects to the end of the building storm drain to receive and convey the contents to a public storm sewer, combined sewer, or other approved point of disposal.

Building subdrain Any portion of a drainage system which cannot drain by gravity into the building sewer.

Caulking Any approved method of rendering a joint water and gas tight. For cast iron pipe and fittings with hub joints, the term refers to caulking the joint with lead and oakum.

Code Regulations and their subsequent amendments or any emergency rule or regulation lawfully adopted to control the plumbing work by the administrative authority having jurisdiction.

Combined building sewer A building sewer which receives storm water, sewage and other liquid waste.

Common vent The vertical vent portion serving to vent two fixture drains which are installed at the same level in a vertical stack.

Conductor See Leader.

Continuous waste A drain connecting a single fixture with more than one compartment or other permitted fixtures to a common trap.

Cross connection Any physical connection or arrangement between two separate piping systems, one containing potable water and the other water of unknown or questionable safety.

Dead end A branch leading from a soil, waste or vent pipe, building drain or building sewer which is terminated by a plug or other closed fitting at a developed distance of two feet or more. A dead end is also classified as an extension for future connection, or as an extension of a cleanout for accessibility.

Developed length   The length as measured along the center line of the pipe and fittings.

Diameter   The nominal diameter of a pipe or fitting as designed commercially, unless specifically stated otherwise.

Downspout   See Leader.

Drain   Any pipe which carries liquid, waste water or other water-borne wastes in a building drainage system to an approved point of disposal.

Drainage system   All the piping within a public or a private premises that conveys sewage, rain water, or other types of liquid wastes to a legal point of disposal.

Drainage well   Any drilled, driven or natural cavity which taps the underground water and into which surface waters, waste waters, industrial waste or sewage is placed.

Durham system   An all-threaded pipe system of rigid construction, using recessed drainage fittings to correspond to the types of piping being used.

Effective opening   The minimum cross-sectional area of the diameter of a circle at the point of water supply discharge.

Effluent   The liquid waste as it flows from the septic tank and into the drainfield.

Fixture branch   The drain from the trap of a fixture to the junction of that drain with a vent. Some codes refer to a fixture branch as a "fixture drain."

Fixture drain   The drain from the fixture branch to the junction of any other drain pipe, referred to in some codes as a "fixture branch."

Fixture unit   A design factor to determine the load-producing value of the different plumbing fixtures. For instance, the unit flow rate from fixtures is determined to be one cubic foot, or 7.5 gallons of water per minute.

Fire lines   The complete wet standpipe system of the building, including the water service, standpipe, roof manifold, Siamese connections and pumps.

Flood level rim   The top edge of a plumbing fixture or other receptacle from which water or other liquids will overflow.

Floor drain   An opening or receptacle located at approximately floor level connected to a trap to receive the discharge from indirect waste and floor drainage.

Floor sink   An opening or receptacle usually made of enameled cast iron located at approximately floor level which is connected to a trap, to receive the discharge from indirect waste and floor drainage. A floor sink is more sanitary and easier to clean than a regular floor drain, and is usually used for restaurant and hospital installations.

Flushometer valve   A device actuated by direct water pressure which discharges a predetermined quantity of water to fixtures for flushing purposes.

Grade   The slope or pitch, known as "the fall," usually expressed in drainage piping as a fraction of an inch per foot.

Horizontal pipe   Any pipe or fitting which makes an angle of more than 45 degrees with the vertical.

Horizontal branch   A drain pipe extending laterally from a soil or waste stack or building drain. May or may not have vertical sections or branches.

Indirect wastes   A waste pipe charged to convey liquid wastes (other than body wastes) by discharging them into an open plumbing fixture or receptacle such as floor drain or floor sink. The overflow point of such fixture or receptacle is at a lower elevation than the item drained.

Industrial waste Liquid waste, free of body waste, resulting from the processes used in industrial establishments.

Insanitary Contrary to sanitary principles, injurious to health.

Interceptor A device designed and installed to separate and retain deleterious, hazardous, or undesirable matter from normal wastes and permit normal sewage or liquid wastes to discharge by gravity into the disposal terminal or sewer.

Leader The vertical water conductor or downspout from the roof to the building storm drain, combined building sewer, or other approved means of disposal.

Liquid waste The discharge from any fixture, appliance or appurtenance that connects to a plumbing system which does not receive body waste.

Load factor The percentage of the total connected fixture unit flow rate which is likely to occur at any point with the probability factor of simultaneous use. It varies with the type of occupancy, the total flow unit above this point being considered.

Loop or circuit waste and vent A combination of plumbing fixtures on the same floor level in the same or adjacent rooms connected to a common horizontal branch soil or waste pipe.

Main The principal artery of *any system* of continuous piping, to which branches may be connected.

Main vent The principal artery of the venting system, to which vent branches may be connected.

May The word "may" as used in the code book is a term of permission.

Mezzanine An intermediate floor placed in any story or room. When the total area of any such mezzanine floor exceeds 33⅓ percent of the total floor area in that room or story, it is considered an additional story rather than a mezzanine.

Pitch "Grade," also referred to as "slope."

Plumbing Includes any or all of the following: (1) the materials including pipe, fittings, valves, fixtures and appliances attached to and a part of a system for the purpose of creating and maintaining sanitary conditions in buildings, camps and swimming pools on private property where people live, work, play, assemble or travel; (2) that part of a water supply and sewage and drainage system extending from either the public water supply mains or private water supply to the public sanitary, storm or combined sanitary and storm sewers, or to a private sewage disposal plant, septic tank, disposal field, pit, box filter bed or any other receptacle or into any natural or artificial body of water or watercourse upon public or private property; (3) the design, installation or contracting for installation, removal and replacement, repair or remodeling of all or any part of the materials, appurtenances or devices attached to and forming a part of a plumbing system, including the installation of any fixture, appurtenance or devices used for cooking, washing, drinking, cleaning, fire fighting, mechanical or manufacturing purposes.

Plumbing fixtures Receptacles, devices, or appliances which are supplied with water or which receive or discharge liquids or liquid borne waste, with or without discharge, into the drainage system with which they may be directly or indirectly connected.

Plumbing official inspector The chief administrative officer charged with the administration, enforcement and application of the plumbing code and all its amendments.

Plumbing system The drainage system, water supply, water supply distribution pipes, plumbing fixtures, traps, soil pipes, waste pipes, vent pipes, building drains, building sewers, building storm drain, building storm sewer, liquid waste piping, water treating, water using equipment, sewage treatment, sewage treatment equipment, fire standpipes, fire sprinklers and related appliances and appurtenances, including their respective connections and devices, within the private property lines of a premises.

Potable water Water that is satisfactory for drinking, culinary and domestic purposes and meets the requirements of the health authority having jurisdiction.

Private property For the purposes of the code, all property except streets or roads dedicated to the public, and easements (excluding easements between private parties).

Private or *private use* In relation to plumbing fixtures: in residences and apartments, and in private bathrooms of hotels and similar installations where the fixtures are intended for the use of a family or an individual.

Private sewer A sewer privately owned and not directly controlled by public authority.

Public or *public use* In relation to plumbing fixtures: in commercial and industrial establishments, in restaurants, bars, public buildings, comfort stations, schools, gymnasiums, railroad stations or places to which the public is invited or which are frequented by the public without special permission or special invitation, and other installations (whether paid or free) where a number of fixtures are installed so that their use is similarly unrestricted.

Public sewer A common sewer directly controlled by public authority.

Public swimming pool A pool together with its buildings and appurtenances where the public is allowed to bathe or which is open to the public for bathing purposes by consent of the owner.

Relief vent A vent, the primary function of which is to provide circulation of air between drainage and vent systems.

Rim In code usage, an unobstructed open edge at the overflow point of a fixture.

Rock drainfield Three-quarter inch drainfield rock 100 percent passing a one inch screen and a maximum of ten percent passing a one-half inch screen.

Roof drain An outlet installed to receive water collecting on the surface of a roof which discharges into the leader or downspout.

Roughing-in The installation of all parts of the plumbing system that can be completed prior to the installation of plumbing fixtures; includes drainage, water supply, vent piping, and the necessary fixture supports.

Sanitary sewer A pipe which carries sewage and excludes storm, surface and ground water.

Second hand A term applied to material or plumbing equipment which has been installed and used, or removed.

Septic tank A watertight receptacle which receives the discharge of a drainage system or part thereof, so designed and constructed as to separate solids from liquid, digest organic matter through a period of detention, and allow the liquids to discharge into the soil outside the tank through a sub-surface system of open-joint or perforated piping, or other approved methods.

Sewage Any liquid waste containing animal, mineral or vegetable matter in suspension or solution. May include liquids containing chemicals in solution.

Shall As used in the code, a term meaning that compliance is mandatory and that the procedure or condition specified must be performed without deviation.

Slope See Grade.

Soil pipe Any pipe which conveys the discharge of water closets or fixtures having similar functions, with or without the discharge from other fixtures, to the building drain or building sewer.

Stack The vertical pipe of a system of soil, waste or vent piping.

Stack vent (Sometimes called a waste vent or soil vent) the extension of a soil or waste stack above the highest horizontal drain connected to the stack.

Standpipe system A system of piping installed for fire protection purposes having a primary water supply constantly or automatically available at each hose outlet.

Storm sewer A sewer used for conveying rain water and/or surface water.

Subsurface drain A drain which receives only subsurface or seepage water and conveys it to a place of disposal.

Sump A tank or pit which receives sewage or liquid waste, located below the normal grade of the gravity system and which must be emptied by mechanical means.

Supports (Also known as "hangers" or "anchors.") Devices for supporting and securing pipe and fixtures to walls, ceilings, floors or structural members.

Supply well Any artificial opening in the ground designed to conduct water from a source bed through the surface when water from such well is used for public, semi-public or private use.

Trap A fitting or device so designed and constructed as to provide a liquid seal which will prevent the back passage of air without materially affecting the flow of sewage or waste water through it.

Trap seal The maximum vertical depth of liquid that a trap will retain, measured between the crown weir and the top of the dip of the trap.

Vent stack A vertical vent pipe installed primarily for the purpose of providing circulation of air to and from any part of the drainage system.

Vent system A pipe or pipes installed to provide a flow of air to or from a drainage system or to provide a circulation of air within such system.

Vertical pipe Any pipe or fitting which is installed in a vertical position or which makes an angle of not more than 45 degrees with the vertical.

Waste pipe Any pipe which receives the discharge of any fixture, except water closets or fixtures having similar functions, and conveys it to the building drain or to the soil or waste stack.

Water-distributing pipe A pipe which conveys water from the water service pipe to the plumbing fixtures, appliances and other water outlets.

Water main A water supply pipe for public or community use.

Water outlet As used in connection with the water-distributing system, the discharge opening for the water (1) to a fixture, (2) to atmospheric pressure (except into an open tank which is part of the water supply system), (3) to a boiler or heating system, (4) to any water-operated device or equipment requiring water to operate, but not a part of the plumbing system.

Water service pipe The pipe from the water main or other source of water supply to the building served.

Water supply system Consists of the water service pipe, the water-distributing pipes, standpipe system and the necessary connecting pipes, fittings, control valves and all appurtenances in or on private property.

Wet vent A waste pipe which serves to vent and convey waste from fixtures other than water closets.

Yoke vent A pipe connecting upward from a soil or waste stack for the purpose of preventing pressure changes in the stacks.

Abbreviations

The abbreviations here are often found on blueprints (building plans) and in plumbing reference books (including the code) to identify plumbing fixtures, pipes, valves and nationally-recognized associations.

A	area
AD	area drain
AGA	American Gas Association
AISI	American Iron and Steel Institute
ASA	American Standard Association
ASCE	American Society of Civil Engineering
ASHRAE	American Society of Heating, Refrigeration and Air Conditioning Engineers
ASME	American Society of Mechanical Engineers
ASSE	American Society of Sanitary Engineering
ASTM	American Society for Testing Materials
AWWA	American Water Works Association
B.S.	bar sink
B	bidet
B. T.	bathtub
B.t.u.	British Thermal Unit
C to C	center to center
CI	cast iron
CISPI	Cast Iron Soil Pipe Institute
C	condensate line
C. O.	cleanout
C. W.	cold water
cu. ft.	cubic feet
cu. in.	cubic inches
C. W. M.	clothes washing machine
C. V.	check valve
D. F.	drinking fountain
D. W.	dish washer
E to C	end to center
EWC	electric water cooler
°F	degrees Fahrenheit
F	Fahrenheit
F. B.	foot bath
F. F.	finish floor
F. C. O.	floor cleanout
F. D.	floor drain
F. D. C.	fire department connection
F. E. C.	fire extinguisher cabinet
F. G.	finish grade
F. H. C.	fire hose cabinet
F. L.	fire line
F. P.	fire plug
F. S. P.	fire standpipe
F. U.	fixture unit
GAL.	gallons
gpm or Gal. per Min.	gallons per minute
Galv.	galvanized
G. S.	glass sink
G. V.	gate valve
G. P. D.	gallons per day
H. B.	hose bibb
Hd or H.D.	head
H. W.	hot water
H. W. R.	hot water return
HWT	hot water tank
in.	inch
I. D.	inside diameter
I W	indirect waste
IPS	iron pipe size
K. S.	kitchen sink
L. or LAV.	lavatory
L. T.	laundry tray
L	length
lb.	pound
Max.	maximum
Mfr.	manufacturer
Min.	minimum
M. H.	manhole
NAPHCC	National Association of Plumbing Heating and Cooling Contractors
NBFU	National Board of Fire Underwriters
NBS	National Bureau of Standards
NFPA	National Fire Protection Association
NPS	nominal pipe size
O	oxygen
O. D.	outside diameter
Oz.	ounce
P. D.	planter drain
P. P.	pool piping
PSI	pounds per square inch
Rad.	radius
R. D.	roof drain
red.	reducer
R. L.	roof leader
San.	sanitary
SH.	Shower
Spec.	specification
Sq.	square
S. B.	Sitz bath

Sq. Ft.	square feet	T	temperature
S. P.	swimming pool	U or Urn	urinal
SS	service sink	V	volume
Std.	Standard	Vtr	vent through roof
SV	service	W	waste
SW	service weight	W.C.	water closet
S & W	soil and waste	W.H.	water heater
		XH	extra heavy

Symbols

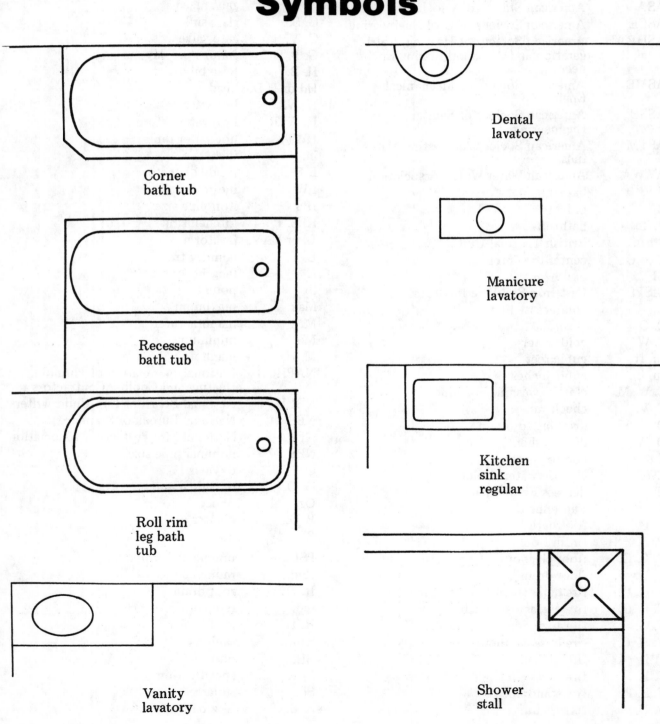

Corner bath tub

Recessed bath tub

Roll rim leg bath tub

Vanity lavatory

Dental lavatory

Manicure lavatory

Kitchen sink regular

Shower stall

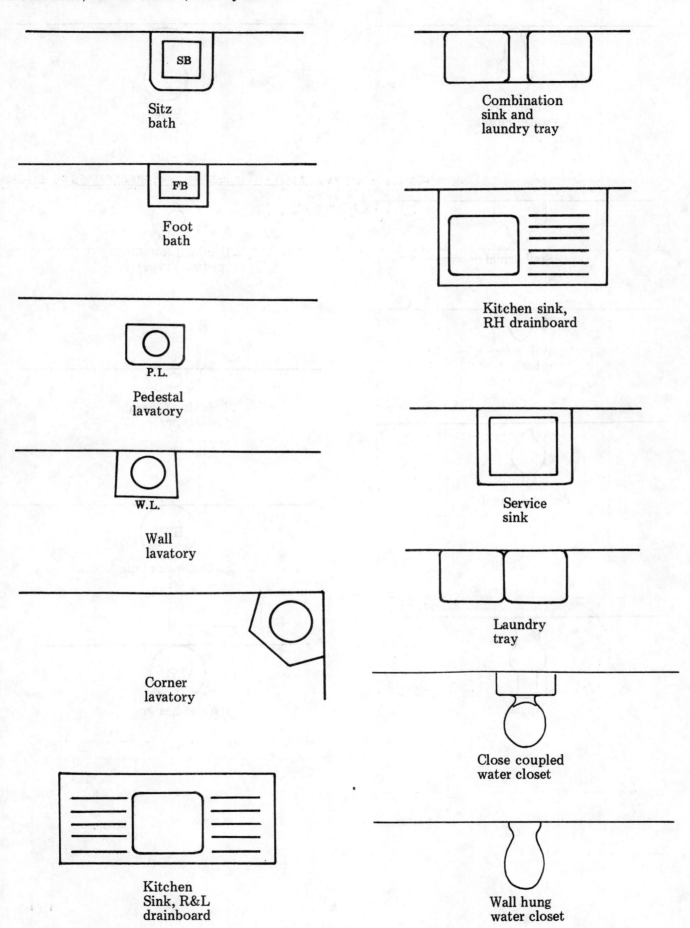

Sitz
bath

Foot
bath

P.L.

Pedestal
lavatory

W.L.

Wall
lavatory

Corner
lavatory

Kitchen
Sink, R&L
drainboard

Combination
sink and
laundry tray

Kitchen sink,
RH drainboard

Service
sink

Laundry
tray

Close coupled
water closet

Wall hung
water closet

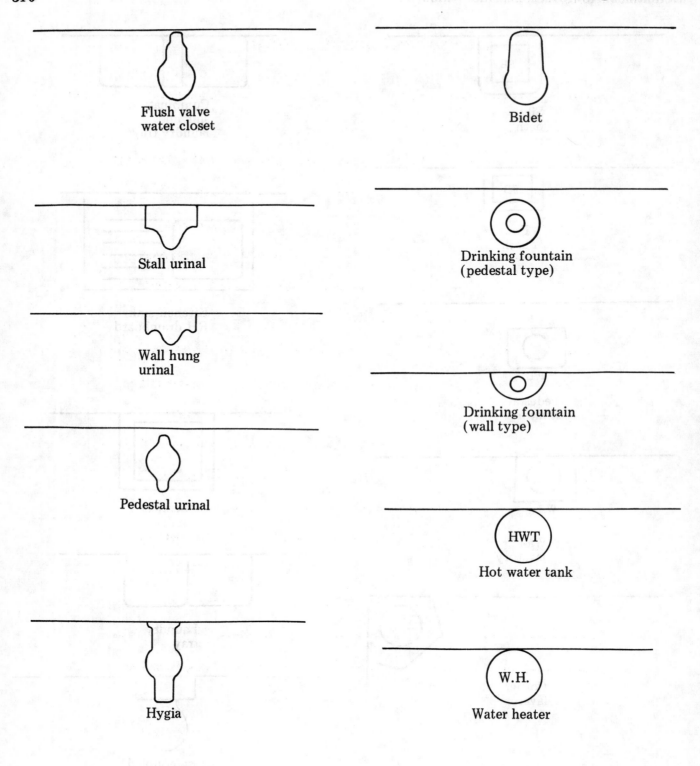

Flush valve
water closet

Bidet

Stall urinal

Drinking fountain
(pedestal type)

Wall hung
urinal

Drinking fountain
(wall type)

Pedestal urinal

HWT

Hot water tank

Hygia

W.H.

Water heater

PLUMBING SYMBOL LEGEND

Symbol		Description
——— — — — · —	C.W.	—— — — · —— Cold water
——— ·· — ·· —	H.W.	——— ·· — ·· — Hot water
—— · — · — · —	H.W.R.	—— · — · — · — Hot water return
———————	W.L.	——————— Waste line
— — — — — —	V.L.	— — — — — — Vent line
———————	S S	——————— Sanitary sewer
———————	C	——————— Condensate line
———————	S D	——————— Storm drain
———————	R.W.L.	——————— Rain water leader
→——→——→	I.W.	→——→——→ Indirect waste
———————	F	——————— Fire line
———————	G	——————— Gas line
	Gate valve
	Globe valve
	Check valve
———————	R	——————— Relief line
	P&T relief valve
	F.C.O.	Floor cleanout
	F.D.	Floor drain
	P.D.	Planter drain
	R.D.	Roof drain
	H.B.	Hose bibb
	A.D.	Area drain

FITTING SYMBOLS

Bell-and-spigot ends

Cap

Plug

Union, screwed

Coupling

Expansion joint, sliding

Sleeve

Expansion joint, bellows

Bushing

Reducer

Eccentric reducer

Reducing flange

Union, flanged

Return bend

Tee

Cross

True Y- wye

Y- Wye single

Combination wye and 1/8 bend

Tee union

Y- Wye double

Double combination wye and 1/8 bend

Elbow, 90 degrees

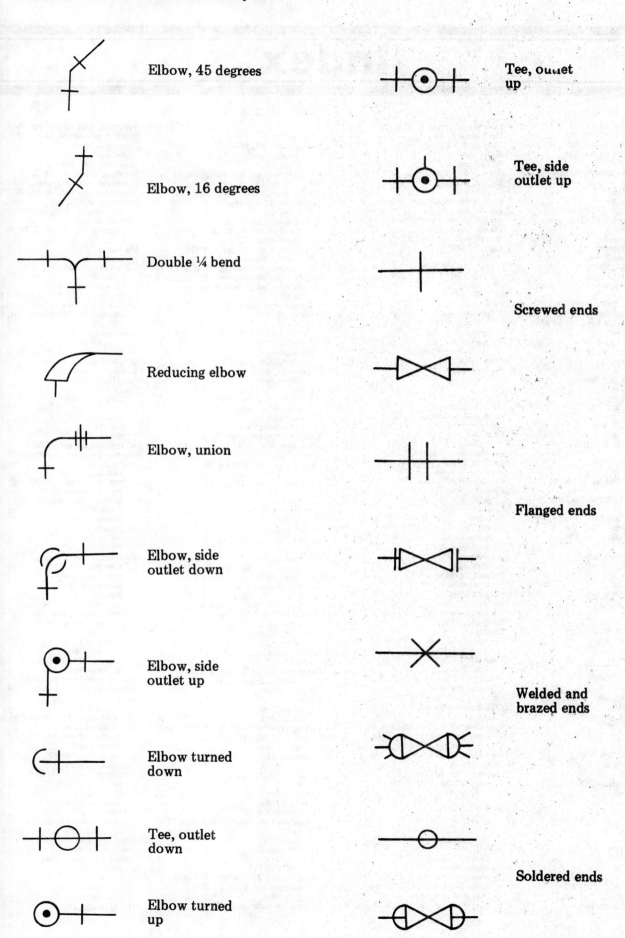

Elbow, 45 degrees

Elbow, 16 degrees

Double ¼ bend

Reducing elbow

Elbow, union

Elbow, side outlet down

Elbow, side outlet up

Elbow turned down

Tee, outlet down

Elbow turned up

Tee, outlet up

Tee, side outlet up

Screwed ends

Flanged ends

Welded and brazed ends

Soldered ends

Index

A

Abbreviations, fixtures, 9 307-308
ABS, 77, 93, 145, 149, 229, 299
　def., 82
Absorption areas, drainfields, 102,
　103-106, 109
Absorption, def., 301
Access, cleanout, 86
Adapters, 86, 92, *145, 146*, 167
Additions, sizing for, 27, 31
Aerators, clogged, 247
Air chambers, 152, *153*, 168
Air conditioning drains, 96
Air gap
　assembly, 230, *232*
　def., 301
　minimum, *154*, 156
　pools, *292*
Air relief device, 298
Aluminum tubing, 136
American Standards Association, 288
Anaerobic, def., 301
Angle valve, 122, *123*, 126
Angles, isometric, 16, *17*
Anti-siphon valve, 224
Antifreeze, 135
Appliances, gas, 281, 286
　connecting, 289
　ventilation, 289, 290
Approved, def., 301
Area drain, def., 301
Asbestos flue pipe, 290
ASTM, 145
Auger, closet, 244, *245*, 246

B

Back pressure, 7, 33, 39, 52, *54*, 72
Back vent, 38
Backfill, 78, *86*, 149, 150, 184, 285, 299
　def., 301
Back flow, def., 301-302
Back flow connection, def., 302
Back flow preventers, 149, 156
　def., 302
Back-siphonage, 156, 226
　def., 302
Back-ups, waste, 26
Backwash, 296, 297
Bacteria, septic tank, 101, 102, 109
Ball, flush, 274, 275
　guide, 276
Ball valve, 254
Ballcock, 226, *271-273*, 274, 275
Base, def., 302
Basin wrench, 221
Basket strainer, 266
Bath valve, *262*
Bathrooms
　back to back, 43
　building drains, 31
　design, 62, 71
　floor plans, 62, *63-70*, 71
Bathtubs, 9, *206-212*
　fixture units, 26
　rough-in, *199*
　slip-resistant, 206, *209*
　stoppages, 244-*245*
　trap sizes, 26
　waste and overflow, *206-212*
Battery of fixtures, 302
Bell traps, 45, *46*
Bells, soil pipe, *78*
Bends, *19, 20*, 72, *73*, 79, *81*, 82
　closet, *20*, 79, 81
　low-hub, *81*
Bibb
　faucet, 124, *125*
　screw, lost, 249
　washer, 124
Bidets, 194, *195*, 202
Bituminous coating, 103
Block or cradle drains, 108

Blockage, old method, 57
Blueprints, 7, 62
Body waste, def., 23
Boiler blow-off, 302
Bottled gas, 280
Branch, def., 302
Branch, fixture, 24
　def., 23
Branch, horizontal, def., 24
Branch, sanitary tee, *80*
Branch, Y-, *81*
Branch interval, def., 302
Branch vent, def., 24, 302
Btu, 129, 281
Building
　department, local, 5
　drain, 23, 24, 31
　def., 302
　DWV in, 93, 94
　line, 58
　plot plans, 31
　sewer, size, 27
　sewer, def., 302
　storm drain, def., 302
　subdrain, def., 302
Buried pipe, 141
Bushings, 143-144

C

Cable, cleaning, 57, 58
Capacities, water heater, 127-128
Capacity, waste piping, 26
Caps, 143, *144*, 145
　vent, 290
Capillary attraction, *51*, 52
Carriers, fixture, 195, *196-197*
Cartridge faucets, 258, *263*
Cartridge filters, pool, 298
Cast-iron pipe, 76, 77, *78*
　cutting, 86, *87*, 88
　tees, 79, *80*
Caulking
　def., 302
　joints, *89-90*
　lavatories, 221
　sleeving, 93
　tools, *89*
Cement flue pipe, 290
Cement mixture, 103
Cement, plastic, 92, 166, 167, 221, 224
Cemented joints, 77, 86, 166-168
Center-to-center measure, *156*, 157
Cesspools, 101
Chain vise, 158
Change of direction, 72-76, 79, *80, 81*
　cleanouts, 58
Channelling, *286*
Chases, 287
Check valve, 122, *123*, 126, 129, 177
Chemical drains, 82
Chemicals, septic tank, 110
Circuit vent, def., 24
Clamp, pipe, 277
Clamp assembly, lavatory, 218, *219*
Cleaning cable, 57 240, 241, *244*
Cleaning fixtures, 250-252
Cleanouts, 50, 57-61, 72, 240, 241
Cleanout tees, *59*, 79
Clearance, 93, 94, 149, 152
　cleanout, 59
　fixture, 194
　trench, 94
Closet
　auger, *245, 246*
　bend, *20*, 79, *81*
　bolts, 223-224
　bowl, 224
　flange, floor, *223*
　flange, offset, 195
　tanks, leaking, 278
　tanks, repairing, 269-270, *274-278*
　tank washers, 269
Clothes washing machines, 9, 26, 247

Code
　def., 302
　gas, 280, 281, 285
　National Plumbing, 26, 27
　sizing, 127
　Southern Florida Building, 26, 27, 34,
　　38-39, 40, 102
　Southern Standard Plumbing, 34
　Standard Plumbing, 26, 27, 38, 39,
　　40, 104
Codes, plumbing, 5
　compared, 26, 27, 34, 39, 40, 127
Collectors, solar, *132-134*
Combination faucet, 124, *125*, 219
Combination lock nut wrench, 266
Combination wye and eighth bend, 79, 81
Combined building sewer, 302
Commercial fixture requirements,
　203-205
Common vents, 38
　def., 24, 302
Compound
　pipe, 162, 214, 226, 240, 275, 277
　sealing, *219*
Compression
　faucets, 122, *124, 255-257*
　fittings, 230
　valve repair, 253-255
Concrete, 93, 149, 152, 184, 196, 213,
　　214, 223, 285, 287
　channelling, *286*
　leaks and, 277
　septic tanks, 103
Condensate drains, 96
Conductor, def., 302
Conduit, *286*
Continuous vent, 38, 41
　def., 24
Continuous waste, 228, 229
　def., 302
Construction, 11
Copper
　cutting, 163
　fittings, *145-146*
　pipe, 77, 127, 144, 148-149, 162-166,
　　285
　reaming, 163
　tubing, 135, 144, 148-149, 162, 285
Corrosion, 82, 118, 119, 141, 145
Countertop lavatories, *217*, 218
Couplings, 143, *145, 146*, 183, 277
CPVC plastic pipe, 127, 146, 151, 152
Cradle or block drains, 108
Crawl spaces, 152
Cross, 82, *142*, 143
　sanitary, *20, 21*
Cross-connection, 101, 149, *153*, 155-156
　def., 302
Cross installation, kitchen sink, 31
Crown vented traps, 45, *46*
Crown weir, 39
Cutter
　soil pipe, *87*
　tubing, *163*
Cutting pipe, 158, *159*, 161, 163, 166-168
　cast-iron, 86, *87*, 88
　copper, 163
　plastic, 88

D

Dead end, 58
　def., 302-303
Definitions, 301-307
Definitions and isometric drawings, 16-25
Demand
　gas, 282
　hot water, 127
　water, 116, 117
Demand load, 113
Department of Environmental Resource
　Management, 11, 101, 172
Deposits, 82
Design, hot water, 127-128

Designing rough plumbing, 16-25
Design temperature, 127
Developed length, 27
　def., 303
Diameter, def., 302
Diaphragm faucets, *256-257*
Diaphragm pressure tanks, 178-*180*
Diatomaceous earth filters, 297-298
Dielectric fittings, 299
Dimensions, roughing-in, *199-202*
Directional
　changes, 72, *73, 74, 75*, 76, 79
　tees, 229, *230*
Discharge
　lines, 137
　valves, air, 137
　vents, 137
Dishwasher cabinet, *231*
Dishwashers, 205, *229-230, 231-232*
Dishwasher Y branch, 230
Distribution box, drainfield, 104, *10[*
　107, 108
Diverter valve, 212
Dope, pipe, 162
Double hub fittings, 72, *76*
Double trapping, 47, *48*
Downspout, def., *303*
Draft hood, 290
Drain
　area, def., 301
　building, def., 23, 302
　cock, 130, 137
　def., 303
　fixture, def., 23
　hose connections, *233*
Drain and waste pipes, sizing, 26
Drainage
　pattern, def., 79
　pattern, fittings, 72, *76*
　piping, cleanouts, 58
　piping, protection, 94
　pools, 294
　system, 7, 26
　system, def., 303
　test holes, 104-105
　wells, def., 303
Drainfields, 101-112
　absorption, 102
　percolation rate, 105
　plastic, 108-109
　plot plans, 11
　swimming pool, 294
　replacement, 109
　reservoir-type, 108
　site, 104
　sizing, 103-105
　tile, *105-108*
　trench, 107, 108, *110*
　underdrains, 104
Draining, and venting, 38-39
Drain/overflow, installation, *210-2[*
Drains
　building, 24
　building, and bathrooms, 31
　building, sizing, 31
　building storm, 302
　clogged, 239
　condensate, 96-97
　fixture, 23, 24
　fixture, sizing, 31, 38-39
　lavatory, *219-220*
　pop-up, 208
　stoppages, 244-*245*
　swimming pool, 294
Drain-waste-vent system, 7
Draw period, 128, 129
Dresser coupling, 277
Drilling, piping, 72
Drinking fountains, 204
Drip pipes, 130
　gas, *287*
Drum traps, 50, *51*
　maintenance, 245
Dry wells, 97, 110

Durham system, 303
DWV
  building, 93-94
  code, 72
  direction changes, 72, *73, 74, 75, 76*
  materials and installation, 72-100
  pipe supports, *94*
  plastic, 82-86, 91-93
  sizing, 72
  underground, 93-94

E

Easements, 11
Effective opening, def., 303
Effluent, 102
  def., 303
Eighth bend, *20*
Elastomeric joints, 77
Elbows, *142, 143, 145, 146,* 299
  IPS, 183
  parts, 269, *270*
Electrical connections, 232, *234*
Electrical service, 11
Ells, *142,* 143, *145, 146,* 299
  IPS, 183
  parts, 269, *270*
Enclosures, shower, 215, *216*
End-to-center measure, *156*
End-to-end measure, *156*
Energy conservation law, 298, 299
Evaporation, trap seals, 52
Excavating for hub, 86
Excavation, *78,* 93, 103
Extension fittings, *143*
Extensions, vent, 94-*96*

F

Face-to-face measure, 156, 157
Face-to-end measure, *156,* 157
Fall per foot, 27, 31
Faucets, 122, 124, *125-126*
  cartridge, 258, 263
  combination, 219
  compression, 122, *124*
  def., 122
  diaphragm, *256-257*
  installing, 221
  laundry, 228
  lavatory, 218-*219, 263,* 268
  maintenance, *261*
  mixing, 124, 257-*263*
  repairing, 255-263
  self closing, 218-219
  sink, *264*
  washers, 248
Fifth bends, 72, *81*
Fit-all, 266
Filter
  aid, 297
  material, drainfield, 108
  piping, 297
  tanks, 297, 298
Filters, swimming pool, 296-298
  flushing, 293-294
Filtration, 292, 296
  rate, 297, 298
Fire lines, def., 303
Fittings
  adaptors, plastic, 86
  air gap, 230, *232*
  change of direction, 72, *73, 74, 75*
  compression, 165-166, 230
  copper, *145, 146*
  double hub, 72, *76*
  drainage, 77
  flare, 165, 285
  galvanized steel, *142-144*
  gas, 288
  high loop, 230
  isometric drawings, 16-22
  layout, 62
  molded, 86
  no-hub, *16-22*
  plastic, 72, 77, 82, *84, 85,* 86, 145-146, 299
  pool supply, 295-*296*
  pressure rated, 146
  recessed, drainage, 77
  soil pipe, 78, 79, *80, 81, 83*
  solar piping, 136-137
  vacuum, 296, *297*
  vent, 290
  water supply, 141-147

G

Galvanized steel pipe, 127, 141, 148, 150, 176, 285
Garbage disposers, 229
Gas
  corrosive, 285
  fuel, 280
  hydrogen sulphide, 118, 145
  sewer, 33, 45, 50, 52, 72, 101, 205
Gas appliances, 281, 286, 289
Gas code, 280, 281, 285
Gas demand, 282
Gas heaters, pool, 299
Gas piping, 141, 280-284
  fittings, 288
  installation, 285-291
  isometrics, *282-283*
  main, 281
  materials, 285
  meter, 280
  service pipe, 280-281
  sizing, 281-282
  water heaters, *235*
Gas supply, 247
Gastight joint, 77

Fixture branch, 24
  def., 23, 303
  developed length, *39*
  re-vent, *40*
  branch sizes, 26
Fixture drains, 24
  def., 23, 303
  size of, 38-39
  sizing, 31
Fixtures, 7, *203-238*
  abbreviations, 9
  battery, def., 302
  cleaning, 250
  clearances, 194
  pump discharge, 31
  minimum requirements, 203-205
  overflows, 205-*206*
Fixture traps, 7
  sizes, various fixtures, 26
  materials, 50
Fixture unit load, def., 26
Fixture units, 26, 27, 31
  def., 26, 303
Flanges, 195, *223,* 267
Flapper, 274
Flare fittings, *145, 146,* 165, 285
Flaring, *165*
Flashing, *96,* 134
Flat, wet vent, 42, *43*
Flat system, 22, *63, 64, 65, 66*
Float
  ball, 274, 275
  rod, 226, 274
Flood-level rim, 96, *97*
  def., 303
Floor
  drain, def., 303
  flange, closet, *223*
  plans, 62, *63-70,* 71
  sink, def., 303
Flow
  control, 122, 124
  pattern, 72, *76*
  rate, 113
  sewage, 45
Flues, 290
Flush ball, 247
  guide, 274, 276
Flush elbows, 269
Flush head, 184
Flushing, pool filter, 293-294
Flushing mechanisms, repair, 271-276
Flushometers, *225-226,* 227
  def., 303
Flush rim, 246
Flush spray heads, 181
  patterns, *182*
  plan, 182-183
Flush tank, *269*
Flush valve, 225, 274, 275-*276*
Footage, pipe, 62
Force cups, 243, *244,* 245-*246*
Foreign substances, 45
Foundations and piping, 93
Fountains, 204
Freeboard, 297
Friction, pipe, 114
Frost, venting and, 33
Frost proofing, 95, 120

Gate valves, 122, *123,* 126, 178, 184, 255
Glass, collector plate, 132
Globe valves, 122, *123,* 126, *254*
  repair, *253-255*
Grade, 287, 303
Grate, drain, 294
Ground joint union, *142, 143,* 288
Ground roughing-in, 7

H

Hacksaw, 88
Hammer, water, 152, 249
Hammer and chisel, 86, *87*
Handicapped, provisions, 205
Hangers, pipe, *151*
Head, testing, 72
Heads, sprinkler, 181
Health department, 11
Heaters, swimming pool, 298, 299
Heat exchangers, 135, 136
Heat transfer, plastic, 82
Heating frozen pipes, 249-*250*
High loop fitting, 230
Hi-lo fitting, *229*
Hook hanger, *151*
Horizontal
  branch, def., 24, 303
  pipe, def., 303
  twin tap tee, 16, *21,* 80
  wet vent, 23, *43*
Hose, garden, *153,* 155-156, 292
Hose bibb, 124, *125,* 155, 178, 292
Hosing, dishwasher, 230
Hot tubs, 298
Hot water
  design, 127-128
  dishwasher, 231
  systems, 127-140
Houses, layouts, 62, *67-70,* 71
Hub-and-spigot, 76
  joints, 88-90
Hubbed tees, 79
Hydrogen sulphide gas, 145, 285
Hydropneumatic pressure tanks, 178-179
Hydrostatic testing, 298
  pressure, 137, 138

I

Ice, 95, 120, 133, 141, 248, 250, 285, 287
Increasers, 33, 82, *83*
Indirect waste, def., 303
Individual vent, *43*
Industrial waste, def., 304
Inlets, recirculation, 295-296
Inlet tees, 102
Insanitary, def., 304
Inspections, 5-7, 27, 72, 92, 137, 172, 194, 214
Installation
  closet tanks, 226
  drainfields, *105-110*
  drain/overflows, *210-211*
  DWV, 72-100
  dishwashers, 230, 231-232
  gas piping, 285
  hot water supply, 231
  lavatories, *218*
  lavatory faucets, 221
  lavatory waste, 221
  pool piping, 298
  plastic building sewers, 86
  shower enclosures, 215, *216*
  shower pans, 214
  solar collectors, 133-134
  solar systems, 136
  urinals, 226-227
  water closets, 223-*227*
  water heaters, 230-232, *234-235*
  water systems, 148-171
Insulation
  pipes, 119, 120, 152, 285
  solar collectors, 132
  water closets, 270
  water heaters, 127
Interceptors, 295, *296,* 304
Inverts, 103
Irrigation wells, 172, 175, 176, 183
Island sink, venting, *40*
Isometric angles, 16, *17*
Isometric drawings, 8
  and definitions, 16-25
  fittings, *16-22*
  layouts, 62
  making, 16, *18*
  venting, 40

J

J bends, 267
Jet pumps, 181, 186-188
Joining pipe, 162, 163-168
Joint compound, 162, 214
Joint runner, *89*
Joints
  brazed, 285
  cast-iron, *88-90*
  caulked, *89-90*
  cemented, 77, 166-168
  compression, 165-166
  elastomeric, 77
  flare, 138, 165
  gas, 288
  gastight, 77
  ground union, *142, 143*
  hub and spigot, *88-90*
  no-hub, 77, *89-90*
  plastic, *91-93,* 166-168
  sealers, 92
  soldered, 138, 163-164, 285
  sweat, 163
  tightening, 92
  watertight, 77

K

Kafer fittings, 82, *83*
Kitchen sinks, 9, 228-229
  clearances, 194
  fixture unit, 26
  piping, 233
  trap sizes, 26
  waste and vent combinations, 34
  waste pipe sizing, 31

L

Lateral, sewer, 22
Lath nails, 152
Laundry sinks, 156, 227-228
Laundry trays
  clearances, 194
  fixture unit, 26
  traps, 47
  trap sizes, 26
Lavatories, 9, 204, 215, 217-219
  clearances, 194, *195*
  drains, 219-220
  faucets, 124, 125, 218-*219, 263,* 268
  fixture units, 26
  pop-ups, 267-269
  roughing-in dimensions, *200*
  stoppages, 244-*245*
  trim, 221
  venting, 39
Lavatory arm, 23
Lawn sprinklers, 149, 181-185
Laying pipe, 148, 285
Layouts, 27, 62, *63-70,* 71
Leaching beds, 108
Lead, caulked joints, 76-77, *88-90*
Lead shower pans, 213
Leaks, 220, 226, 277-278
  faucets, 255, 256, 257
  gas, 289
  plastic joints, 167
  locating, 277-278
  water closet, 269-270
Length, developed, def., 303
Level, 216, 226
Lift wire, 274
LP gas, 280

M

Mack washer, 267
Main
  building drain, 23
  def., 304
  gas, 281
  sewer, 22
  water, 113
Main drain, pool, 294
Maintenance, 239-252
  drains, 239
  faucets, 259-260, *261,* 262
  water heaters, 247
  water supply, 118-*120*
Main vent, def., 304
Materials
  drainfields, 105
  DWV, 72-100
  fixture traps, 50

gas piping, 285
hot water piping, 127
pool piping, 298-299
solar systems, 136
vent extensions, 77
water pipe and fittings, 141
May, def., 304
Measurements, roughing-in, 194-195, *197, 199-202*
Measuring
offsets, 157
pipe, 156-157, 162
Meteoric water, 172
Meters
gas, 280
water, 113
Methane, 280
Mezzanine, def., 304
Mineral oil, traps, 52
Minerals, 172
Mixing faucets, 124, 257
repair, *257-263*
Municipal sewer, def., 22

N

National Plumbing Code, 26, 27
National Sanitation Foundation, 299
Negative action, 47
Negative pressure, 33, 52
Nipples, 143, *144*, 184
No-hub pipe, *16-22*, 77, *90-91*

O

Oakum joints, 76-77, 88-90
Offset fittings, 39, 82, 156, 195
Offsets, measuring, *157*
Oil
mineral, 52
penetrating, 245
thread-cutting, 162, 163
O rings, 255, 257
Outhouses, 101
Outlet tees, 102
Overflow rims, *155*, 156
pool, 292
Overflows, fixtures, 205-*206*
Overflow tubes, 225, 274-275
Owl fittings, 16

P

Packing washer, 124
Packing, faucet, 255, 263, 265
Paint, 118, 214
Painting pipe, 285
Pans, shower, *213-214*
Paper, 152
drainfields, *104*, 105-*106*, 108, 109
Partitions, piping in, 152
Patching, drainage pipe, 57
Percolation rate, 105, 109, 110
Permits, 5, 138, 299
Pipe
ABS, 299
cast-iron, 76, 77, *78*, 79, *80*, 86, *87*, 88
chases, 287
clamps, 277
compounds, 92, 240
copper, 77, 127, 144, 148-149, 162-166, 285
covering, 120
cutters, soil, *87, 160*, 161
cutting, 158, *159*, 161, 163, 166-168
dope, 162
drip, 130, *287*
friction, 114
galvanized steel, 127, 141, 148, 150, 176, 285
gas service, 280-281
hangers, *151*, 290
insulation, 119, 120
laying, 148
leaks, 220, 226, 277-278
measuring, 156-157, 162
paint, 118
plastic, 47, 50, 77, *82, 84, 85*, 86, 88, 93, 96, 145-146, 149, 166-168, 176, 184, 199, 299
polyethylene, 299
PVC, 77, 82, 93, 145, 149, 176, 184, 199, 299
reaming, 158, *159*, 161, 163

sizes, 115
sizing, 26
straps, 150, 151, 290
supports, *92, 94*, 148
tapping and drilling, 72
thawing, 248, 249-*250*
thermoplastic, 299
threading, *160, 161*-162
threads, 156, 288
vise, 157, *158, 159*, 161
wrenches, 162
water service, 148
weights, 77, 141
Piping
diagrams, 8, *9*
direction changes, 72
dishwasher, 233
filter, 297
foundations and, 93
gas, 141, 235, 280-281
installation, 72-100, 136, 148-171, 210-211, 221, 231, 285, 298-299
partitions, *152*
slabs, 96-97
swimming pool, *293, 294*, 298-299
vent, 33-44
vent, gas, 290
water distributing, 141-147, 150-155
water heater, *130, 131, 134, 135, 136, 234-235*
Piping, water, 141-147
Pitch, 27, 86, 287
def., 304
pan, 134
Plans, plot, 11, *12, 13, 14*, 15
Plans examiners, 11
Plastic, leaks in, 167
Plastic chemical drains, 82
Plastic pipe and fittings, 47, 72, 77, 82, 84, 85, 86, 96, 136, 145-146, 166-168
building sewers, *86*
cutting, *88*
joints, 91-93
pressure rated, 146
restrictions, *93*
roughing-in, dimensions, 86
sizes, 86
tailpieces, 47
traps, 47, 50
Plastic tubing, drainfields, 108-109
Plot plans, 11, *12, 13, 14*, 15
Plugs
cleanout, *58, 59, 60*, 61
drain, *219*, 220
pipe, 143, *144*
Plumbing
cost, 7
fixtures, 203-238
permits, 5-6
system components, 7
maintenance, 239-252
procedures, 156-168
repairs, 253-279
Plumbing, def., 304
fixtures, 304
official inspector, 304
system, 304
Polyethylene pipe, 299
Polyethylene tubing, *183*, 184
Pools, swimming, 292-300
Popoff lines, 129
Pop-up drains, 208, *220*, 267-269
Positive pressure, 33, 52, *54*
Potable water, def., 305
Potable water supply system, 7
Pot traps, 45, *46*
P traps, *19, 21*, 45
Pressure, 113, 114, 137, 178, 181
drop, 173
filter tanks, 297, 298
gauge, 297, 298
gauge, gas, 288-289
hydrostatic, 137, 138
positive and negative, 33
problems, 247-249
rated, 146, 176
reducing valve, 113
regulator, 137
tanks, 178-*179, 180*
testing, 288-289, 297
Private, defs., 305
Private or building sewer, def., 22-23
Private water systems, 8, 172-181, 184-185

Property line, sewer, 22
Public, defs., 305
Public sewer, def., 22
Public water system, 8
Pullout plugs, *219*
Pulsating sprinkler heads, 181
Pump discharge, drain, 230
Pump discharge fixture, 31
Pumps
comparing, 183
diaphragm, 179, *180*
dishwasher, 230
hot water, 134, 135
installation, 179-181
sizing, 183
swimming pool, 292, 298
troubleshooting, 185-193
well, 174, 176, 177, *178*, 183
wiring, 179
Putty, 221, 267
PVC, 77, 82, 93, 145, 149, 176, 184, 199

Q

Quarter bend, *19*, 72, *73*
Questions, 10, 15, 25, 32, 44, 55, 61, 71, 97, 111, 121, 126, 138, 146, 168-171, 184-185, 198, 236, 251-252, 278-279, 290-291, 299-300

R

Rain birds, 181
Rain collars, 134
Rain water, 172
Reamer, pipe, *161*
Reaming pipe, 158, *159*, 161, 163
Recirculation inlets, 295-296
Recovery rate, 128
Reducer fittings, 79, 82, *83*, 142, 143
Regulator, pressure, 137
Relief device, filtration, 298
Relief lines, 130
Relief pipe, 231
Relief valves, 129-130, 137, 231, *234-235*
Relief vent, def., 24
Relief vents, 24, 32, 39, 40
Repairing
faucets, 255-263
leaks, 277-278
valves, *253-255*
Repairs, plumbing, 253-279
Residential fixture requirements, 203
Residential waste piping sizing, 27-32
Re-vent, 24, 32, 39, 40
Reverse trap closets, 222
Roughing-in, 7, 62, 86, 194-202
clearances, 194
def., 194
designing, 16-25
measurements, 194-*195, 199-202*
Rubber hose, 230
Running threads, 162
Running traps, 45, 46

S

Sand filters, 296-297
Sanitary crosses, *20, 21*, 82
Sanitary sewer, 23, 305
Sanitary system, 7
Sanitary tees, 16, *19, 21*, 39, 72, *73*, 79, 80
Scale, 119
Schedule 40, 77, 86, 145, 176, 184, 299
Schedule 160, 184
Screens, clogged, 247
Seals, trap, 7, 50, 52, *53*
Seal, setting, 270
Seal washer, 267
Seat dressing, 253, 256
Sediment interceptor, *296*
Seepage pit, 110
Self-siphonage, 47, 49, 50-51
Septic tanks, 101-112
plot plans, 11
stoppages, 240
Septic Tank Association, 101
Service tees, 143
Setbacks, plot plans, 11
Setting seal, 269-*270*
Sewage, 7, 45, 101
system, 22
Sewer cables, 57, 240-241, *244*
Sewer gases, 33, 45, 50, 52, 72, 101, 205
Sewers, 11, 22-23, 27, 86, 302

Shall, def., 305
Shank extenders, *271-273*
Short sweeps, *19*, 79
Shower
baths, 212
enclosures, 215, *216*
heads, 246-247
pans, *213-214*
rods, 215
stalls, *212*, 214
stalls, fixture units, 26
stalls, trap sizes, 26
fixture units, 26
trap sizes, 26
valves, *260, 265*
Showers, 9, 204
clearances, 194, *195*
stoppages, 244
tile, 212-213
Shutoff valves, 129
gas, 287-288
Sight glass, 297
Sill cocks, *153*
Sink arm, 23
Sinks, 9, 156, 204, 205, 227-229
island, venting, *40*
kitchen, clearance, 199
rough-in dimensions, *199*
trapping, 47
stoppages, 241, *243*, 244
Siphon action, 52, *53*
Siphon action closets, 222, 223
Siphonage, 7, 39, 52, *53*, 72, 95
Siphon jet closets, *222*
Site, drainfield, 104
Sizes, building sewer, 27
pipe, 115
pipe, plastic, 86
pipe, soil, 78
Sizing, 26
accessory buildings, 31
code, 127
drain and waste pipes, 26
drains, 31
DWV, 72
gas systems, 281-282
hot water systems, 115, 127
pressure tanks, 178-179
pumps, 183
PVC and polyethylene, 184
lawn sprinklers, 181
room additions, 27
septic tanks, 102
sewer pipe, 27, 32
vent system, 33-34
waste pipes, 27-32
water piping, *113-118*
water service pipe, 115
Skimmers, pool, 298
Slabs, 96-97, 150, 196
Sleeving, 93, 152
Slip joints, *271-273*
Slope, 108, 109, 305
Sludge, septic tank, 109
Soil pipe, 24, 305
cutter, *87*
stack, def., 24
tees, 79, *80*
Solar collectors, *132-134*
Solar water systems, 131-138
Soldered joints, 163-*164*, 285
Solder fittings, *145*
Solvent welded fittings, 86
Southern Standard Plumbing Code
South Florida Building Code, 26, 27, 38-39, 40, 102
Spacers, vent, *288*
Spacing, filter elements, 298
Spas, 298
Split ring hangers, *151*
Spray heads, sprinklers, 181-185
layout pattern, *182*
Sprinkler systems, 149, 181, *182-18*
Spud washers, 269
Square cut washers, 265
Stack, 24, 305
Stacked system, 22, *63, 64, 65, 66*
Stacks, vent, 23, 24, 33
as cleanouts, 59, 61
Stack vent, 24, 305
Stack venting, 34, *35*
Standard Plumbing Code, 26, 27, 38, 40, 104
Static head, 114

d waste valve, 122, *123*, 126
ges, 26, 72
ng, 240-241, *244-247*
ting, 239-240, *241-243*
s
n, 245
up, *220*
out, *219*, 220
hot water, 128-129, 131, 132,
135, 136-138
rain, building, 302
ewer, def., 306
ge cross, 82
runs, 156
tees, 79
rs, 206, 207, 213, 266
45, 52, *53*
pipe, 151
lbows, *142*, 143
ees, 143
23-224
ards, 152, 287
n, building, 302
sible pumps, 189-193
ace drain, def., 306
lines, 11
s, 176, 177
def., 306
fittings, 295-*296*
tube, 249
well, def., 306
s pipe, *92*, *94*, 97, 148, 151, 285,
290, 306
s, plumbing, 308-313
ttings, *234*
g, water closets, 270
oints, 163
short, *19*, 81
ng pools
, 296-298
pment, 294-298
rs, 293-294, 296-298
ags, 295-*296*
ers, 298
ng, *293*, *294*
heck valves, 123

ces, *219*, *220*, 228, 229
tic, 47
her, 266
, side, 79

water, 128-129, 131, *134*, *135*,
138
izing, 128-129
r, *297*, 298
ing, 278
sure, 178-*179*
airing, 269-270, *274-278*
hers, 269
er closet, 224-225, 269
sanitary cross, 20
tees, 79, *80*
g, 72
ing, 76, 77
nches, *80*
9, *80*, *142*, 143, *145*, *146*
out, *59*, 79
tional, 229, *230*
102
t, 102
ary, 16, *19*, 21, 39, 72, *73*, 79
c tanks, 103
ipe, 79, *80*
57, 79
long turn, 39

Test holes, drainage, 104-105
Testing, 72, 137-138, 166, 168, 267,
288-289, 297-299
Test tee, *57*, 79
Thawing frozen pipes, 249-*250*
Thermoplastic pipe, 299
Thermosyphon, 134-135
Thread-cutting oil, 162, 163
Threader, pipe, *161*, 162
Threads, 156
gas pipe, 288
leaking, 277
running, 162
Tightening joints, 92
Tile
drainfields, *105-108*
showers, 212-213
Tip-toe drain, 206-207
Toilets, 203-205
Topping out, 6
Trap
def., 306
horizontal distance from vent, 38
parts, 52, *55*
Trap arms, 92
Trapping, double, 47, *48*
Trapping, vertical drop, 49, 50
Traps, 7, 45-61 229
bell, 45, *46*
building, 50, *51*
crown vented, 45, *46*
drum, 50, *51*, 245
fixture, *45-50*, 205
laundry, 47, 228
materials, 50
P-, *19*, *21*, 45
plastic, 47, 50
pot, 45, *46*
prohibited, 45, *46*, *53*, 54
running, 45
S-, *45*, 52, *53*
seals, 33, 50, 52, *53*, 72, 306
separate, 47
sinks, 47
siphonage, 52, *53*
sizes, 26
shower, 213
urinal, 227
water closets, 47, 50
Trap washers, 263
Trenches, 72, 86, 93, 107, 108, 122,
148-149, 184, 285
Trench loads, 78, 141
Trim, cleaning, 221, 250-251
Trip lever, 274
Trip waste, 207-208
Troubleshooting pumps, 185-193
Tube
lavatory supply, 221, 226
overflow, 225, 274, 275
Tubing, 141
aluminum, 132, 136
copper, 144, 148-149, 162, 285
cutter, *163*
polyethylene, *183*, 184
solar collector, 132, 133
Tub valves, *260*, 265
Tunnelling, 72, 148
Two-way cleanout, *58*
Types K, L, M copper, 144, 145

U

U-hangers, *151*
UL approved, 138, 290
Underdrains, 104, *108*
Underground DWV, 93-94
Underwriters Laboratories, 138, 290
Uniform Solar Energy Code, 136

Unions, *142*, 143, *145*, *146*, 164, 288
Urinals, 23, 204, 226-227
roughing-in, 202
U.S. Commercial Standard, 299
U.S. Department of Commerce, 38

V

Vacuum breaker, 149, *150*, *153*, 154
Vacuum fitting, 296, *297*
Vacuum gauge, 297, 298
Valves, 122, *123*, 126, 137, 230, 253-255
air discharge, 137
anti-siphon, 224
ball, *254*
ballcock, 226
bath, 262
check, 129, 177
compression, 165
diverter, 212
flush, 225, 274, 275-276
flushometer, *225-226*, 227
gas, 287-288
gate, 178, 184, 253, *254*, 255
globe, *253-255*
locating leaks, 277-278
pressure reducing, 113
relief, 129-*130*, 137, 231, *234-235*
shutoff, 129, *232*, 255
tub and shower, *260*, 265
water fill, *232*
water control, 149-150
water meter, 149
Vent, branch, def., 24, 302
Vent, circuit, def., 24
Vent caps, 290
Vent extensions, materials, 77
Ventilation, gas appliances, 289-290
Vents
back, 38
cleanouts, 61
cold weather, 95
common, 24, 38, 308
continuous, 24, 38, *41*
discharge, 137
distance from trap, 38
flat wet, *42*
gas, *286*, 289-290
horizontal wet, 23, *43*
individual, 24, 38, *43*
relief, 24, 32, 39, *40*
re-, 24, 32, 39, *40*
stack, 24, 34, *35*, 41
stack wet, *41*
vertical, 23, 24
water heater, 128, 235
wet, 23, *42*
Vent spacers, *288*
Vent stacks, 23, 24, 31, 32, 33, 59, 61, 82
Vent terminals, 92, 94-*96*, 289
Vent system, 7, 24, *28*, 31, 33, 34, 35-38,
39, 40, *41-43*, 97, 289, 290
Vermin, 45, 50
Vertical drop, trapping, 49, 50
Vestibules, toilet, 204, 205
Vise, pipe, 157, *158*

W

Washdown closets, *222*
Washers, 124, 248, 253, 255-256, 263,
265, 267, 269, 275
Washing machines, 247
Wash-up sinks, 204
Waste, food disposers, 229
Waste, trip, 207-208
Waste opening, 212-213, 229
Waste outlets, 221
Waste piping, 23-24, 26, 27-32, 34, *36*
assemblies, installing, 221

continuous, 302
laundry, 228
overflows, *205-212*
sink, 229
stack, 24
urinal, 227
Water, demand, 116, 117
Water, hot, 127-140
Water, loss, 248
Water, meteoric, 172
Water, rain, 172
Water closet elbows, *270*
Water closets, 9, *196-197*, 204, 221-226,
269-278
carriers, 195, 196-*197*
clearances, 194, *195*
fixture units, 26
hardware, 197
installation, 223-227
leaks, 269-270
roughing-in, 195, *200-201*
seats, 276-277
stoppages, *245-246*
trapping, 47, 50
venting, 33, 38
washers, 269, 275
Water control valve, 149-150
Water distributing pipe, 7-8, 113-121,
150-*155*
Water fill valve, *232*
Water hammer, 152, 168, 249
Water heaters, 9, *127-131*, 230-232,
234-235, 247, 289
solar, 128, 131-138
swimming pool, 299
Water main, 113
Water meter, 11, 113, 277
valves, 149
Water pipe, 141-148
Water piping diagram, 8
Water pressure, 114, 137, 178, 181,
247-249
Waterproofing, shower, 213
Water seal, traps, 45-50, 54
Water service pipe, 7-8, 11, 113, 115, *148*
Water softener, 119, *173*
Water supply systems, 7-8, *113-121*,
148-181, 184-185
swimming pools, 292-293
Water supply tube, 221, 226
Watertight joints, 77
Waste water, pools, 293-294
Wax, 224
Weights, pipe, 78, 141
Well points, *175*
Well pumps, 174, 176, 177, *178*, 183,
185-193
Wells, 122, 156, 172-181, 184-185,
292-293
drainage, def., 303
dry, 97
plot plans, 11
pumps, 174, 176, 177, *178*, 183,
185-193
suction lines, *176-177*
Well water, 172, 293
Wet venting, 28, 31, 34, 36, 306
Wet vents, 23, *41*, *42*, *43*
Wind effect, trap seals, 50, 52
Wiring, pump, 179
Woolfelt, *120*
Wrenches, pipe, 162, *221*, 253, 255, 266,
269
Wyes, *20*, 79, *81*

XYZ

Yarning, *89-90*
Yoke vent, 306
Yoke vise, 158

Practical References for Builders
Construction Estimating Guides

Estimating & Bidding for Builders & Remodelers

New and more profitable ways to estimate and bid any type of construction. Shows how to take off labor and material, select the most profitable jobs for your company, estimate with a computer (FREE estimating disk enclosed), fine-tune your markup, and learn from your competition. Includes Estimate Writer, an estimating program with a 30,000-item database on a 5¼" high-density disk when you buy the book. (If your computer can't use high-density disks, add $10 for 5¼" or 3½" double-density disks.) **272 pages, 8½ x 11, $29.75**

National Plumbing & HVAC Estimator

Manhours, labor and material costs for all common plumbing and HVAC work in residential, commercial, and industrial buildings. You can quickly work up a reliable estimate based on the pipe, fittings and equipment required. Every plumbing and HVAC estimator can use the cost estimates in this practical manual. Sample estimating and bidding forms and contracts also included. Explains how to handle change orders, letters of intent, and warranties. Describes the right way to process submittals, deal with suppliers and subcontract specialty work. Includes *FREE* estimating disk with all the cost estimates in the book plus a handy program called Estimate Writer that makes it easy to write plumbing and HVAC estimates. Estimate Writer is free on 5¼" high density (1.2Mb) DOS disk. Add $10 for 5¼" (360K) or 3½" (720K) disks. **288 pages, 8½ x 11, $32.25. Revised annually.**

National Repair & Remodeling Estimator

The complete pricing guide for dwelling reconstruction costs. Reliable, specific data you can apply on every repair and remodeling job. Up-to-date material costs and

labor figures based on thousands of jobs across the country. Provides recommended crew sizes; average production rates; exact material, equipment, and labor costs; a total unit cost and a total price including overhead and profit. Separate listings for high- and low-volume builders, so prices shown are accurate for any

size business. Estimating tips specific to repair and remodeling work to make your bids complete, realistic, and profitable. *Repair & Remodeling Estimate Writer FREE on a 5¼" high-density (1.2 Mb) disk when you buy the book.* (Add $10 for *Repair & Remodeling Estimate Writer* on extra 5¼" double density 360K disks or 3½" 720K disks.) **320 pages, 11 x 8½, $29.50. Revised annually.**

Estimating Home Building Costs

Estimate every phase of residential construction from site costs to the profit margin you include in your bid. Shows how to keep track of manhours and make accurate labor cost estimates for footings, foundations, framing and sheathing finishes, electrical, plumbing, and more. Provides and explains sample cost estimate worksheets with complete instructions for each job phase. **320 pages, 5½ x 8½, $17.00**

National Construction Estimator

Current building costs for residential, commercial, and industrial construction. Estimated prices for every common building material. Manhours, recommended crew, and labor cost for installation. Includes Estimate Writer, an electronic version of the book on computer disk, with a stand-alone estimating program — *FREE* on 5¼" high density (1.2Mb) disk. The National Construction Estimator and Estimate Writer on 1.2Mb disk cost $31.50. (Add $10 if you want Estimate Writer on 5¼" double density 360K disks or 3½" 720K disks.) **592 pages, 8½ x 11, $31.50. Revised annually**

Building Cost Manual

Square foot costs for residential, commercial, industrial, and farm buildings. Quickly work up a reliable budget estimate based on actual materials and design features, area, shape, wall height, number of floors, and support requirements. Includes all the important variables that can make any building unique from a cost standpoint. **240 pages, 8½ x 11, $16.50. Revised annually**

Estimating Tables for Home Building

Produce accurate estimates for nearly any residence in just minutes. This handy manual has tables you need to find the quantity of materials and labor for most residential construction. Includes overhead and profit, how to develop unit costs for labor and materials, and how to be sure you've considered every cost in the job. **336 pages, 8½ x 11, $21.50**

Construction Estimating Reference Data

Provides the 300 most useful estimating reference tables. Labor requirements for nearly every type of construction, including: sitework, concrete work, masonry, steel, carpentry, thermal and moisture protection, doors and windows, finishes, mechanical and electrical. Each section details the work being estimated and gives the appropriate crew size and equipment needed. **368 pages, 8½ x 11, $30.00**

Professional Remodeling Guides

Manual of Professional Remodeling

The practical manual of professional remodeling that shows how to evaluate a job so you avoid 30-minute jobs that take all day, what to fix and what to leave alone, and what to watch for in dealing with subcontractors. Includes how to calculate space requirements; repair structural defects; remodel kitchens, baths, walls, ceilings, doors, windows, floors and roofs; install fireplaces and chimneys (including built-ins), skylights, and exterior siding. Includes blank forms, checklists, sample contracts, and proposals you can copy and use. **400 pages, 8½ x 11, $23.75**

Remodeler's Handbook

The complete manual of home improvement contracting: evaluting and planning the job, estimating, doing the work, running your company, and making profits. Pages of sample forms, contracts, documents, clear illutrations, and examples. Includes rehabilitation, kitchens, bathrooms, adding living area, reflooirng, residing, reroofing, replacing windows and doors, installing new wall and ceiling cover, repainting, upgrading insulation, combating moisture damage, selling your services, and bookkeeping for remodelers. **416 pages, 8 ½ x 11, $27.00**

Practical Plumbing Guides

Plumber's Handbook Revised

This new edition shows what will and won't pass inspection in drainage, vent, and waste piping, septic tanks, water supply, fire protection, and gas piping systems. All tables, standards, and specifications completely up-to-date with recent plumbing code changes. Covers common layouts for residential work, how to size piping, selecting and hanging fixtures, practical recommendations, and trade tips. The approved reference for the plumbing contractor's exam in many states. **240 pages, 8½ x 11, $18.00**

Planning Drain, Waste & Vent Systems

How to design plumbing systems in residential, commercial, and industrial buildings. Covers designing systems that meet code requirements for homes, commercial buildings, private sewage disposal systems, and even mobile home parks. Includes relevant code sections and many illustrations to guide you through what the code requires in designing drainage, waste, and vent systems. **192 pages, 8½ x 11, $19.25**

Plumber's Exam Preparation Guide

Hundreds of questions and answers to help you pass the apprentice, journeyman, or master plumber's exam. Questions are in the style of the actual exam. Gives answers for both the Standard and Uniform plumbing codes. Includes tips on studying for the exam and the best way to prepare yourself for examination day. **320 pages, 8½ x 11, $21.00**

Estimating Plumbing Costs

Offers a basic procedure for estimating materials, labor, and direct and indirect costs for residential and commercial plumbing jobs. Explains how to read and understand plot plans, design drainage, waste, and vent systems, meet code requirements, and make an accurate take-off for materials and labor. Includes sample cost sheets, manhour production tables, complete illustrations, and all the practical information you need. **224 pages, 8½ x 11, $22.50**

Practical Construction References

Roof Framing

Shows how to frame any type of roof in common use today, even if you've never framed a roof before. Includes using a pocket calculator to figure any common, hip, valley, or jack rafter length in seconds. Over 400 illustrations cover every measurement and every cut on each type of roof: gable, hip, Dutch, Tudor, gambrel, shed, gazebo, and more. **480 pages, 5½ x 8½, $22.00**

Paint Contractor's Manual

How to start and run a profitable paint contracting company: getting set up and organized to handle volume work, avoiding mistakes, squeezing top production from your crews and the most value from your advertising dollar. Shows how to estimate all prep and painting. Loaded with manhour estimates, sample forms, contracts, charts, tables and examples you can use. **224 pages, 8½ x 11, $19.25**

Stair Builders Handbook

If you know the floor-to-floor rise, this handbook gives you everything else: number and dimension of treads and risers, total run, correct well hole opening, angle of incline, and quantity of materials and settings for your framing square for over 3,500 code-approved rise and run combinations — several for every 1/8-inch interval from a 3 foot to a 12 foot floor-to-floor rise. **416 pages, 5½ x 8½, $15.50**

Masonry & Concrete Construction

Every aspect of masonry construction is covered, from laying out the building with a transit to constructing chimneys and fireplaces. Explains footing construction, building foundations, laying out a block wall, reinforcing masonry, pouring slabs and sidewalks, coloring concrete, selecting and maintaining forms, using the Jahn Forming System and steel ply forms, and much more. Dozens of photos, illustrations, charts, and tables. **224 pages, 8½ x 11, $17.25**

Construction Superintending

Includes every job phase from taking bids to project completion on both heavy and light construction: excavation, foundations, pilings, steelwork, concrete and masonry, carpentry, plumbing, and electrical. Explains scheduling, preparing estimates, record keeping, dealing with subcontractors, and change orders. Includes the charts, forms, and established guidelines you need. **240 pages, 8½ x 11, $22.00**

Rafter Length Manual

Complete rafter length tables and the "how to" of roof framing. Shows how to use the tables to find the actual length of common, hip, valley, and jack rafters. Explains how to measure, mark, cut and erect the rafters; find the drop of the hip; shorten jack rafters; mark the ridge and much more. Loaded with explanations and illustrations. **369 pages, 5½ x 8½, $14.25**

Building Layout

Shows how to use a transit to locate a building correctly on the lot, plan proper grades with minimum excavation, find utility lines and easements, establish correct elevations, lay out accurate foundations, and set correct floor heights. Explains how to plan sewer connections, level a foundation that's out of level, use a story pole and batterboards, work on steep sites, and minimize excavation costs. **240 pages, 5½ x 8½, $11.75**

Concrete Construction & Estimating

Explains how to estimate the quantity of labor and materials needed, plan the job, erect fiberglass, steel, or prefabricated forms, install shores and scaffolding, handle the concrete into place, set joints, finish and cure the concrete. Full of practical reference data, cost estimates, and examples. **571 pages, 5½ x 8½, $20.50**

Construction Surveying & Layout

A practical guide to simplified construction surveying. How to divide land, use a transit and tape to find a known point, draw an accurate survey map from your field notes, use topographic surveys, and the right way to level and set grade. You'll learn how to make a survey for any residential or commercial lot, driveway, road, or bridge — including how to figure cuts and fills and calculate excavation quantities. Use this guide to make your own surveys, or just read and verify the accuracy of surveys made by others. **256 pages, 5½ x 8½, $19.25**

Practical Construction References

Rough Carpentry

All rough carpentry is covered in detail: sills, girders, columns, joists, sheathing, ceiling, roof and wall framing, roof trusses, dormers, bay windows, furring and grounds, stairs, and insulation. Explains practical code-approved methods for saving lumber and time. Chapters on columns, headers, rafters, joists, and girders show how to use simple engineering principles to select the right lumber dimension for any species and grade. **288 pages, 8½ x 11, $17.00**

Contractor's Survival Manual

How to survive hard times and succeed during the up cycles. Shows what to do when the bills can't be paid, finding money and buying time, transferring debt, and all the alternatives to bankruptcy. Explains how to build profits, avoid problems in zoning and permits, taxes, time-keeping, and payroll. Unconventional advice on how to invest in inflation, get high appraisals, trade and postpone income, and stay hip-deep in profitable work. **160 pages, 8½ x 11, $16.75**

Wood-Frame House Construction

Step-by-step construction details, from the layout of the outer walls, excavation and formwork, to finish carpentry and painting. Contains all new, clear illustrations and explanations updated for construction in the '90s. Everything you need to know about frame, roofing, siding, interior finishings, floor covering and stairs — your complete book of wood-frame homebuilding. **304 pages, 8½ x 11, $19.75. Revised edition**

Residential Electrical Design

Shows how to draw up an electrical plan from blueprints, including the service entrance; grounding; lighting requirements for kitchen, bedroom and bath; and how to lay them out. Explains how to plan electrical heating systems and what equipment you'll need, how to plan outdoor lighting, and much more. If you're a builder who sometimes has to plan an electrical system, you should have this book. **194 pages, 8½ x 11, $11.50**

Contractor's Guide to the Building Code Revised

This completely revised edition explains in plain English exactly what the Uniform Building Code requires. Based on the 1991 code, the most recent, it covers many changes made since then. Also covers the Uniform Mechanical Code and the Uniform Plumbing Code. Shows how to design and construct residential and light commercial buildings that'll pass inspection the first time. Suggests how to work with an inspector to minimize construction costs, what common building shortcuts are likely to be cited, and where exceptions are granted. **544 pages, 5½ x 8½, $28.00**

Builder's Guide to Accounting Revised

Step-by-step, easy-to-follow guidelines for setting up and maintaining records for your building business. A practical, newly-revised guide to all accounting methods showing how to meet state and federal accounting requirements and the new depreciation rules. Explains what the 1986 Tax Reform Act can mean to your business. Full of charts, diagrams, blank forms, simple directions and examples. **304 pages, 8½ x 11, $22.50**

Craftsman Book Company
6058 Corte del Cedro
Carlsbad, CA 92018

FAX No. (619) 438-0398

24 hr. order line

1-800-829-8123

☎ **Mail Orders**

We pay shipping when your check covers your order in full.

In a hurry? We accept phone orders charged to your Visa, MasterCard and American Express **Call 1-800-829-8123**

Name (Please print) _____

Company _____

Address _____

City/State/Zip _____

Total enclosed _____ (In California add 7.25% tax)

Use your ☐Visa, ☐MasterCard or ☐Am Exp.

Card # _____

Expiration date _____ Initials _____

10 Day Money Back GUARANTEE

☐ 22.50 Builder's Guide to Accounting Revised
☐ 16.50 Building Cost Manual
☐ 11.75 Building Layout
☐ 20.50 Concrete Construction & Estimating
☐ 30.00 Construction Estimating Reference Data
☐ 22.00 Construction Superintendent
☐ 19.25 Construction Surveying & Layout
☐ 28.00 Contractor's Guide to the Build. Code Rev
☐ 16.75 Contractor's Survival Manual
☐ 18.00 Electrical Blueprint Reading Revised
☐ 29.75 Estimating & Bid. for Blders & Rem. with free 5¼" high density *Estimate Writer disk.* *
☐ 17.00 Estimating Home Building Costs
☐ 22.50 Estimating Plumbing Costs
☐ 21.50 Estimating Tables for Home Building
☐ 23.75 Manual of Professional Remodeling
☐ 17.25 Masonry & Concrete Construction
☐ 31.50 National Construction Estimator with free 5¼" high density *Estimate Writer disk.* *

☐ 32.25 National Plumbing & HVAC Estimator with free 5¼" high density *Plumbing & HVAC Estimate Writer disk.* *
☐ 29.50 National Repair & Remodeling Est. with free 5¼" high density *Repair & Remodeling Estimate Writer disk.* *
☐ 19.25 Paint Contractor's Manual
☐ 19.25 Planning Drain, Waste & Vent Systems
☐ 21.00 Plumber's Exam Preparation Guide
☐ 18.00 Plumber's Handbook Revised
☐ 14.25 Rafter Length Manual
☐ 27.00 Remodeler's Handbook
☐ 11.50 Residential Electrical Design
☐ 22.00 Roof Framing
☐ 17.00 Rough Carpentry
☐ 15.50 Stair Builders Handbook
☐ 19.75 Wood Frame House Construction
☐ 22.00 Basic Plumbing with Illustrations
☐ **FREE Full Color Catalog**

* *If you need* ☐ *5¼" or* ☐ *3½" double density disks add $10 extra.*

Craftsman Book Company
6058 Corte del Cedro
Carlsbad, CA 92018
FAX No. (619) 438-0398

24 hr. Order Line Call

1-800-829-8123

☎ **Mail Orders**

We pay shipping when your check covers your order in full.

In a hurry? We accept phone orders charged to your Visa, MasterCard and American Express **Call 1-800-829-8123**

Name (Please print) _____

Company _____

Address _____

City/State/Zip _____

Total enclosed _____ (In California add 7.25% tax)

Use your ☐Visa, ☐MasterCard or ☐American Exp.

Card # _____

10 Day Money Back GUARANTEE

☐ 22.50 Builder's Guide to Accounting Revised
☐ 16.50 Building Cost Manual
☐ 11.75 Building Layout
☐ 20.50 Concrete Construction & Estimating
☐ 30.00 Construction Estimating Reference Data
☐ 22.00 Construction Superintending
☐ 19.25 Construction Surveying & Layout
☐ 28.00 Contractor's Guide to the Build. Code Rev
☐ 16.75 Contractor's Survival Manual
☐ 18.00 Electrical Blueprint Reading Revised
☐ 29.75 Estimating & Bid. for Blders & Rem. with free 5¼" high density *Estimate Writer disk.* *
☐ 17.00 Estimating Home Building Costs
☐ 22.50 Estimating Plumbing Costs
☐ 21.50 Estimating Tables for Home Building
☐ 23.75 Manual of Professional Remodeling
☐ 17.25 Masonry & Concrete Construction
☐ 31.50 National Construction Estimator with free 5¼" high density *Estimate Writer disk.* *

☐ 32.25 National Plumbing & HVAC Estimator withfree 5¼" high density *Plumbing & HVAC Estimate Writer disk.* *
☐ 29.50 National Repair & Remodeling Est. with free 5¼" high density *Repair & Remodeling Estimate Writer disk.* *
☐ 19.25 Paint Contractor's Manual
☐ 19.25 Planning Drain, Waste & Vent Systems
☐ 21.00 Plumber's Exam Preparation Guide
☐ 18.00 Plumber's Handbook Revised
☐ 14.25 Rafter Length Manual
☐ 27.00 Remodeler's Handbook
☐ 11.50 Residential Electrical Design
☐ 22.00 Roof Framing
☐ 17.00 Rough Carpentry
☐ 15.50 Stair Builders Handbook
☐ 19.75 Wood Frame House Construction
☐ 22.00 Basic Plumbing with Illustrations
☐ **FREE Full Color Catalog**

* *If you need* ☐ *5¼" or* ☐ *3½" double density disks add $10 extra.*

Mail This Card Today
For a Free
Full Color Catalog

Over 100 books, videos, and audios at your fingertips with information that can save you time and money. Here you'll find information on carpentry, contracting, estimating, remodeling, electrical work, and plumbing.

All items come with an unconditional 10-day money-back guarantee. If they don't save you money, mail them back for a full refund.

Name _____

Company _____

Address _____

City / State / Zip _____

Craftsman Book Company / 6058 Corte del Cedro / P. O. Box 6500 / Carlsbad, CA 92018

BUSINESS REPLY MAIL
FIRST CLASS MAIL PERMIT NO. 271 CARLSBAD CA

POSTAGE WILL BE PAID BY ADDRESSEE

CRAFTSMAN BOOK COMPANY
PO BOX 6500
CARLSBAD CA 92018-9974

BUSINESS REPLY MAIL
FIRST CLASS MAIL PERMIT NO. 271 CARLSBAD CA

POSTAGE WILL BE PAID BY ADDRESSEE

CRAFTSMAN BOOK COMPANY
PO BOX 6500
CARLSBAD CA 92018-9974

BUSINESS REPLY MAIL
FIRST CLASS MAIL PERMIT NO. 271 CARLSBAD CA

POSTAGE WILL BE PAID BY ADDRESSEE

CRAFTSMAN BOOK COMPANY
PO BOX 6500
CARLSBAD CA 92018-9974